The GREAT SALEM FIRE OF 1914

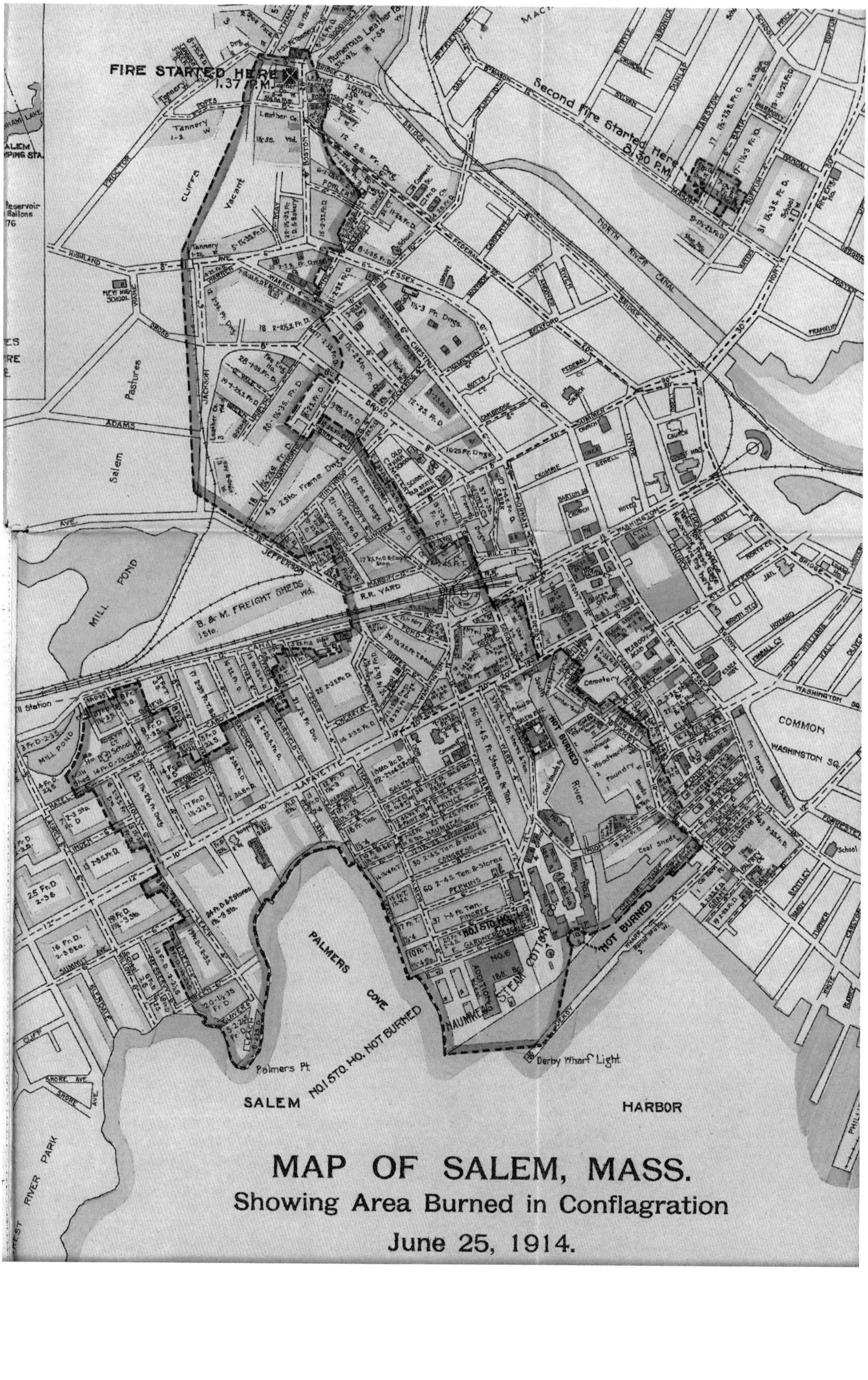

MAP OF SALEM, MASS.

Showing Area Burned in Conflagration

June 25, 1914.

IMAGES FROM THE PHILLIPS LIBRARY COLLECTION

The GREAT SALEM FIRE OF 1914

BARBARA PERO KAMPAS
THE PHILLIPS LIBRARY AT THE PEABODY ESSEX MUSEUM

THE History PRESS

Published by The History Press
Charleston, SC 29403
www.historypress.net

Frontispiece: This map outlines the sections of Salem ravaged by the fire. The fire started on Boston Street in the early afternoon of June 25, 1914, in the northwest section of town known as Blubber Hollow, named because of the whale blubber originally used in tanning leather. Fueled by strong winds, the preponderance of wooden structures and the lack of water pressure, the fire quickly became a conflagration spreading southeast toward the Naumkeag Steam Cotton Company on Congress Street in South Salem, where it was stopped only because it reached the Salem Harbor. The map was published in the Fire Report, Index #12375, submitted by Fire Inspector Charles H. Smith of the Association of Factory Mutual Insurance Companies of Boston after he inspected the Naumkeag Steam Cotton Company Ruins on June 26. (*Neg #32846*)

All images courtesy of the Phillips Library unless otherwise noted.

First published 2008

Manufactured in the United States

ISBN 978.1.59629.471.4

Library of Congress Cataloging-in-Publication Data
Kampas, Barbara Pero.
The great Salem fire of 1914 : images from the Phillips Library collection / Barbara Pero Kampas.
p. cm.
Includes bibliographical references.
ISBN 978-1-59629-471-4
1. Salem (Mass.)--History--20th century--Pictorial works. 2. Fires--Massachusetts--Salem--History--20th century--Pictorial works. 3. Phillips Library (Peabody Essex Museum)--Photograph collections. I. Phillips Library (Peabody Essex Museum) II. Title.
F74.S1K36 2008
974.4'5--dc22
2008034254

Contents

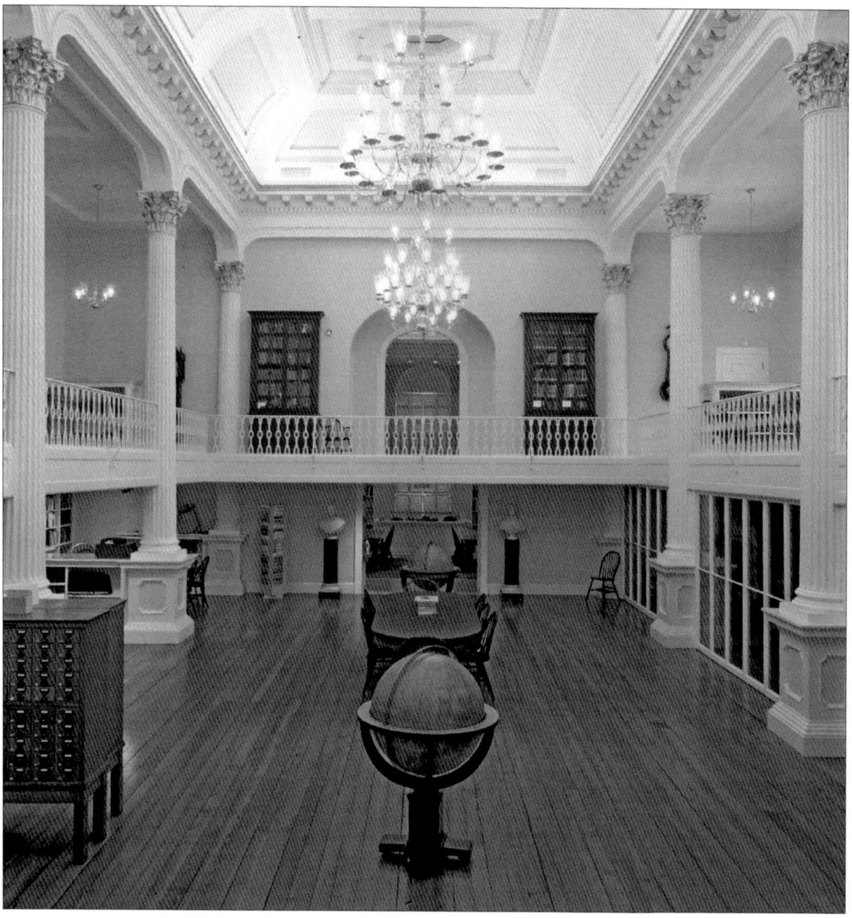

The Peabody Essex Museum (PEM), founded in 1799, is America's oldest continuously operating museum. The Phillips Library, the research and documentation division of PEM, is one of New England's older libraries. The library resides in two historic buildings in Salem—the John Tucker Daland House and Plummer Hall—opposite the museum in East India Square. Scholars use the library to research maritime history and art, American decorative arts, Asian art and culture, Native American history and art, New England life and culture and the art and culture of Oceania. The collection includes original manuscripts, printed volumes, broadsides, photographic archives, historic maps, ships' logs and account books, diaries and Essex County probate and court records. Researchers may visit during public hours or by appointment. *Photographed by Jeff Dykes and Mark Sexton (Plummer Hall).*

Preface

The Great Fire of 1914 had a significant impact on Salem, Massachusetts. Much has been written about the fire and many images were taken to record this historic tragedy. The images come from professional photographers, newspaper journalists and citizens of Salem who wanted to chronicle a historic event. To create this book, images were culled from personal scrapbooks, newspaper clippings, negatives (film and glass), historic documents and photographic prints in the Photo Archives of the Phillips Library at the Peabody Essex Museum.

Whenever possible, the location of the photograph and the name of the photographer have been identified in the captions. The earlier issues of the *Naumkeag Directory for Salem* identify photographers as photograph artists; this designation is apt when one looks at the images in this volume. Viewing pictures of the fire with today's technology made it possible to see details that were not readily discernible when the images were originally printed in 1914. Such details have been extracted from many of the photographs reproduced in this book to show the impact of the fire.

The history of the fire and technical information on how it started and spread were readily available from many sources in the Phillips Library. The impact on the people who survived was not as available from library sources, but attempting to understand their plight was a major goal of this project. Letters to relatives, poetry written immediately after the fire or many years later to reminisce, journal articles, papers archived from the organizations created to provide relief for the homeless families and firsthand accounts in our collection were invaluable sources to ascertain the fire's impact. All sources, primary and secondary, which have been identified in this text can be found in the collection of the Phillips Library. Unless otherwise noted, the events mentioned in the book occurred in 1914.

Captions identifying street locations use names of streets in use at the time of the fire. New streets were added during the reconstruction process, as was a new number system for the addresses on the streets as they were rebuilt and widened after the fire. Every caption in the book provides a negative number or identifying name for the image. All the photographs in this publication are available for reproduction. Please contact the Photo Services Department of the Phillips Library at the Peabody Essex Museum if you are interested in a reproduction of any image in this publication.

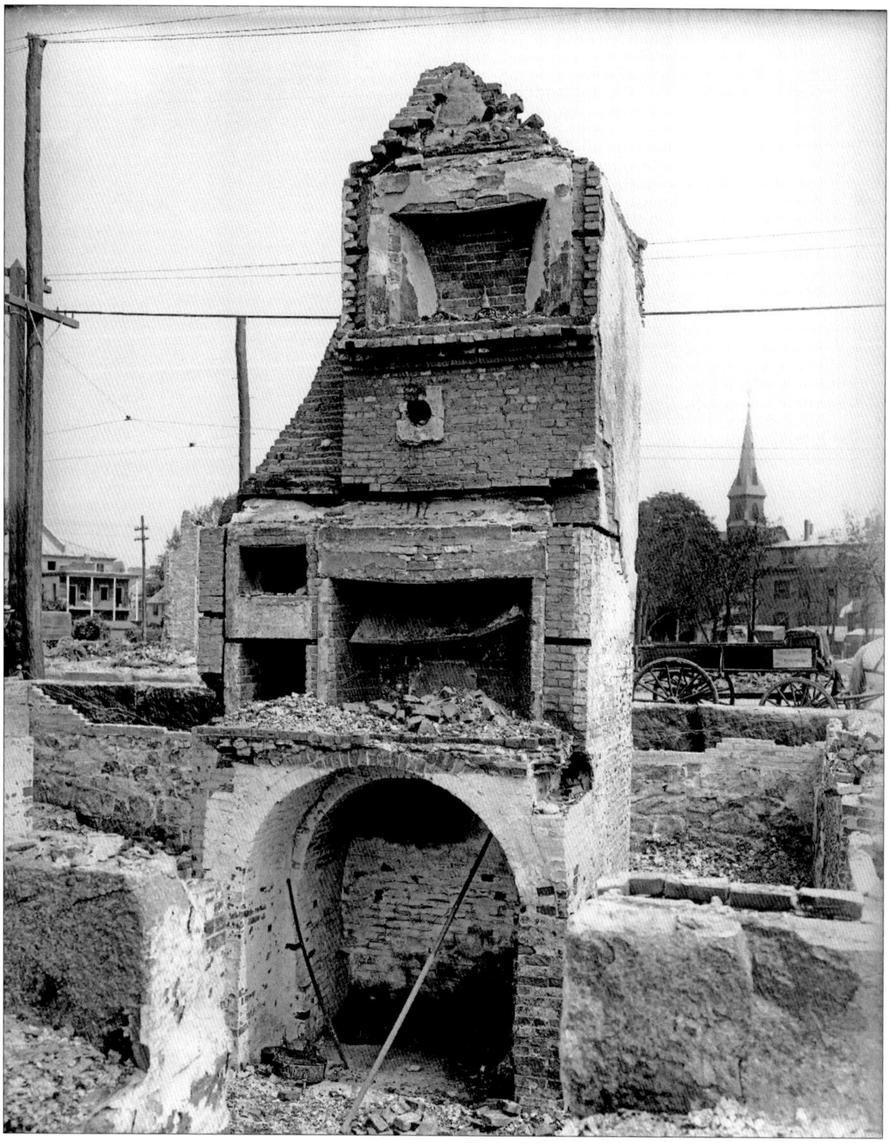

The Phillips Library Photo Archives hold prints and glass negatives of Frank Cousins, one of Salem's most famous photographers. Cousins, active as a photographer from the early 1880s to the early 1920s, specialized in architectural studies. He is well-known for overall views of buildings and street blocks as well as for his photographs of building details and interiors. This image represents one of his detailed views of chimneys and building remains in the ruins of the Salem Fire. This particular chimney was on Boston Street, close to where the fire started. (*Neg #2698*)

Acknowledgements

The mission of a museum library is complex: supporting the activities of the museum as well as supporting the needs of local researchers and international scholars. My thanks and appreciation go to Dan Monroe, Josh Basseches and Lynda Hartigan for supporting the Phillips Library in all of its endeavors. It has been gratifying to work with and learn from Sidney Berger as we strive to achieve his goals for the library. To the staff of the Phillips Library—Irene Axelrod, Roseanna Enos, Kathy Flynn, Tamara Gaydos and Ruth Stewart—thank you for your support and encouragement during the writing of this book. The staff of the IT Department—Sean Pyburn, Isabel Soares and Ben Arlander—was always available to solve the technical problems and to assist me as I learned new software required to complete this project. Christine Michelini's insightful suggestion to delve deeply into the Photo Archives led to the discovery of new images of the Salem Fire, which helped me understand the human impact of the tragedy. Jeff Dykes's visual acuity and ability to view photographs from many perspectives taught me to look deeply into the images and to extract details. His guidance helped to elucidate the connections between the photographs; his expertise with digital imaging brought depth and clarity to the images. The result of this project would have been considerably different without the invaluable assistance of Christine and Jeff. I appreciate the help provided by Yohanny Gonzalez and Michelle Cangemi, the Museum Action Corps (MAC) interns working with Jeff, who helped scan images for this book. To Britta Karlberg, thank you for sharing your love of the library collections before you retired and for moving me from "outside the box" to "there is no box." The writing of this book has been an intriguing journey that greatly enhanced my awe of the collections in the Phillips Library.

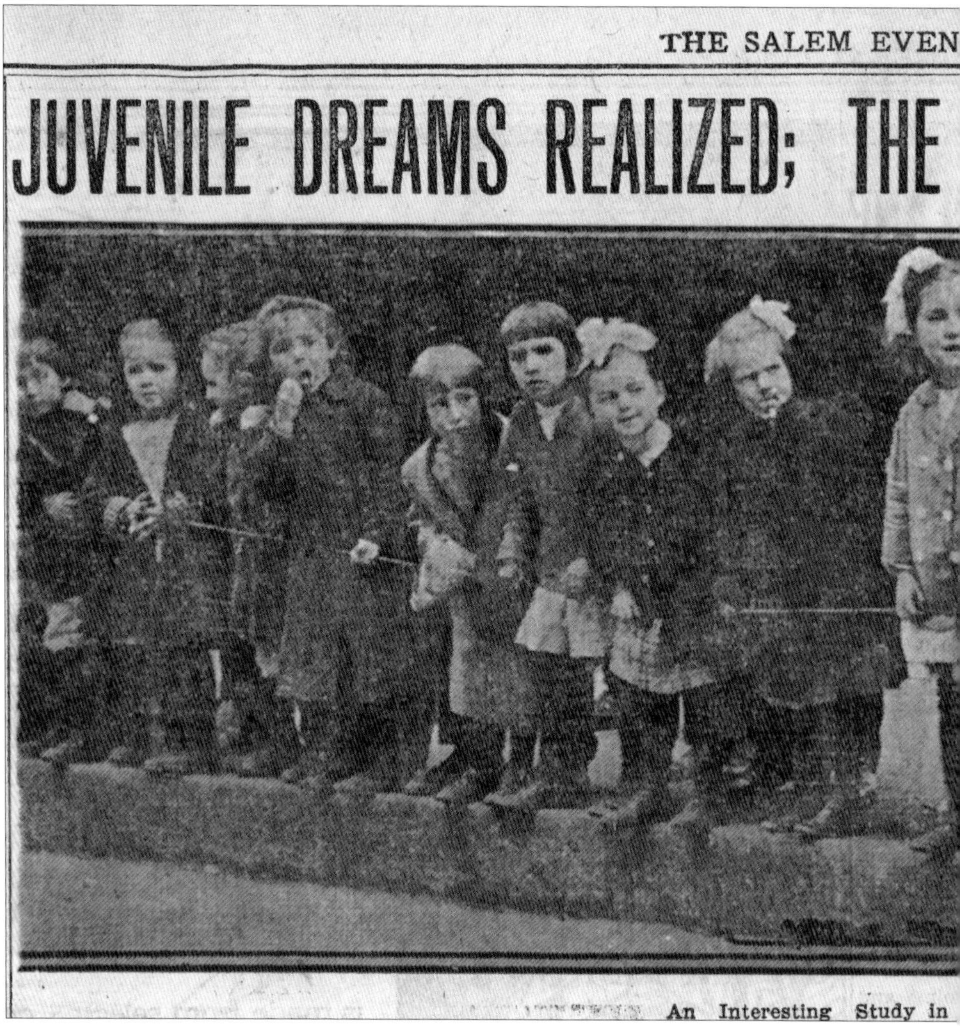
THE SALEM EVEN

JUVENILE DREAMS REALIZED; THE

An Interesting Study in

This photo, published two days before the fire, on June 23 in *The Salem Evening News*, depicts children enraptured by Barnum and Bailey's Greatest Show on Earth circus parade. The caption with the photo reads: "An interesting study in expression caught as the parade passed." The reporter stated that "the street parade was even more than was promised in magnitude, splendor and diversity." The parade was nearly three miles long and included bands, plumed horses, clowns and over forty elephants and ended with the famous calliope, the "steam car of the Muses." (*Neg #32882*)

G NEWS— SALEM, MASS., TUESDAY, JUNE 23, 1914.

CIRCUS CAME TO TOWN THIS MORNING

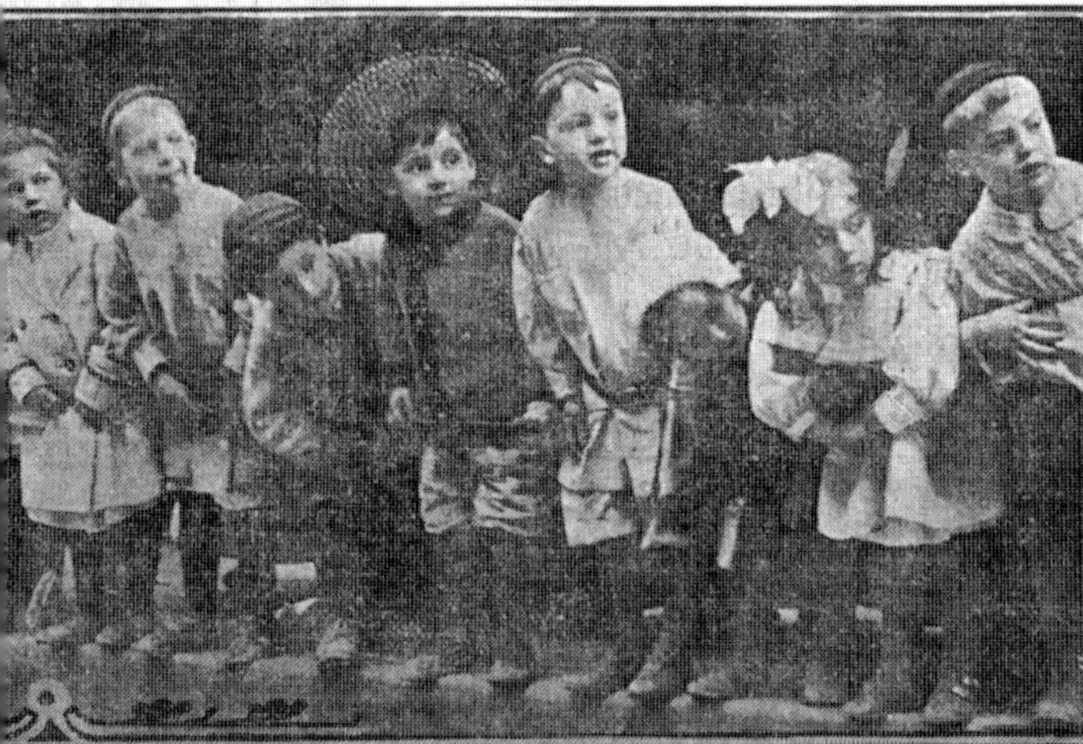

ression Caught as the Parade Passed.

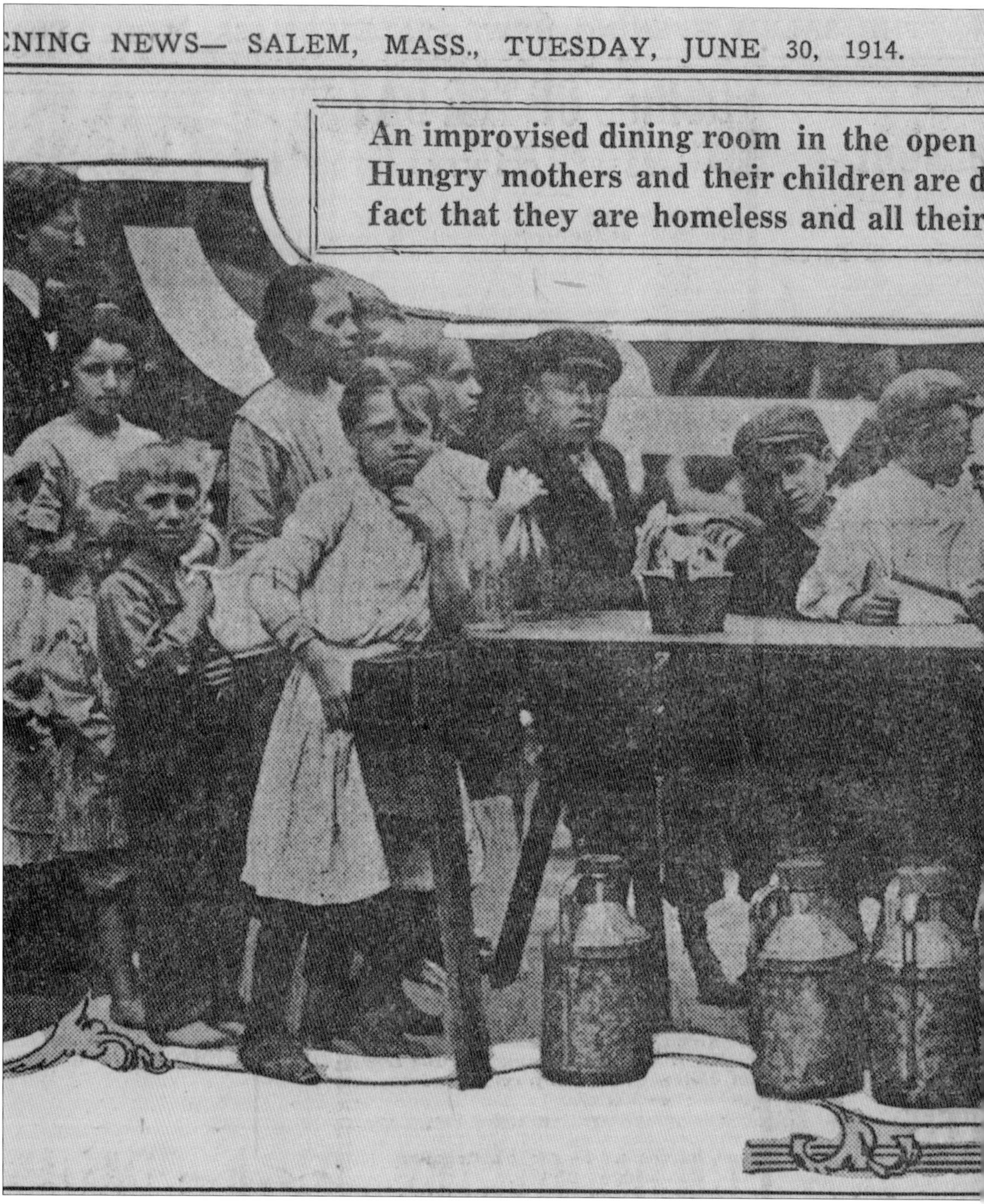

The children in this photo, published by *The Salem Evening News* on June 30, provide a marked contrast to those watching the circus parade several days earlier. These children were among the thousands rendered homeless by the fire. Immediately after the fire an improvised "dining room" was set up outside the State Amory on Essex Street. The caption with the photo reads in part: "Hungry mothers and their children are drinking milk and contentedly munching crackers despite the fact that they are homeless and all their possessions were swept away by the fire." (*Neg #32885*)

air near the Salem bread-line outside the State Armory. rinking milk and contentedly munching crackers despite the possessions were swept away by the fire.

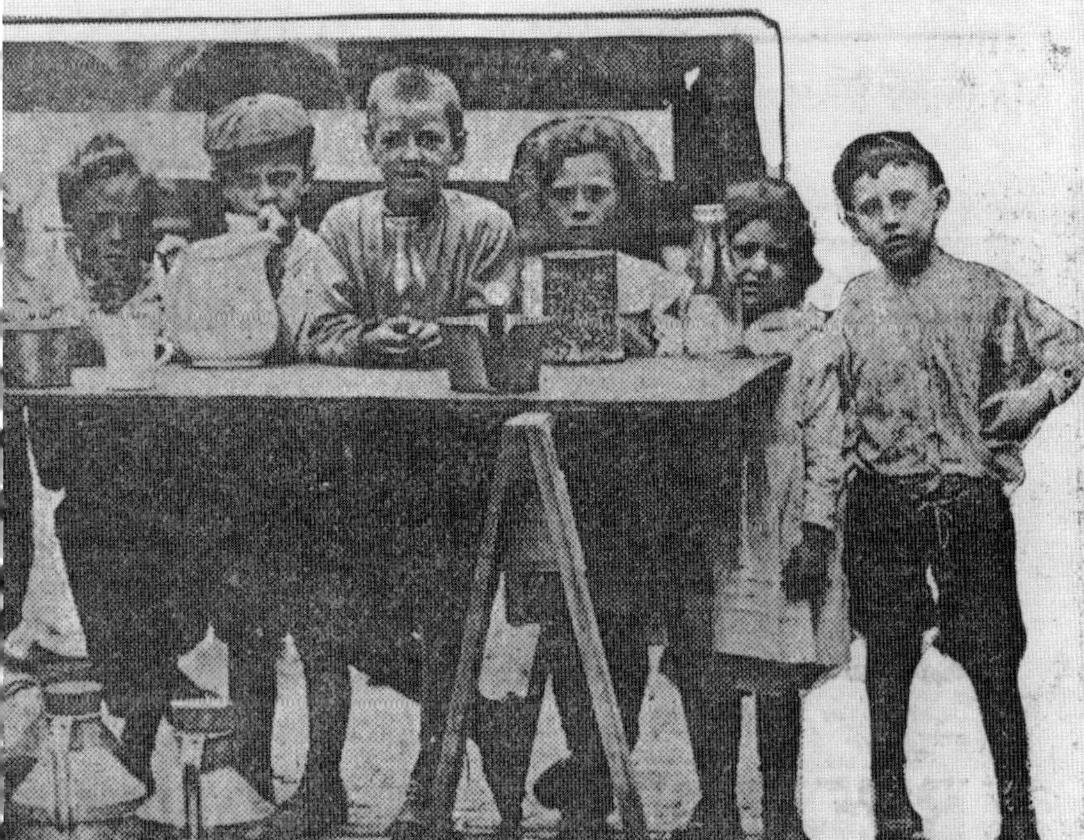

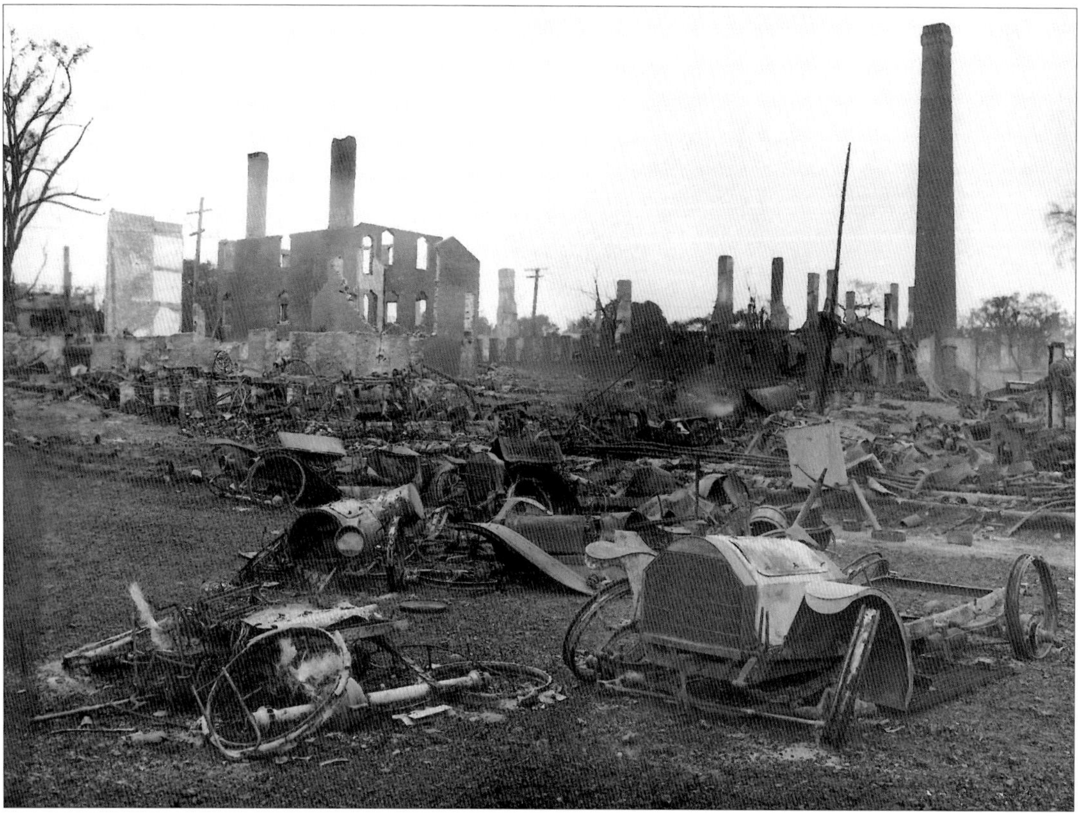

The heat of the fire was intense and burned almost everything in its path. An unidentified newspaper clipping found in one of the many scrapbooks in the library told the story of a man who discovered his automobile on fire. He jumped into the car and drove it directly into the ocean. "Leaping into the car he ran it full tilt down to the water and into Salem bay as far as it could go before the engine stopped. There he left it, and the car burned down to its waterline, leaving the lower parts intact from flames." (*Neg #32785*)

Introduction

In the life of any town, certain things happen that can change the lives of its inhabitants permanently. The catalysts of change could be political, fiscal or natural, sometimes these three overlapping. Usually the incident or situation causes tremendous uproar—physical and emotional pandemonium—and affects the lives of thousands. And the consequences are vast and cataclysmic. The aftermath of such phenomena changes the course of many lives, as well as the life of the town, and it is a central topic of conversation for weeks or years afterward. The more cataclysmic the event, the more it is on the lips of the people affected and the more it is covered in the press.

In time, as years go by and generations pass, the event fades from memory, and it is buried under layers of the subsequent doings of the place. But libraries, fulfilling one of their primary missions, are repositories of memories, both visual and verbal. Such is the case for Salem, Massachusetts.

On June 25, 1914, Salem, Massachusetts, experienced such a cataclysm. In the early afternoon, at the Korn Leather Factory at 57 Boston Street, a mixture of acetone, amalacitate, alcohol and celluloid exploded. The fire alarm was reported at 1:37 p.m. Help to fight the fire was sent out from nearly two dozen nearby cities and towns. There had been a long dry spell and the town was ripe for such a disaster. The factory was making what is known as "tip finish," the varnish that shines patent leather. What happened in the factory is unclear. What we do know is that the explosions that took place set off a conflagration that destroyed more than 250 acres and 1,376 buildings (including the homes of twenty thousand people), and left ten thousand people out of work.

Reports of the fire read like a novel. The July 1, 1914 edition of the *New York Times* printed an article with the headline "Mrs. Belknap Asks Aid For Salem Poor." Mrs. Belknap had come to Manhattan a few days after the fire to seek assistance for the victims of the fire. Mrs. Belknap's mother, formerly of New York, lost her home, the Endicott mansion, during the fire, along with many priceless relics. Passages from the article appear below:

> *My husband and I were visiting my mother-in-law at Hartford at the time of the fire…It was very hot, and, seeking a cool place, we had taken the children in our car down to the lighthouse on Marblehead point, about seven miles from Salem. The wind at midday was blowing hard from the direction of Salem.*
>
> *Suddenly, at about 1:00 o'clock, a great column of smoke shot into the air from the direction of Salem. Almost immediately afterward it seemed the air about us was filled with bits of charred leather and of board, which kept falling about us. We hurriedly finished our luncheon and started*

back, for we could see flames at many points. We went as fast as we could in our auto. As we got in front of 235 Lafayette Street, houses in the rear of us started to blaze, brands having been carried to them by the wind.

By this time the wind had veered a little north of west. It was about 2:30 o'clock. As we went on we found house after house on fire. We crossed the railway tracks…and entered a district of substantial homes of workingmen, which were all now ablaze.

We ran up Warren Street, to find it all flaming, and by the time we reached the Endicott mansion the roof of that old building was all on fire, and men were pouring water on it. That was the last house in the block to go, as it was built of brick.

Later she added:

I have been in Salem several times since the fire, and was there on Sunday. The sidewalks in the burned area were then still so hot that you could not walk on them. Several fires have started since, due, it is said, to smoldering embers in the coal packets, which are burning so fiercely that they cannot be put out.

One thing that impressed me while the fire was at its worst was the fact that there was no disorder. The working people of Salem are a self-respecting lot of citizens, and orderly. They deserve any assistance that can be given to them.

Clearly the writer was one of the more affluent residents of the town, for she talked about her car and her family outings and the treasures she was able to save from her home.

The Vandyck and most of the paintings were saved…the furniture had to be dragged into the street. We managed to save a collection of glass which had been in the family for two or three generations. An old clock of King George III's time, which was brought over by a member of the family in the seventeenth century, and the Belknap clock, which stood in the dining room, were carried to safety.

There is a sangfroid in her description of the event. I am certain that the less well-to-do, losing everything they had, would not speak with such imperturbability. The scene was horrendous, as the photographs in this volume will attest. The utter devastation, literal and figurative, must have been unimaginable to the thousands who were displaced and who lost everything they owned.

As Arthur B. Jones, in *The Salem Fire* (Gorham Press, 1914), says of Salem, "The building laws have been very unsatisfactory, but there has been no organized effort to improve them until within the last two years."

He adds that as early as 1910 the citizens of the town knew of the fire danger in the city:

Franklin H. Wentworth, now secretary of the National Board of Fire Underwriters, when councilman in Salem in 1910, introduced an order in regard to new roof coverings, as follows: "Every roof hereafter constructed, whether within or without the fire limits, shall be covered with a non-combustible roof covering satisfactory to the inspector of buildings.

"Whenever a non-combustible roof covering is hereafter replaced in whole or in part, whether within or without the fire limits, the new roof covering shall be non-combustible to the satisfaction of the said inspector.

> *"Whenever roof covering, whether within or without the fire limits, other than non-combustible, is hereafter replaced in whole or in part, the new roof covering shall be noncombustible, to the satisfaction of the said inspector if, as much as two-thirds of the superficial area of the entire roof is repaired."*

Jones mentions ex-Councilman Charles J. Collins, who called for better water pressure in the pipes. And Jones concludes chapter 5 with this statement: "Our water supply is deficient for fire fighting. We know it and have known it for years. What are we going to do about it?"

The shoe factory, as I have noted, was storing many inflammable materials. Jones's book quotes comments from a high state official meeting at the Chief's Club on July 9, 1914:

> *The shed was filled with a quantity of cheap sheep skins, and there were three parts of barrels filled with the highly inflammable substance I have mentioned. How the flames originated I do not know, but I do know that once the fire started it spread to the barrels and the sheep skins, and swept up the open elevator well to the third floor, where there was more of the inflammable compound and a saturated floor.*
>
> *If this factory had been equipped with a sprinkler system, the building might have been saved and the conflagration averted. There was nothing in this building which water would not have extinguished.*

Jones describes the fast spread of the fire, the many efforts of private parties and the dozens of firefighters required to stop the flames, the destruction of one building after another, the "Pathetic Incidents" (chapter 17) and the inevitable and unstoppable devastation. He then talks about the relief efforts of the Red Cross and others, and the generosity of many to the devastated victims.

The day after the fire, June 26, 1914, the U.S. Army set up Tent City at the end of Lafayette Street, where thousands of those who had lost their homes lived. By June 29 there were more than four hundred tents and fifteen hundred people living in Tent City in Forest River Park, reports Jones. By the end of June rebuilding began, tentatively at first, the city council having, on June 29, begun meetings on the problems faced by the city. One major proposal ordered that

> *five local men be appointed to be called the Salem Rebuilding Commission to serve three years without pay, to have charge and control of the construction of all public buildings, take land by right of eminent domain, grant permits to build, pass regulations as to the location, size, material to be used in construction, the space between said buildings, changing of parks, squares, streets, sidewalks, discontinuing, laying out, relocating, altering, widening and repairing of all streets.*

This petition passed both branches July 3 and went to the governor to sign on July 7, according to Jones.

What else was there to do? A vast tract of their city lay before them in ashes. Thousands of people were homeless and jobless. But the sense of despair in the air was beginning to be mitigated by the forceful and persevering citizens, looking to their futures.

Jones, not only a firsthand witness but also one who was there fighting the blaze, concludes his book with this paragraph:

The Salem Rebuilding Commission has a whole lot of work cut out for it, but we know they will meet it in a fair way and for the welfare of the majority. With our fine water front, best of railroad facilities, freedom from labor troubles we have faith in our city's growth as a manufacturing center. As a residential city it is all that can be desired. The splendid spirit of optimism which has so far carried the people over the hard places, we believe will prevail in the future and from the ruins rise a city far better than the Salem of the past.

And, indeed, Salem is again a thriving and lovely city, with great culture, beautiful old homes, good schools and much optimism.

The Phillips Library at the Peabody Essex Museum has a large collection of photographs showing the before, during and after of the fire. Selecting slightly more than two hundred of these has been an immense job, for so many more could have been shown in these pages.

Barbara Kampas, the Phillips Library's Head of Library Collections, must be given full credit for the herculean task of putting this book together, from inception to completion. She has done exceptional research, hunted up and sifted through thousands of photographs, selected the ones in the present volume and even did much of the design and layout of the book. She was ably assisted by Christine Michelini and Jeff Dykes of the library's Photo Services Department and by the editors at The History Press in Salem.

Anticipating the centennial of the Great Salem Fire by six years, we have put together a volume to intrigue, inform and commemorate the city and its thrilling history.

Sidney E. Berger
Ann C. Pingree Director
Phillips Library
Peabody Essex Museum
Salem, Massachusetts

Chapter 1

Salem Before the Fire

Nineteen miles north of Boston, Salem is one of the earliest settlements in America. It was founded at the mouth of the Naumkeag River in 1626 by a company of fishermen from Cape Ann who were led by Roger Conant; it was first called Naumkeag, which meant eel land. This section of the North Shore had originally been inhabited by the Naumkeag Indians. Salem was the second city chartered in Massachusetts and it held the seat of government for the Massachusetts Bay Colony under Governor John Endicott in 1628–29. The town incorporated itself on June 24, 1629, and changed its name to Salem, the name related to the Hebrew word "shalom" and the Arabic word "salam," both meaning peace. Exactly 283 years and a day after incorporation, the Great Fire of Salem occurred.

In the eighteenth century, Salem was a major shipbuilding and fishing port, trading with the Far East, and by 1790 it was the sixth largest city in the United States, with the highest wealth per capita. Salem was incorporated as a city in March 1836. By the late nineteenth century it boasted three libraries and two museums: the Athenaeum, the Salem Public Library and the Essex Institute Library; and the Essex Institute Museum and the Peabody Academy of Science. By the nineteenth century, it was an industrial city with a thriving tanning industry, shoe industry and a cotton textile factory, which were worked by immigrants. By this time, the silting harbor forced the shipping and trade industry into deep-water harbors, such as Boston, changing the economic profile of the city.

The land of the city was fairly level in the manufacturing and business sections of town, with gently rolling hills in the outlying districts. Streets varied in width from 20 to 150 feet. The city to the west and directly east of Blubber Hollow, the leather district, was residential, with houses fairly well separated. The residential section merged into the mercantile district. South and east of the Boston and Maine Railroad was a manufacturing section and beyond this another residential section, the Point, extending to cotton mills and the harbor. Most buildings in this manufacturing district were of frame construction. Shingle roofs were the rule except in the business section.

The water supply before the fire was operated by the municipality, with water flowing by gravity from Lake Wenham, five or six miles north of Salem, from which it was pumped to a twenty-million-gallon distributing reservoir through one mile of thirty-six-inch cast-iron pipe. Pumps from the reservoir were rated at five million gallons per day. Passage of the water was through main and secondary feeders with several dead ends in the service stream. Approximately 40 percent of the water pipes were cement lined.

In July 1907, the National Board of Fire Underwriters reported on Salem, warning the city that its conflagration hazard required immediate attention. The water supply was considered

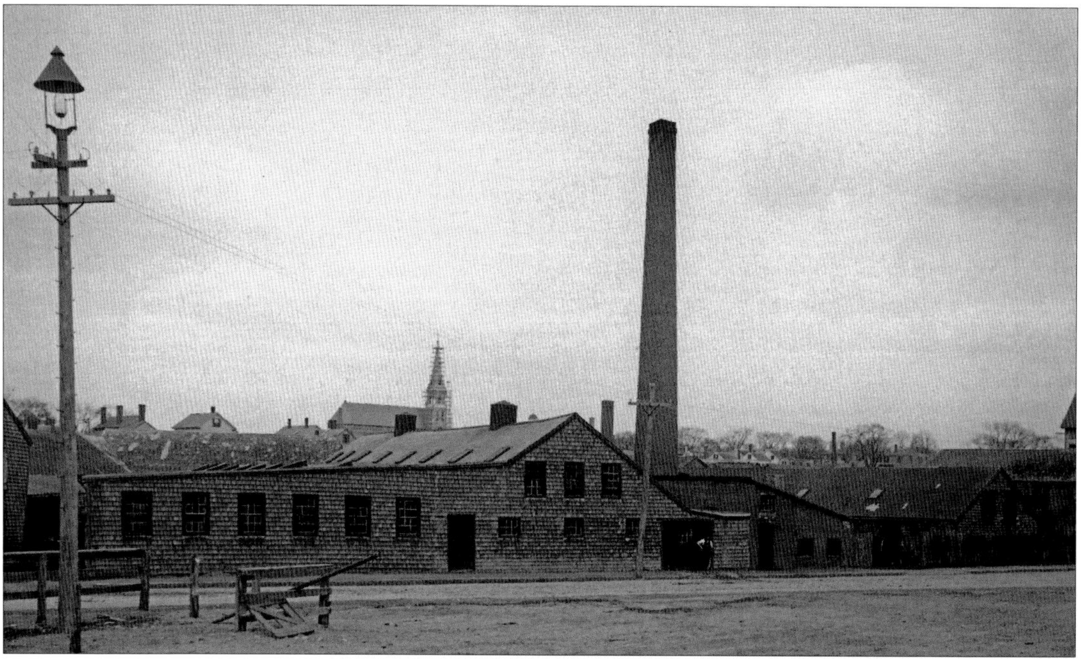

Both of these photographs were taken by Frank Cousins several years before the fire. They depict the thriving leather district along Goodhue Street. The top image shows the George Putnam and the Hamlin Tanneries at 16–18/20 Goodhue Street. (*Neg #809*) The bottom photo, dated 1891, depicts the rear view of the Robson, Hamlin and Putnam Tanneries. (*Neg #810*)

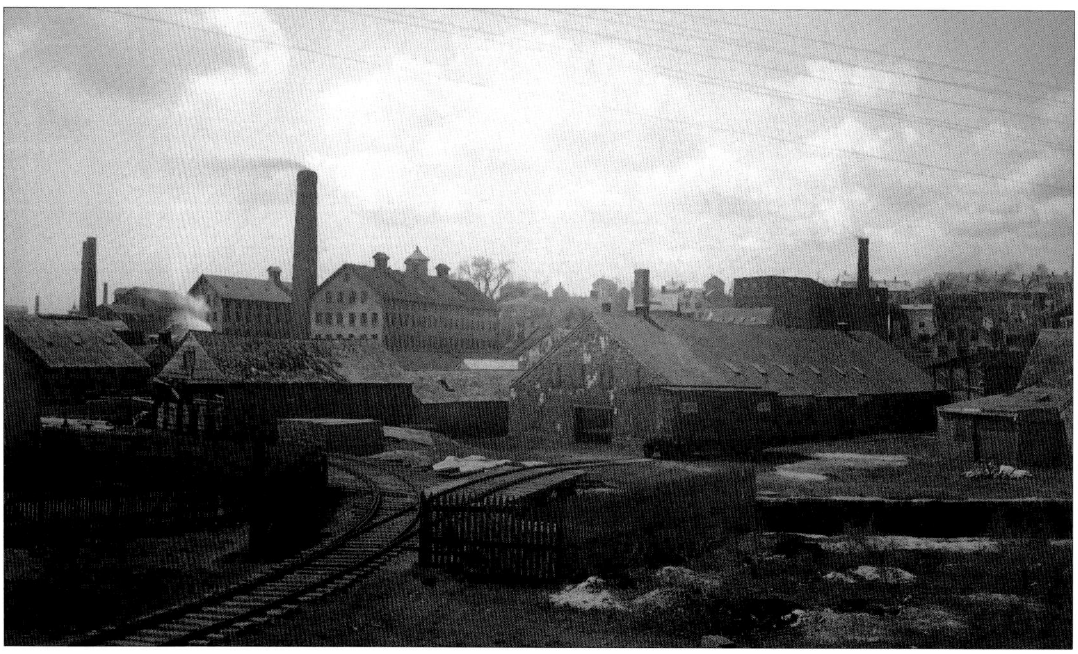

barely sufficient for present use and would have to be increased. Water pressure was satisfactory in the business district but light in some of the residential districts. The fire department, consisting of twenty-three permanent men and eight call men, was weak in men and apparatus. Financial support from the city for the fire department was smaller than that in cities of similar size. The fire department was managed well but considered insufficient because of the shortage of men and lack of training drills to keep the men acquainted with firefighting techniques and equipment. The Underwriters' report considered the building laws inadequate; they did not contain modern methods of construction with requirements for protection to exposures and floor openings, fire escapes, standpipe or hose and sprinklers. Fire limits were extensive enough, but frame construction was allowed within these limits, creating potentially hazardous conditions. The result of the report was that Salem was a conflagration hazard:

> *The features most prominent in creating a severe conflagration hazard in Salem are the large amount of frame construction in and surrounding high value districts; lack of protection in exposed windows; generally narrow streets; a weak fire department; and an unreliable water distribution system.*

The city did not heed the warning.

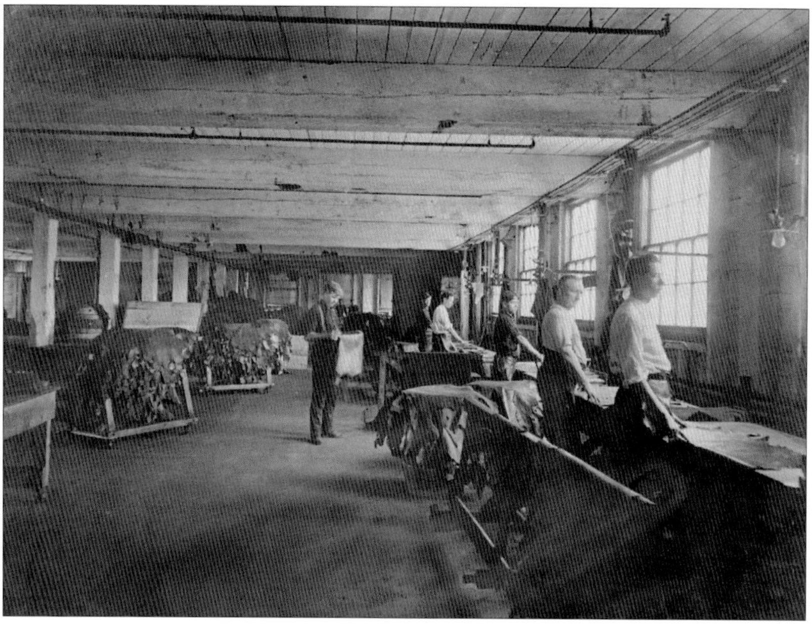

This page, facing and following pages: Salem's shoe and leather industry developed during the 1880s and 1890s, specializing in women's leather shoes. The industry specialized in leather tanning, upper shoe leather and embossing, shoe cut stock and findings and making shoes. Blubber Hollow, Salem's oldest district, was the center of the leather industry; it included eleven streets and forty-two firms, of which the Carr Leather Factory was one. It was on Pope Street and had a thriving business before the fire. The images on this page and the next three pages show employees of the company actively involved in the various steps of the tanning process. (*this page: Neg #31264D, top, and Neg #31264E, bottom; page 23: Neg #31264I, top, and Neg #31264H, bottom; page 24: Neg #31264F, top, and Neg #31264B, bottom; and page 25: Neg #31264G, top, and Neg #31264C, bottom*)

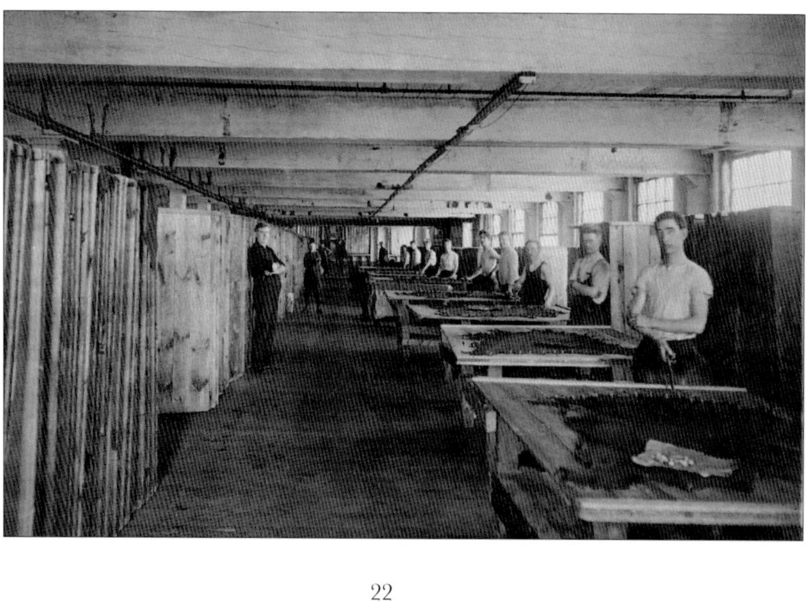

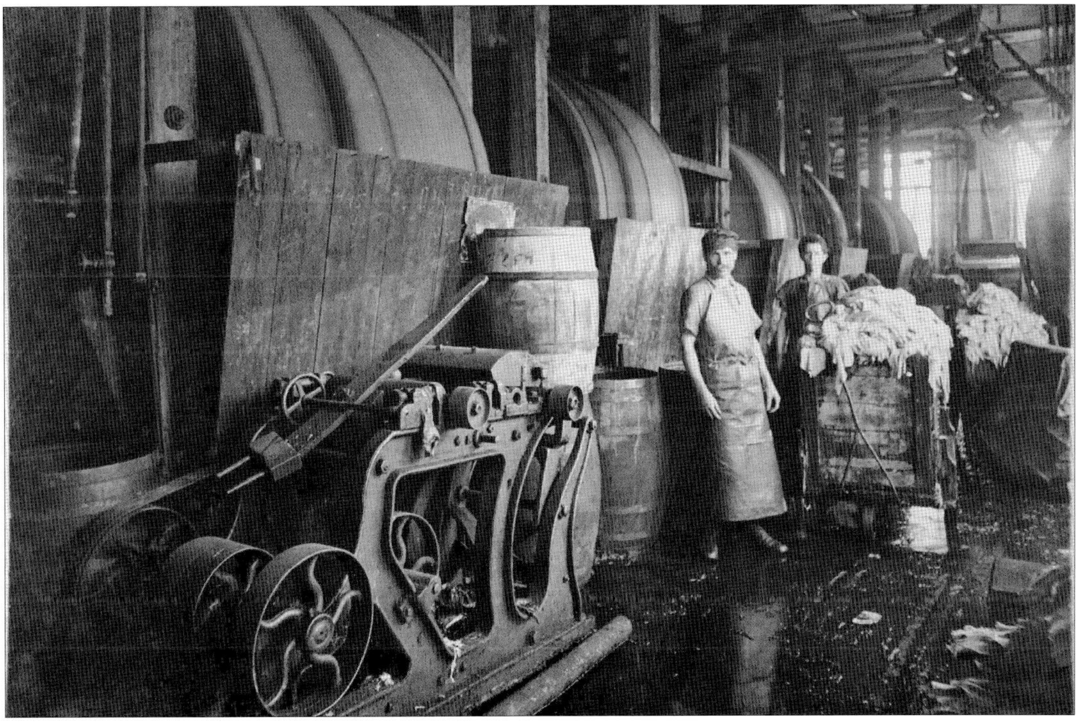

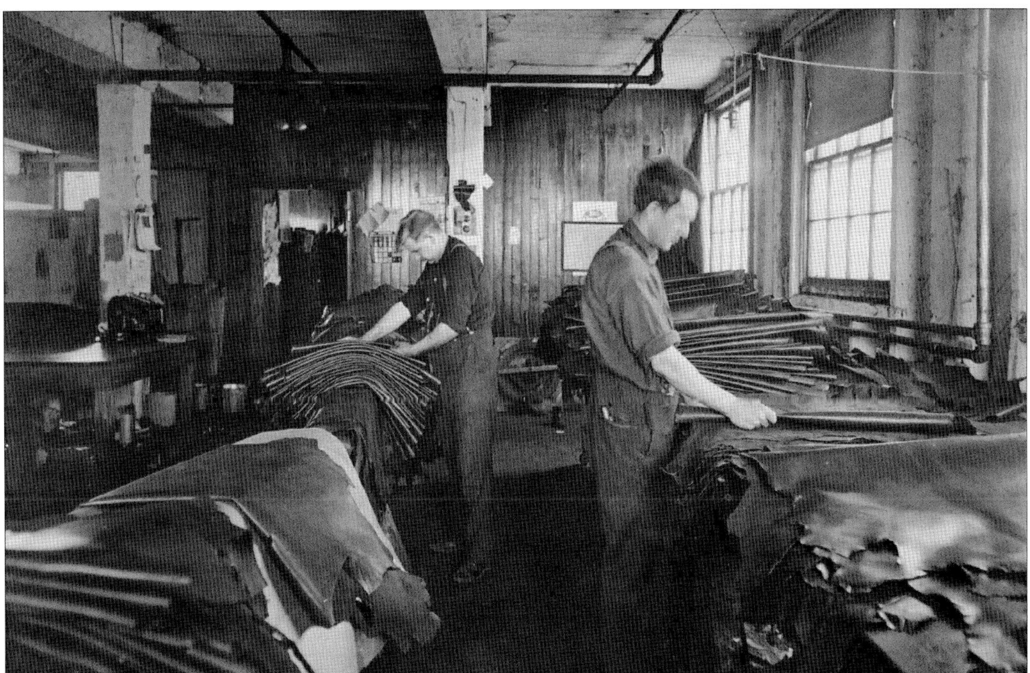

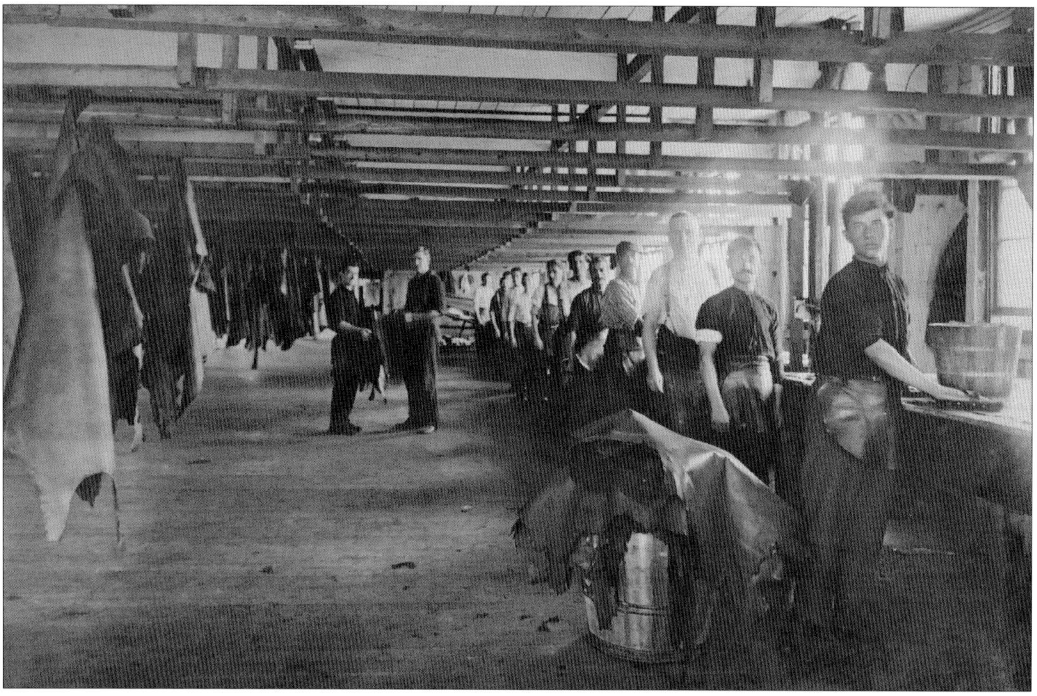

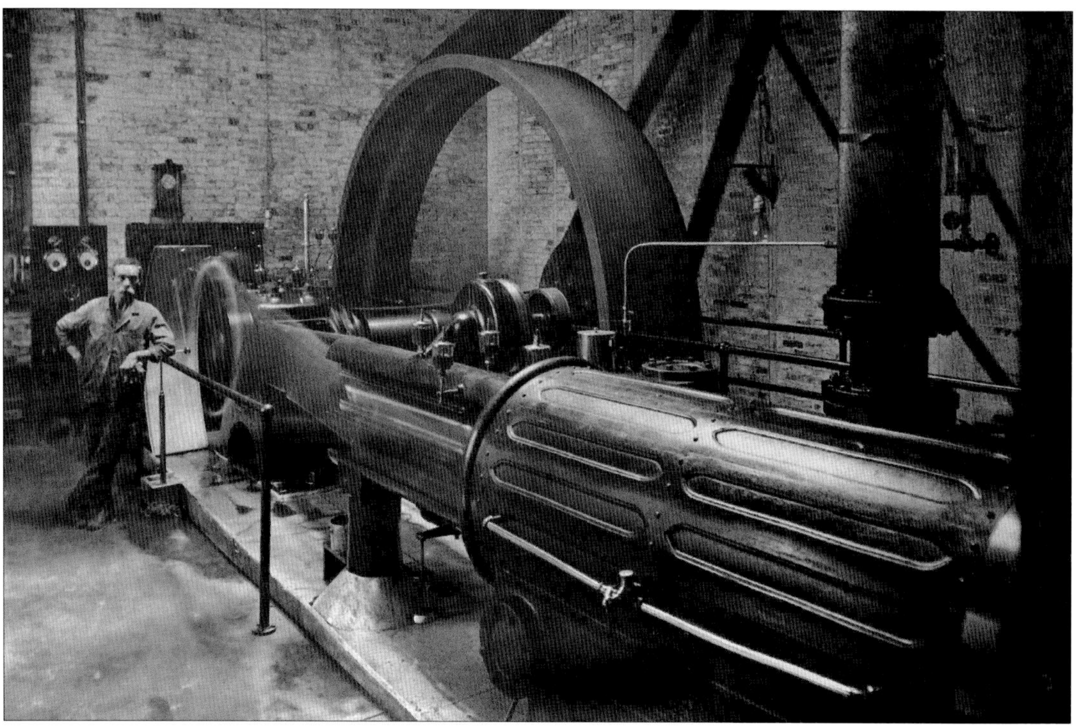

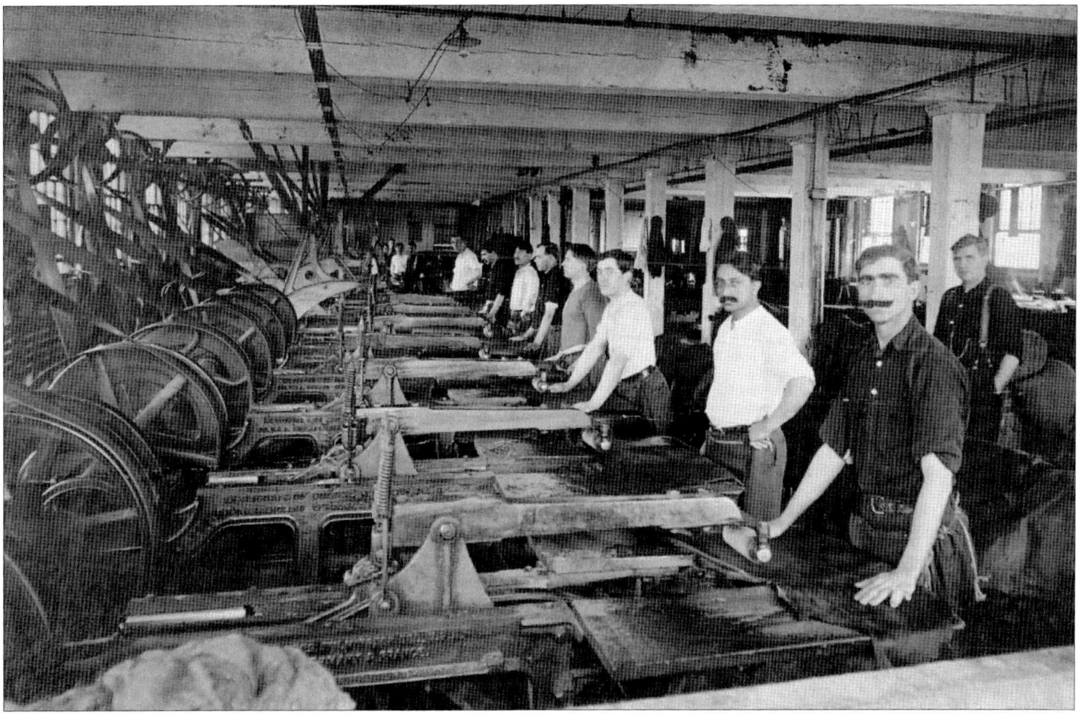

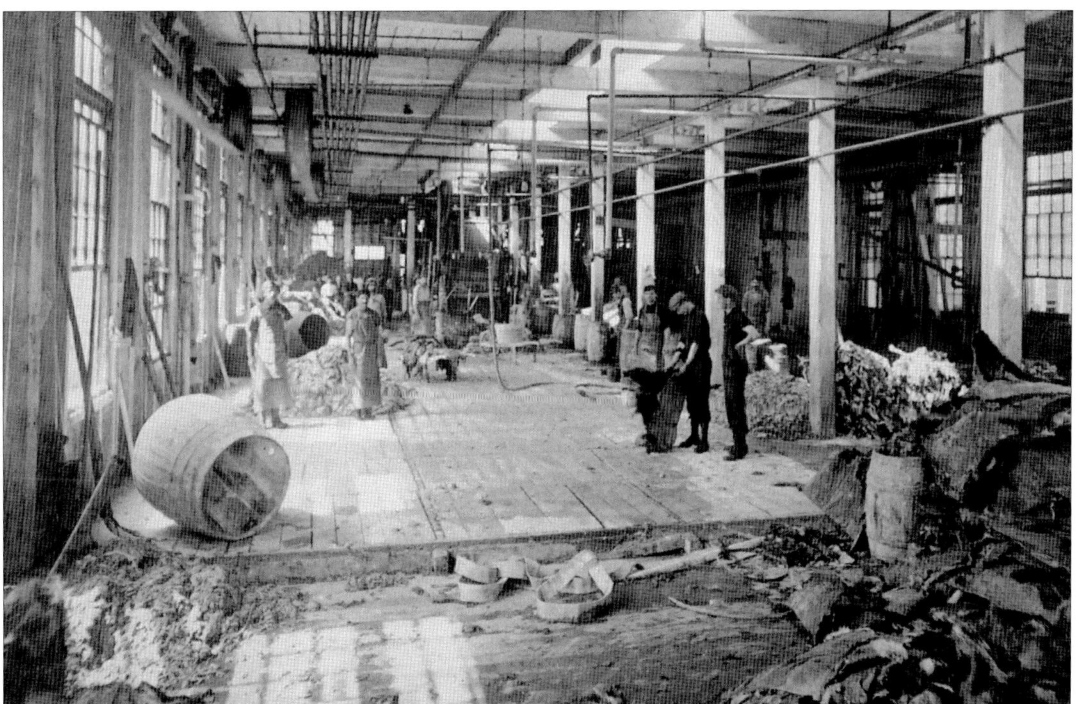

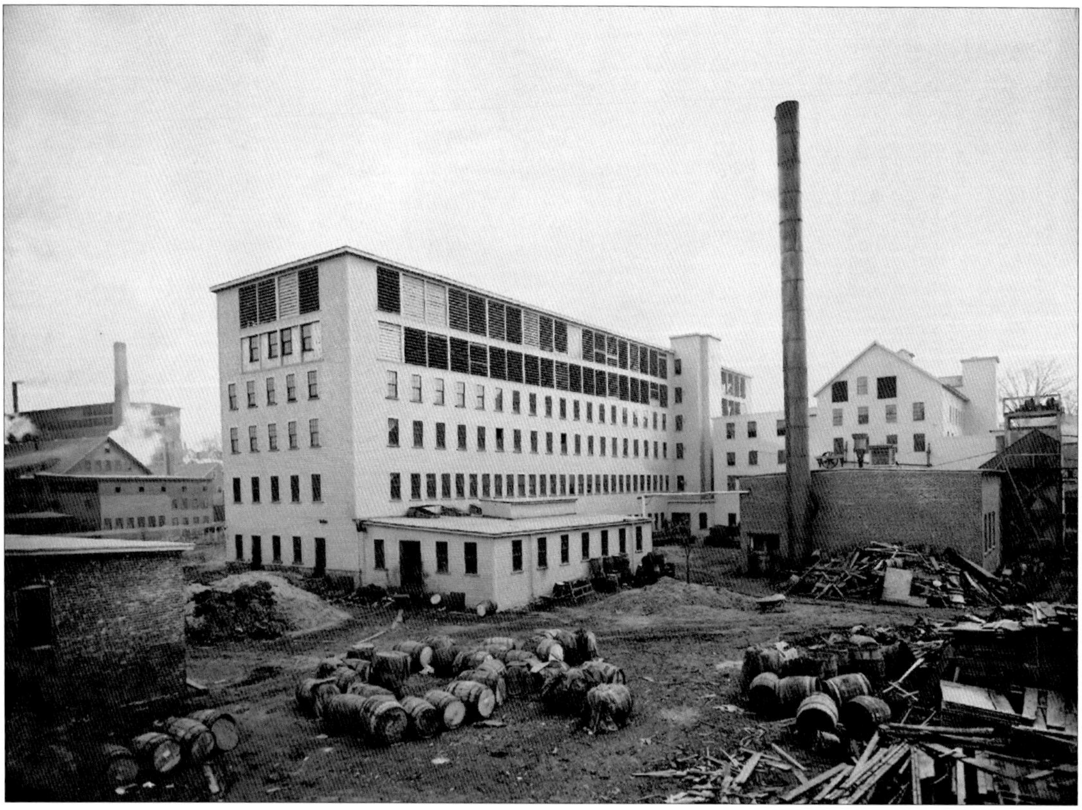

These three photographs show another tannery and a currier business. The top images on both pages depict the front and rear of 23 Boston Street, the home of the Arthur T. Way Leather Company, one of Salem's oldest tanneries. Despite a sprinkler system and a fire wall, the tannery was totally destroyed during the fire. There was a separate shed on the property in which greases and solvents were stored, which exploded during the fire, spreading flames through the entire tannery. (*Neg #8066, this page; Neg #8362, next page*) The bottom photo on the next page, photographed in 1876 by Moulton, shows the rear of 48 Boston Street, the home of G.W. and J.F. Varney Curriers. The currying trade was a vital part of the leather industry. Currying is the process of stretching and finishing tanned leather, making it supple and strong for the use of a saddler or cobbler. (*Neg #7253*)

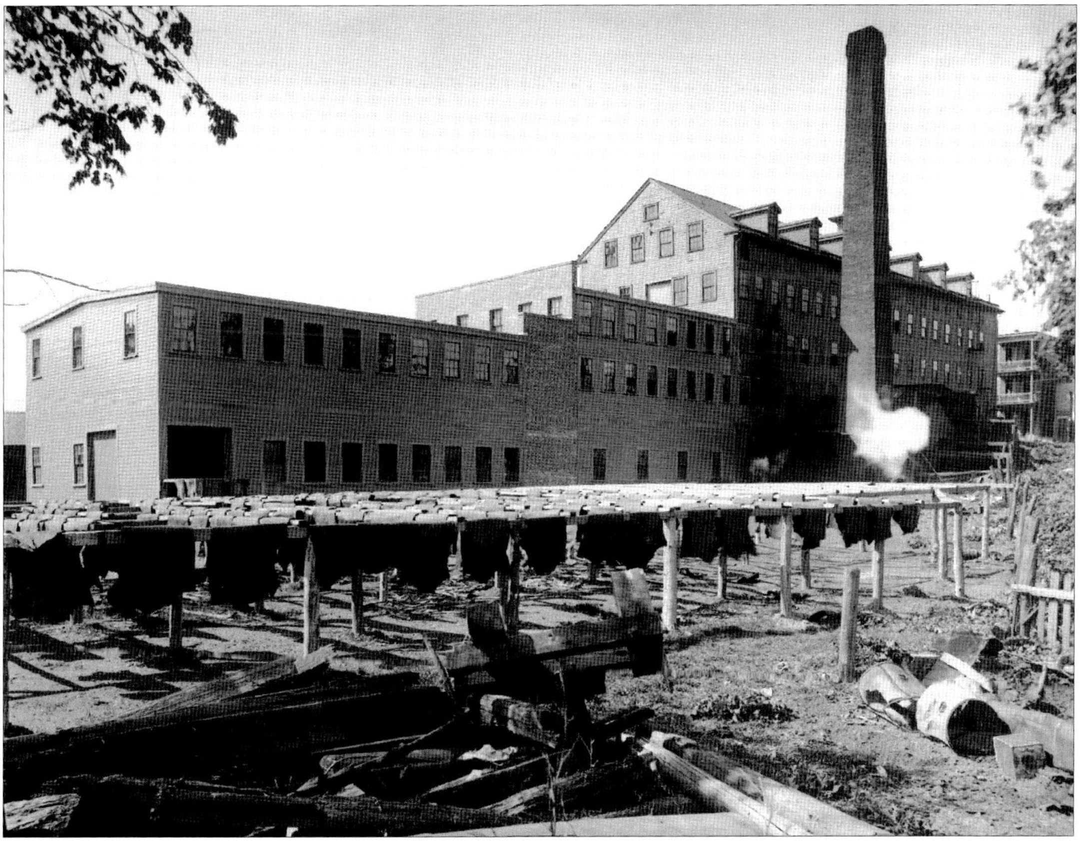

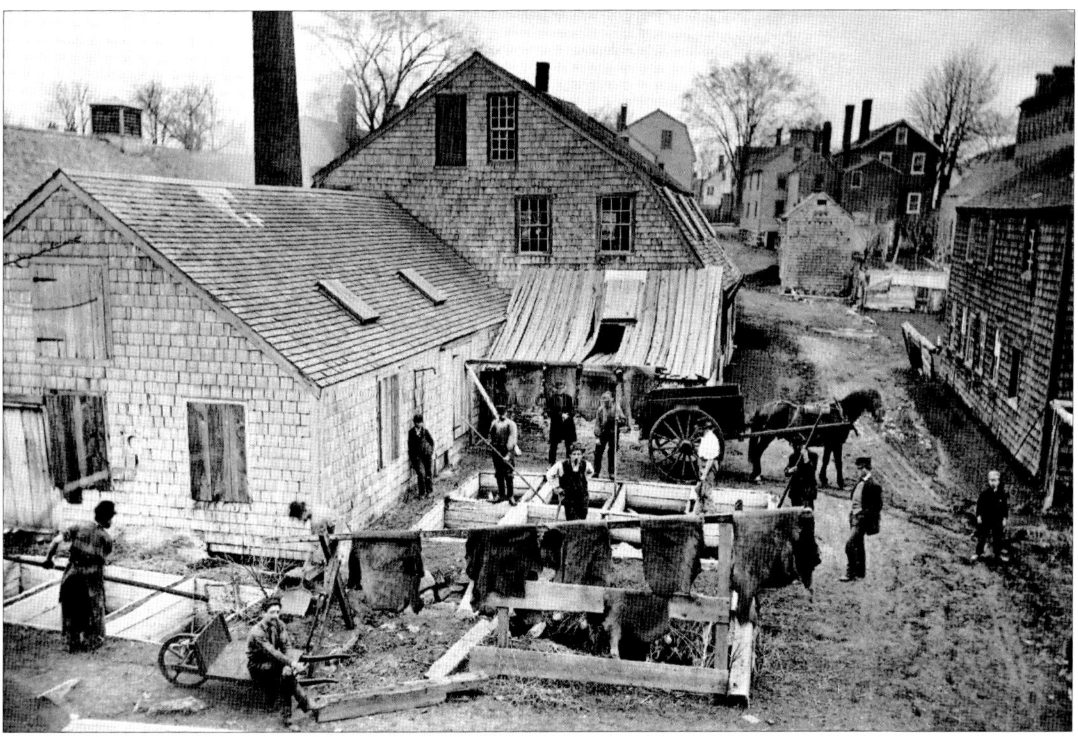

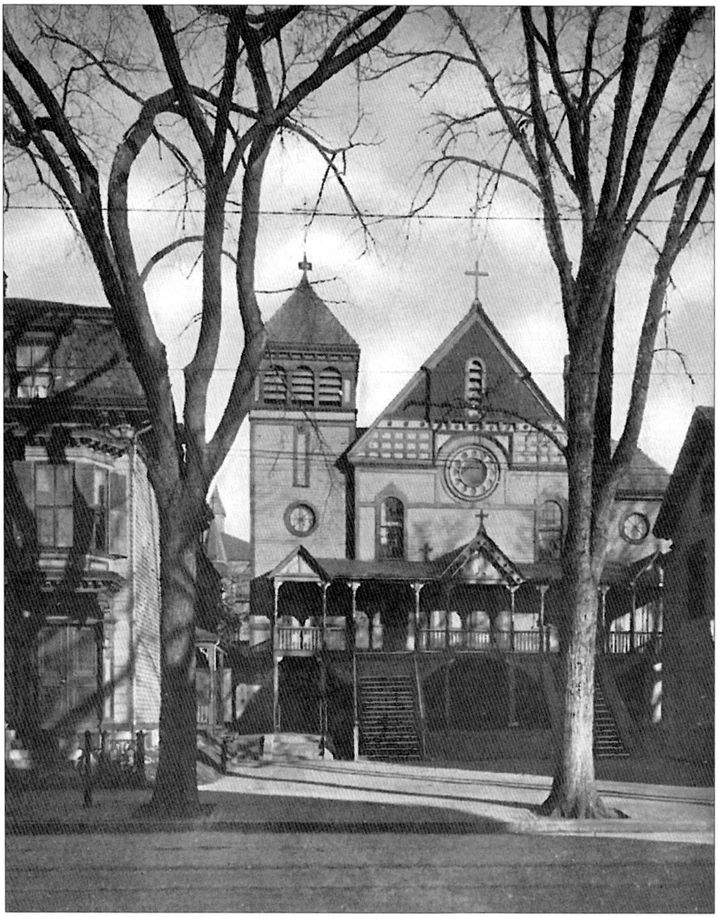

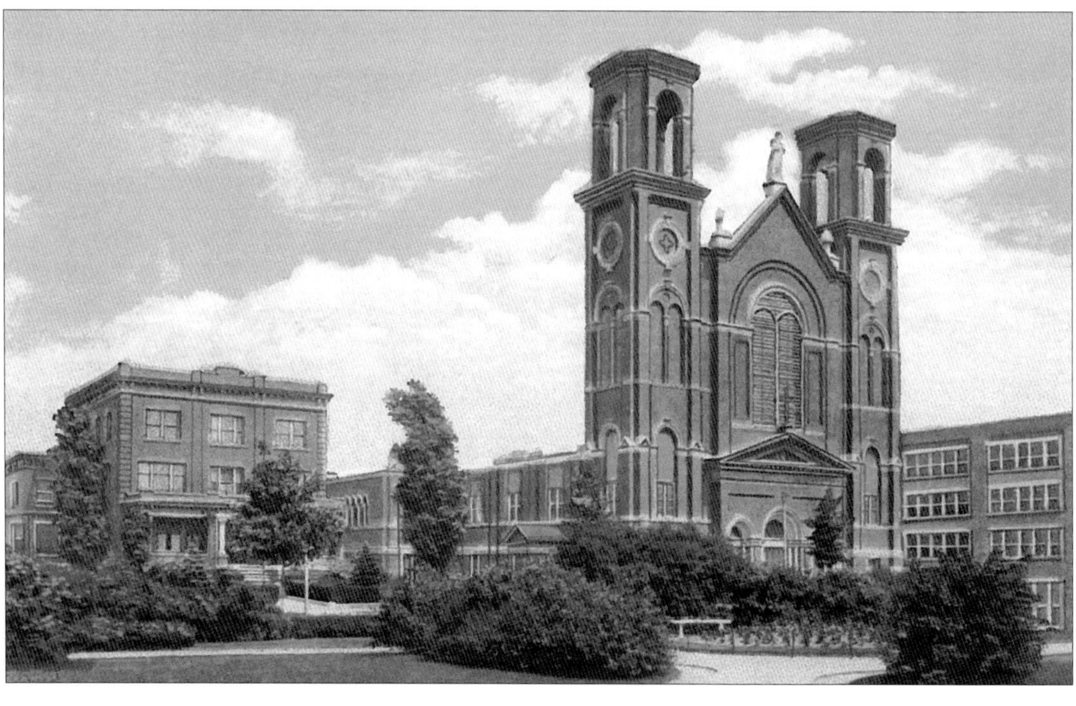

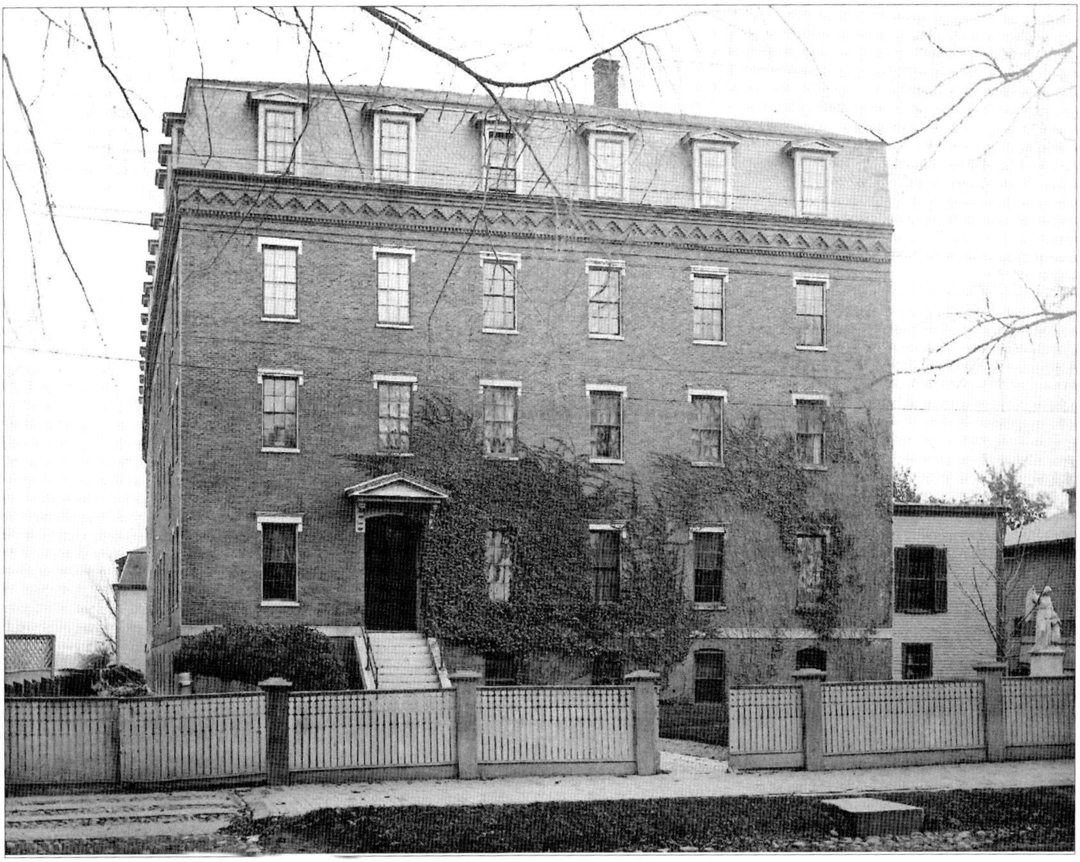

St. Joseph's French Church, 135 Lafayette Street, was built in 1884 and dedicated in 1885. Its parishioners were French Canadians who had immigrated to Salem to work in the Naumkeag Steam Cotton Company. The church housed the organ, which was originally in Mechanics Hall that was built by J.H. Wilcox and Co., and included a grotto dedicated to Our Lady of Lourdes in its basement. (*Neg #32863*) A larger brick church was built between 1910 and 1911 at 133 Lafayette Street, next to which was built the Parish Hall. Two large towers stand guard around the statue of St. Joseph. The church was dedicated on September 7, 1913, one year prior to the fire that severely damaged the church. (*Neg #32864*) The City Orphan Asylum, 215 Lafayette Street, was built in 1875. Before the fire, the Orphan Asylum was home to 22 aged and invalid adults and 125 children cared for by the Sisters of the Grey Nuns, originally from Montreal, and later known as the Sisters of Charity. (*Neg #32847*)

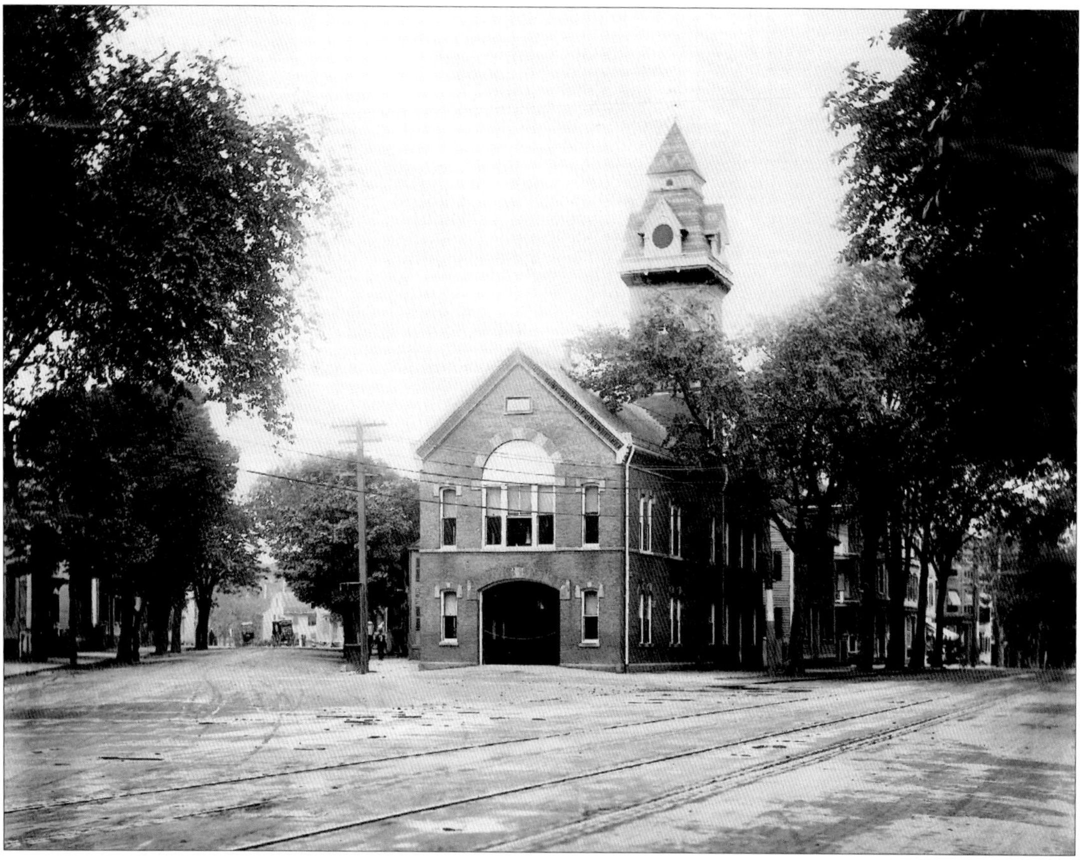

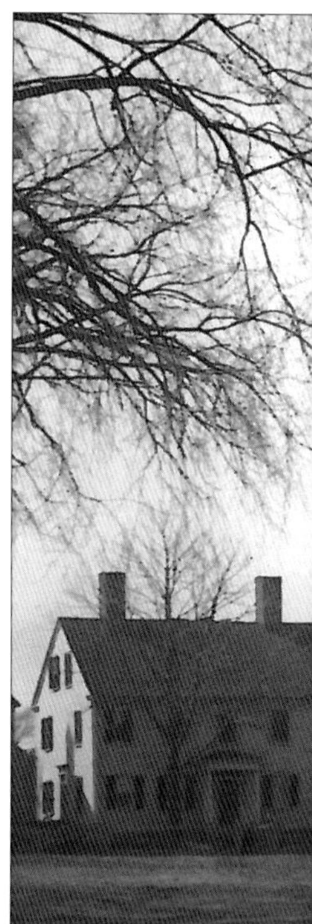

Engine House No. 1, home to Steamer One, was built twenty-five years before the fire at the junction of Washington and Lafayette Streets. Some of the residents of Lafayette Street brought their possessions to the firehouse for safekeeping only to have the firehouse and these possessions destroyed by the conflagration. (*Neg #8128*) The Salem Electric Light Company was constructed in 1889 on Peabody Street. The electrical station survived the fire and was able to produce electricity immediately after the fire on June 26. (*Neg #32943*) The houses in this neighborhood, on the north side of Broad Street, as viewed from Flint Street, are typical of the wooden structures that burned so easily in the fire. Embers from neighboring fires landed on wooden roof shingles, igniting the homes. According to the 1895 edition of the *Visitor's Guide to Salem*, the dwellings of the early Salemites "were plain and prim. The huge chimney stack…and close proximity of these old houses…was the natural imitation of the thickly settled English towns from which they came." The proximity of the houses made it easy for the fire to spread quickly within and between neighborhoods. (*Neg #5158*)

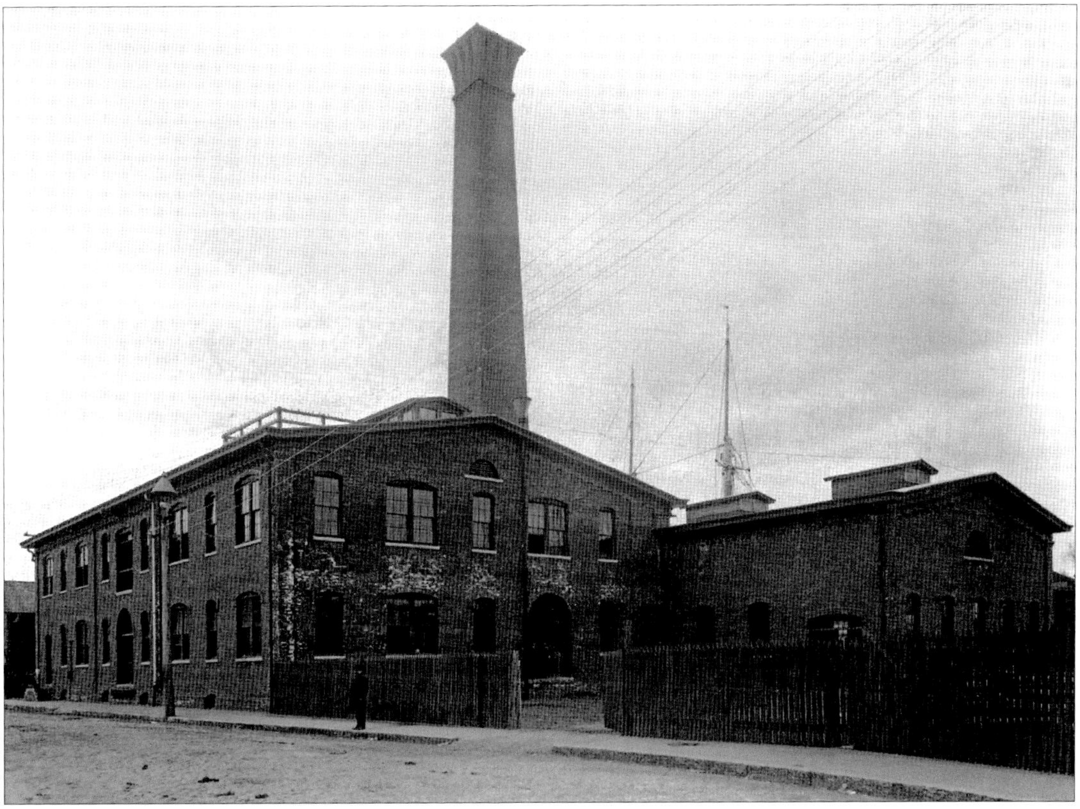

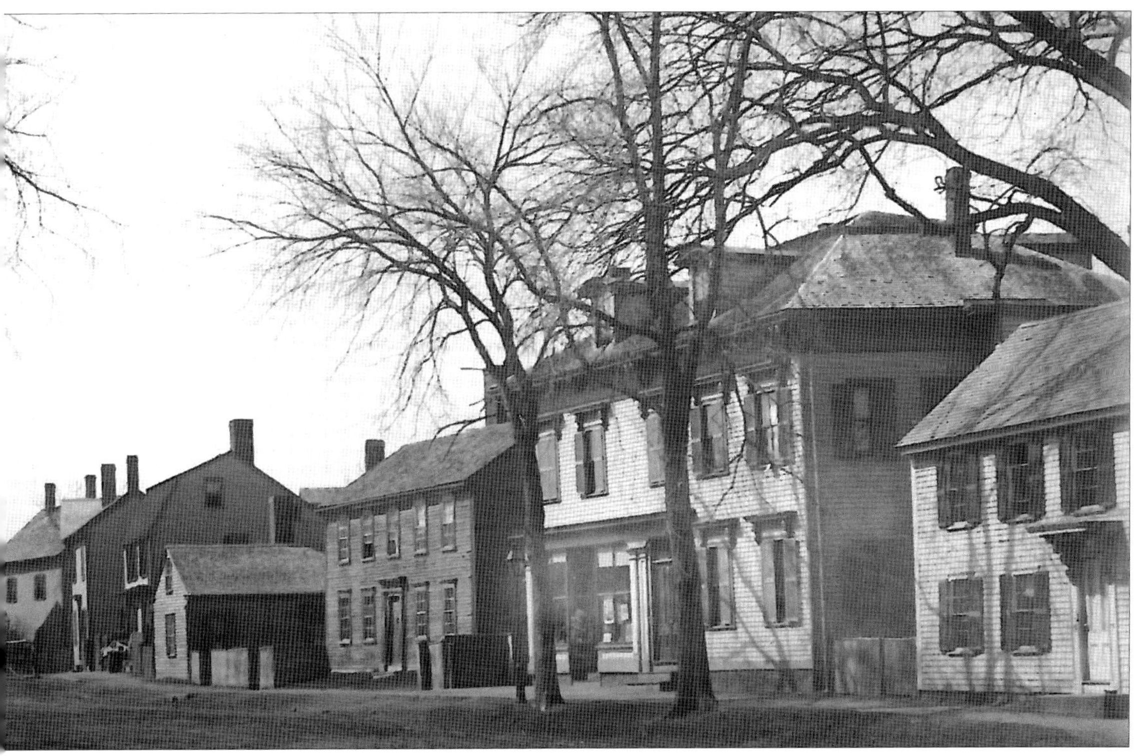

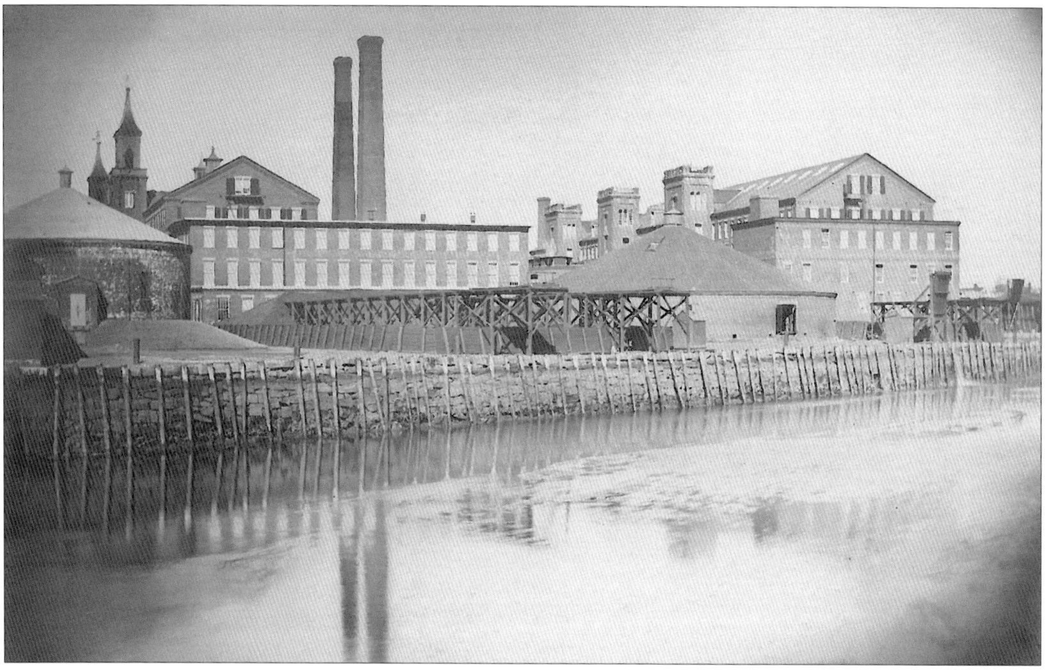

The Naumkeag Steam Cotton Company was one of the oldest industries in Salem. Founded in 1839 by Nathaniel Griffin, a retired shipmaster of Salem, the mills were in the Stage Point section of town where the South River emptied into the harbor. The mill was built in 1845 by Charles James and production of Pequot sheetings began in 1847. Griffin capitalized on the fact that Salem was a port city to which vessels brought supplies of cotton and coal, enabling the mills to use steam power, a distinction from the other textile mills in New England, which were run by water power. The flooding and ebbing tides of the waterfront also supplied the necessary humidity to control the static electricity during the manufacture of cotton yard goods. (*Neg #32850, above, photographed by Lefavour; Neg #32861, below*)

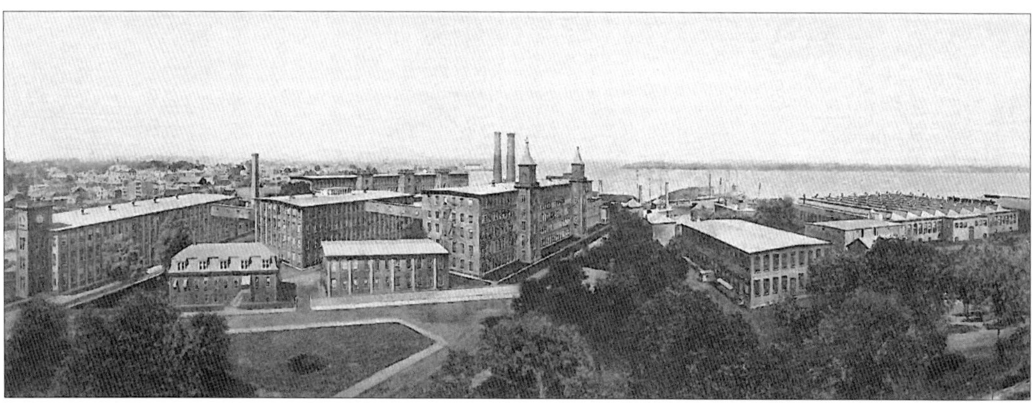

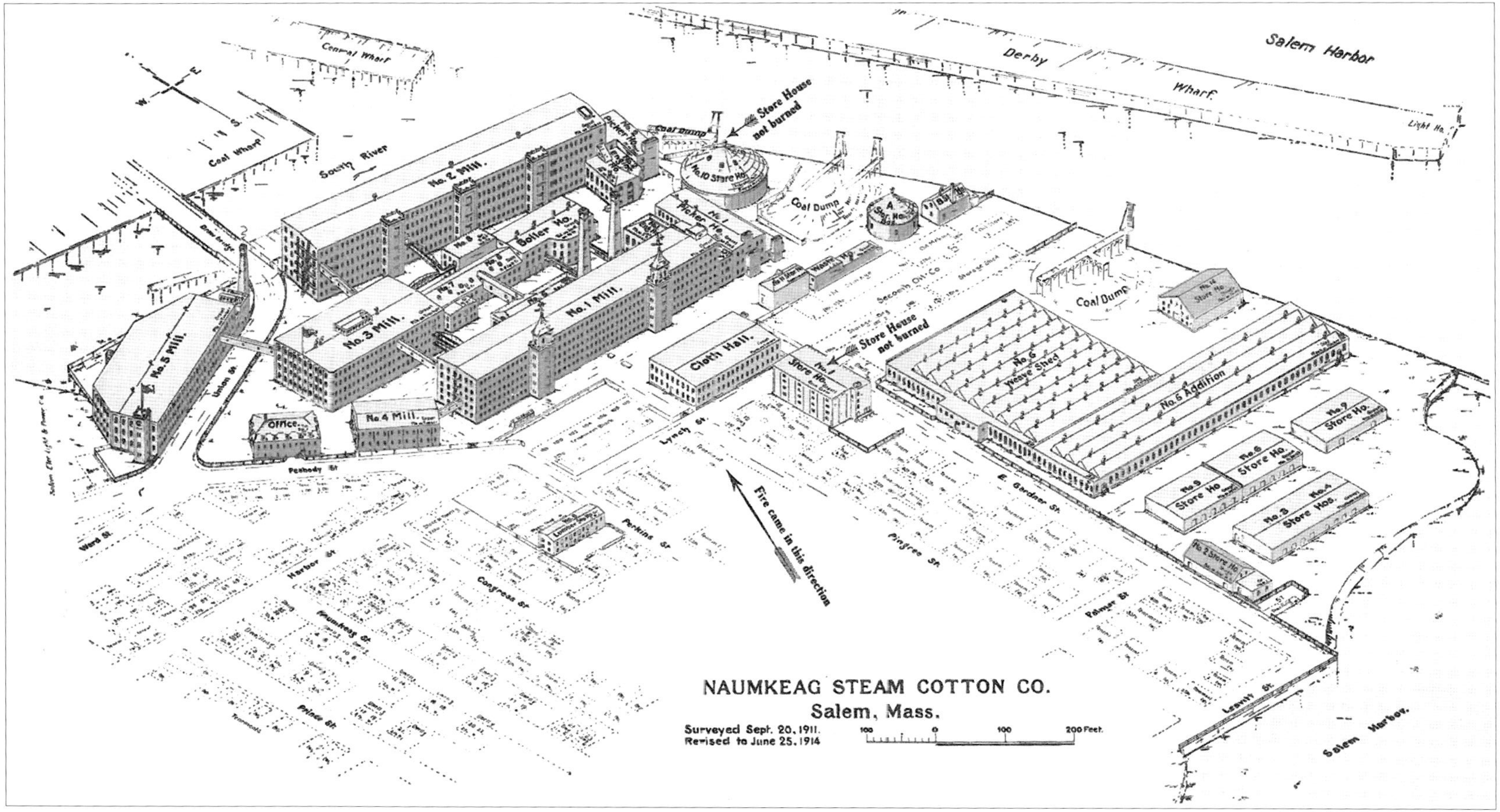

Of the buildings depicted on the map, only two survived the fire due to their fireproof construction—the round concrete structure, Storehouse No. 10, originally known as the Gasometer, located directly on the harbor; and the four-story Storehouse No. 1 on East Gardner Street. This map was published in the same document as the map on the frontispiece. (*Neg #32831*)

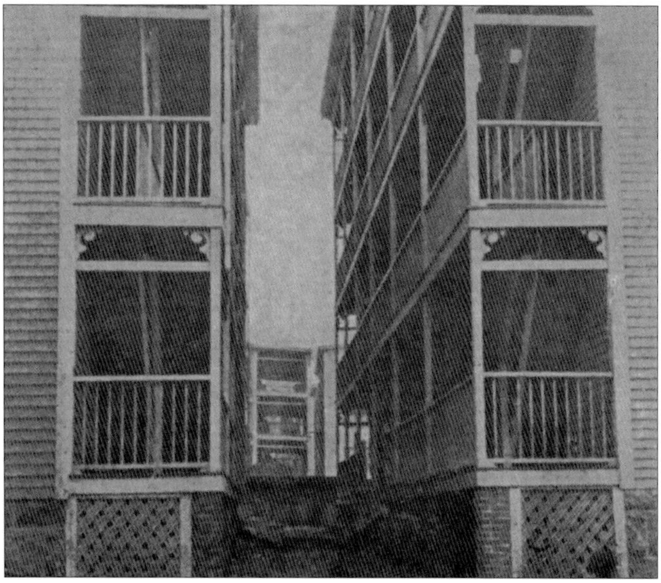

Both of these images were taken in Ward 5, referred to as the Point, which consisted of multifamily tenements, the home to many French Canadians who worked at the Naumkeag Steam Cotton Company. The wooden structures were built close together with backyards and front yards shared by residents. A news story after the fire mentioned that the four-decker homes had been crowded in so thickly that residents could exchange goods from window to window. Such proximity increased the intensity of the heat from the fire, fueling its path of destruction through the area. The Salem Rebuilding Commission printed a report in 1917 that stated that before the fire the city "was fast on the way to become a city of ugly and unsafe wooden three and four deckers." (*Neg #32902, above; Neg #32897, below*)

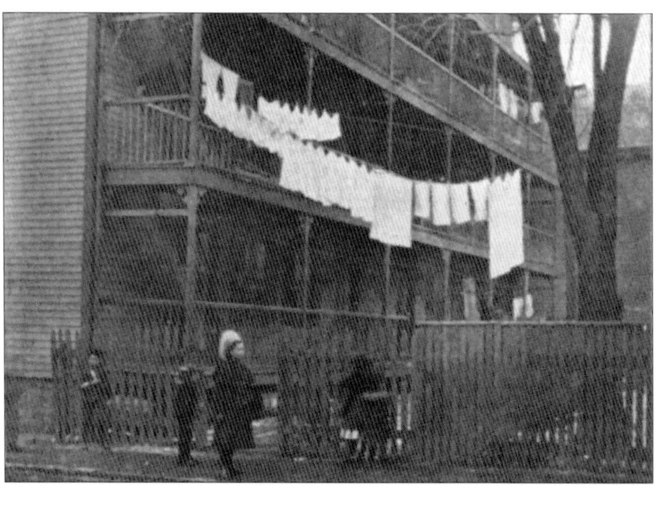

Chapter 2

The Conflagration

While the dictionary defines a conflagration as a large, destructive fire, the National Fire Protection Association applies the term to "sweeping fires that spread beyond control to destroy cities or large areas of built up property." The criteria to distinguish a conflagration from a large-loss fire include the "destruction of considerable numbers of buildings under a diversity of ownership, spread across streets or open spaces from the area of origin to involve other areas, and magnitude beyond the capacity of available fire fighting facilities." Three characteristics distinguish a conflagration: fires originating in hazardous occupancies or fire breeders in congested areas that spread to adjacent properties due to lack of exposure protection; fires occurring in residential areas that spread beyond control due to combustible construction, wood-shingle roofs and inadequate protection facilities; and fires following destructive explosions. The Salem Fire of 1914 resulted from all three characteristics.

Conflagrations spread because they create thermal updrafts. Heated air and gases rise and air is drawn in from the side of the burning area so that a high wind is often reported by observers while the air appears relatively calm a few miles away. The thermal updrafts carry pieces of burning material aloft in several directions, starting additional fires. Conflagrations also spread horizontally through heat waves that raise the temperature of exposed buildings to their ignition point before fire reaches them directly. This fact alone makes exposure protection of utmost importance when one fights fires; firemen find that when the full force of their attention should be focused on a main fire, a large number of fires start over a wide area, causing fire protection to be overtaxed and the chance of preventing a conflagration is gone.

Several years before the fire, Franklin Wentworth, a Salem councilman, tried to introduce a new building ordinance to require noncombustible roofs on all structures. It was felt that his suggestions would only serve the interests of the insurance industry. Wentworth went on to become the secretary of the National Board of Fire Underwriters. After the fire, he submitted a report about the Salem fire indicating that because of the Chelsea fire several years earlier, Salem had had time to know the danger and protect itself from it. "She [Salem] assumed no responsibility to protect the lives and livelihood of her common people, and her property-owning class was content to run its daily risk because it knew it could saddle its fire losses upon the country." According to Wentworth, the fire was slow burning and the wind was light; had that not been the case, the business district would have encountered problems as well, but because the wooden factory buildings without fire protection were strategically placed in the fire district, the fire was continually fed.

The second most frequent contributing factor to create a conflagration is an inadequate water distribution system. Systems like the one in Salem may be adequate for normal water distribution purposes, but they fall short of the demands of the large number of pumping engines required to fight a conflagration. The cement-lined water mains did not burst until after the fire had spent its full force. Knowledge that such a circumstance could happen prevented firefighters from using water from the mains; instead, water was used from steamers supplying their own water. Abundant water supplies from Peabody, Danvers and Beverly were tied into Salem but it was felt the extra pressure would increase the possibility of a water main break and these supplies were called in only when the water pressure dropped significantly.

According to Wentworth, modern fire departments are "agencies of conservation. They make inspections and remove the causes of fire. The Salem fire department [was] not modern. The Salem conflagration should never have spread beyond the Korn Factory." Monies spent on automatic fire protection and regular inspection for hazardous materials could have prevented the fire.

To prevent conflagrations in the future, Wentworth suggested the following: enforce city planning and zoning laws; segregate hazardous manufacturing operations; target hazards of oil and lumber storage; remove dilapidated structures from congested areas; widen streets, create parks to facilitate firefighting and serve as fire breaks; require automatic sprinklers in new buildings; and develop auxiliary sources of water supply with accessible locations for fire pumpers to suction the water and replenish machines. Many of these suggestions were incorporated in the efforts of the Salem Rebuilding Commission and in the establishment of a high-pressure water system in Salem a few years after the fire.

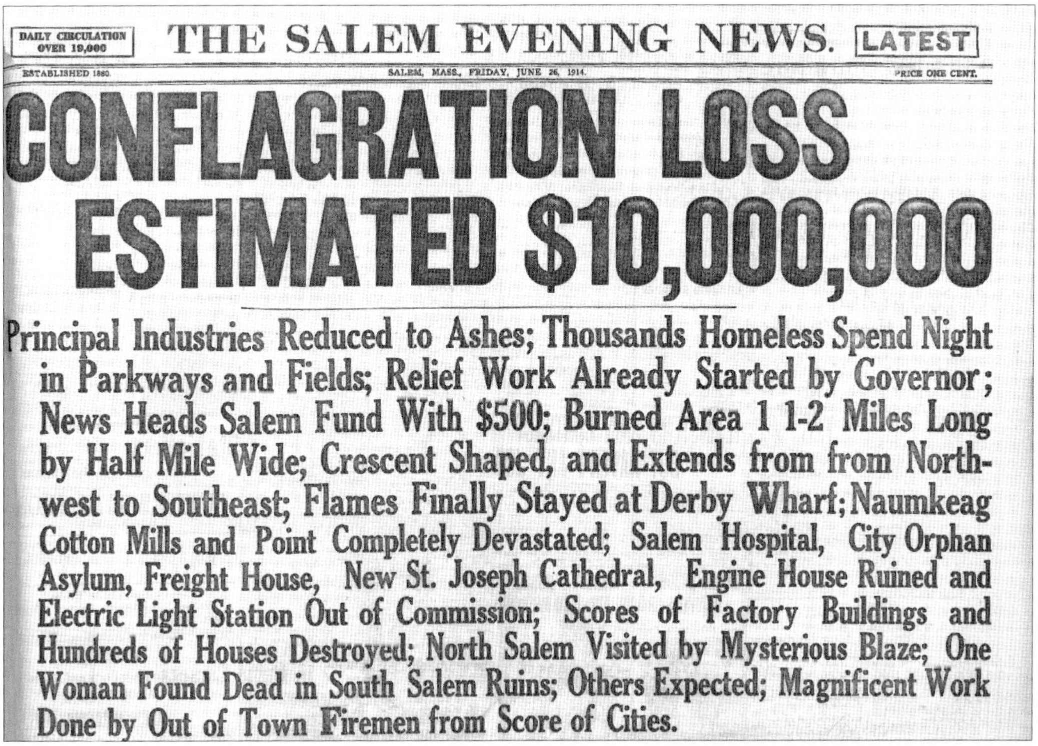

DAILY CIRCULATION OVER 19,000

THE SALEM EVENING NEWS. LATEST

ESTABLISHED 1880. SALEM, MASS., FRIDAY, JUNE 26, 1914. PRICE ONE CENT.

CONFLAGRATION LOSS ESTIMATED $10,000,000

Principal Industries Reduced to Ashes; Thousands Homeless Spend Night in Parkways and Fields; Relief Work Already Started by Governor; News Heads Salem Fund With $500; Burned Area 1 1-2 Miles Long by Half Mile Wide; Crescent Shaped, and Extends from from Northwest to Southeast; Flames Finally Stayed at Derby Wharf; Naumkeag Cotton Mills and Point Completely Devastated; Salem Hospital, City Orphan Asylum, Freight House, New St. Joseph Cathedral, Engine House Ruined and Electric Light Station Out of Commission; Scores of Factory Buildings and Hundreds of Houses Destroyed; North Salem Visited by Mysterious Blaze; One Woman Found Dead in South Salem Ruins; Others Expected; Magnificent Work Done by Out of Town Firemen from Score of Cities.

The headline from *The Salem Evening News* on Friday, June 26, reported the extent of the devastation of the fire and its impact on the residents of Salem, and it acknowledged the help from neighboring fire departments. The height of the blaze, photographed by the Tichnor Brothers of Boston, depicts the inferno as seen from the Salem High School at 2:30 p.m. on the afternoon of June 25, 1914. (*Neg #32884, above; Neg #32858, below*)

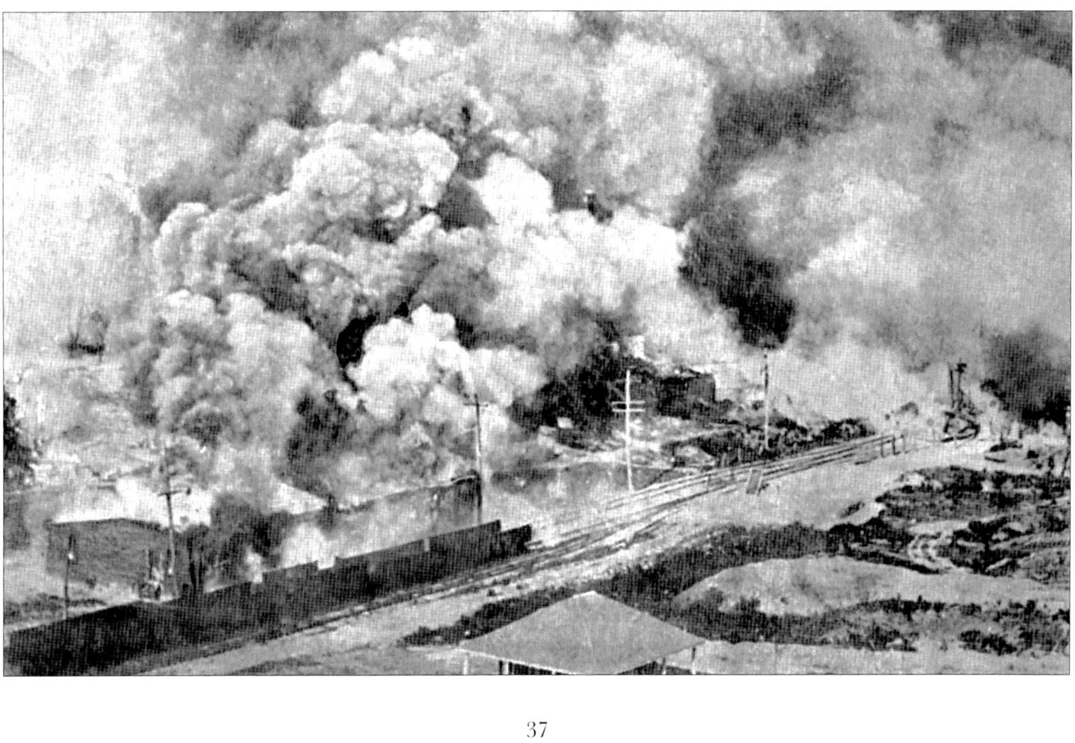

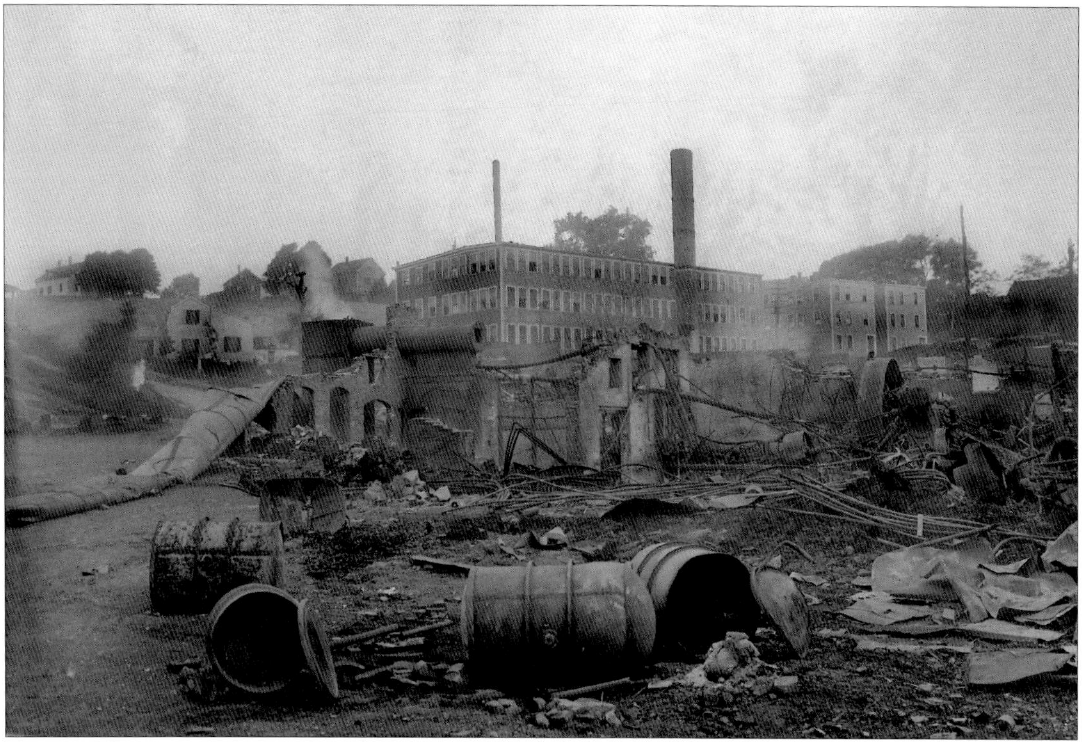

The fire started in an unsprinklered leather factory, the Korn Leather Company, one of the buildings owned by the Creedon Building Association. The plant was at 55–65 Boston Street, at the corner of Proctor Street in the Blubber Hollow district of Salem, the northwest section of the city. A series of explosions occurred in the shed on the property, which housed barrels of inflammable chemicals. The Salem Tip Finish Company occupied the top floor of the Korn factory. Chemicals used in the compound to give the patent leather finish to shoe tips exploded just after the start of the fire. According to *The Salem Evening News*, celluloid was one of the chemicals involved in these explosions; the same chemical was in a box consigned to the Korn Leather Company at the Merritt Express Stable on Barr Street, across the North River, in Ward 2. This box exploded later that evening, starting another fire, which was less destructive but destroyed several buildings plus the stable and the horses inside. Charles Haywood's book *General Alarm* accredits the assistance of the steamers and hose wagons from Lawrence with the success of battling the North Salem fire; their equipment, one of which used a deck gun, capable of receiving three lines that merged into one powerful stream, carried water to every part of a burning structure, successfully stopping the fire. On July 2, *The Salem Evening News* reported that Max Korn, the president of the leather company, had received several threats because people held him responsible for the fire; he rebuilt his factory in Peabody rather than in Salem. (*Neg #32787*)

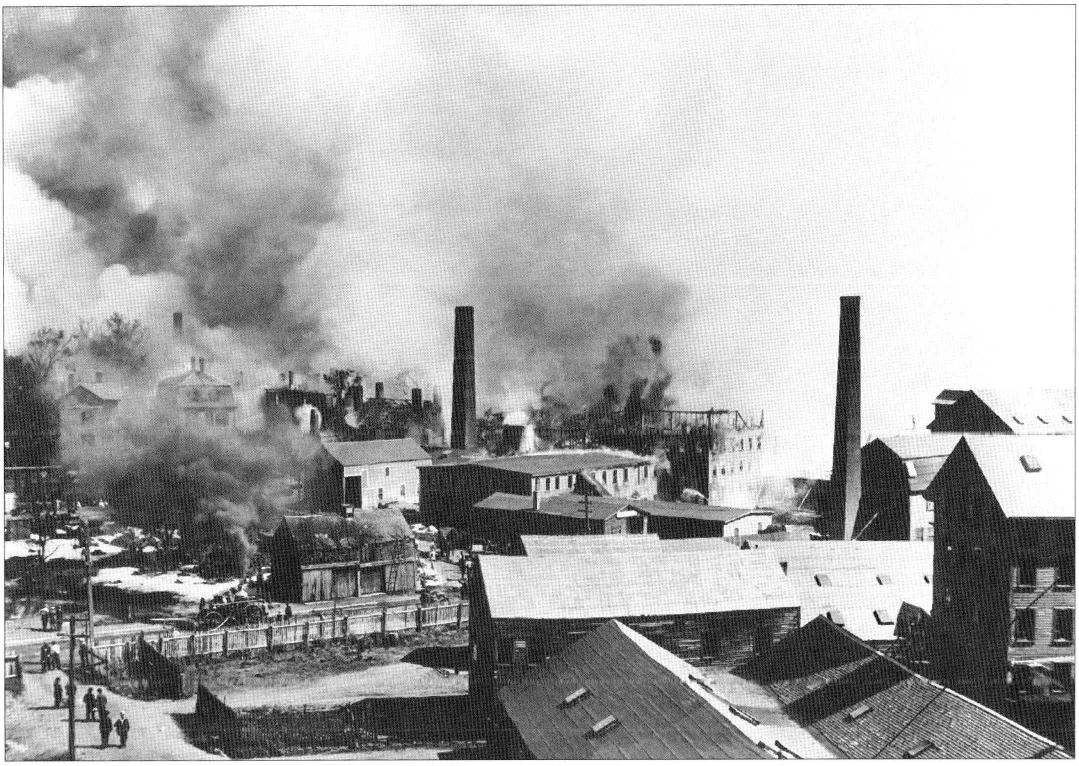

The fire spread quickly across the street to Bridge Street near the corner of Boston and Goodhue Streets. (*Neg #15155, above*) The leather tanning industry in Blubber Hollow consisted of multistory buildings, many of which had sprinklers. However, since many of them were of wood-frame construction, the sprinklers were not sufficient to save the buildings. The leather district in Salem never fully recovered after the destruction of the fire—31 shoe and leather plants were damaged and 1,450 jobs in the industry were lost. (*Neg #32828, below*)

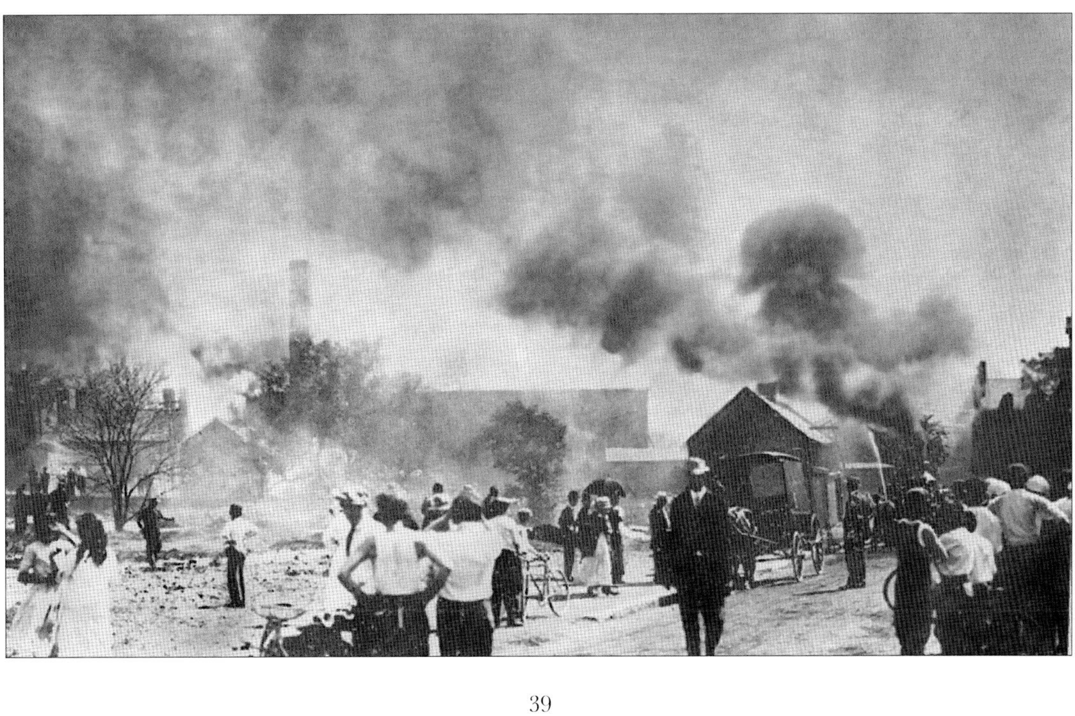

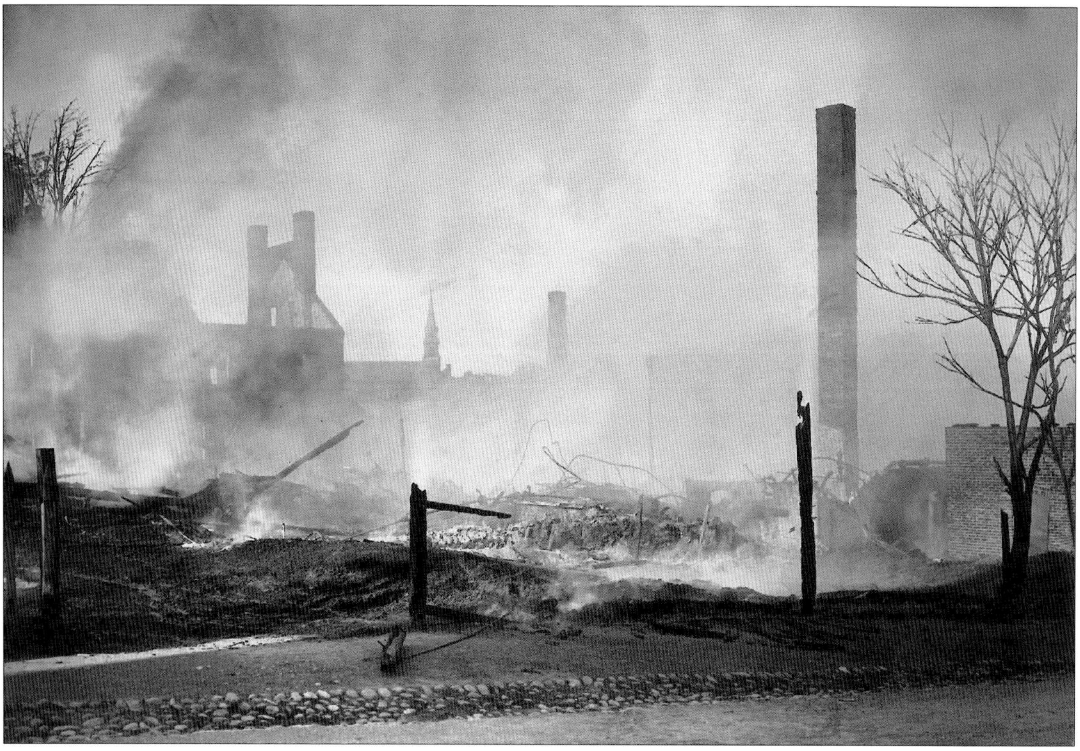

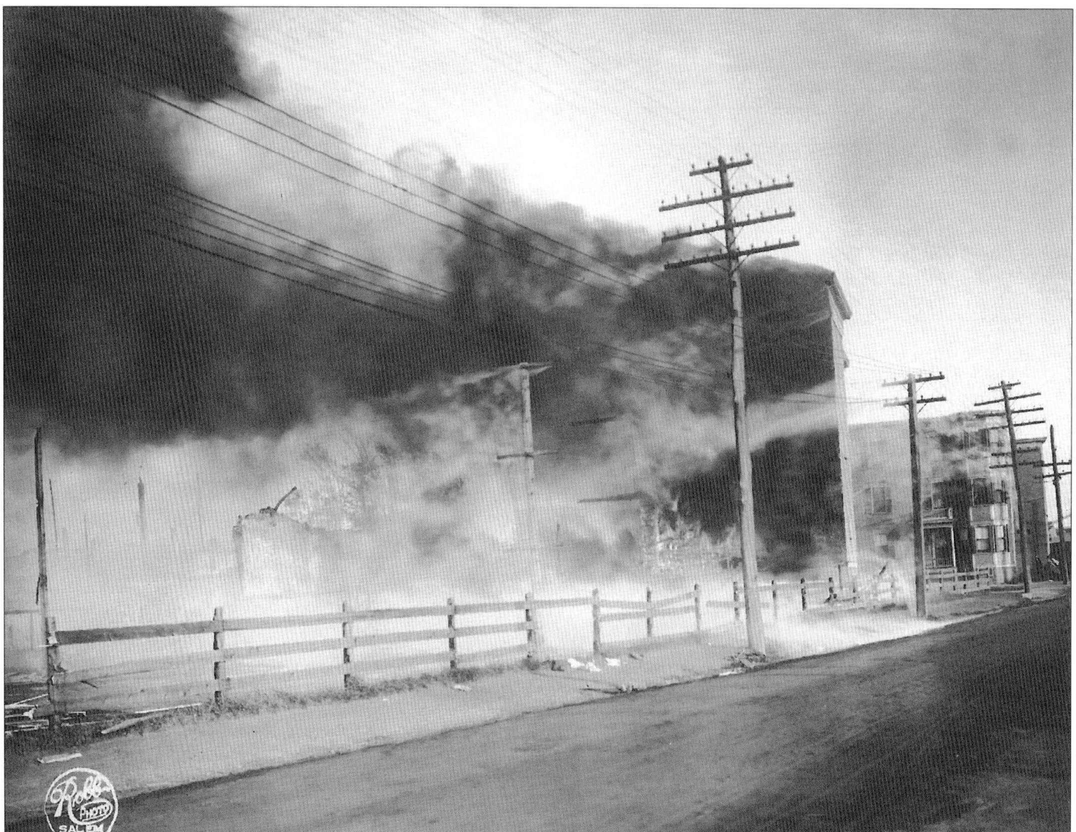
Robb
PHOTO
SALEM
MASS

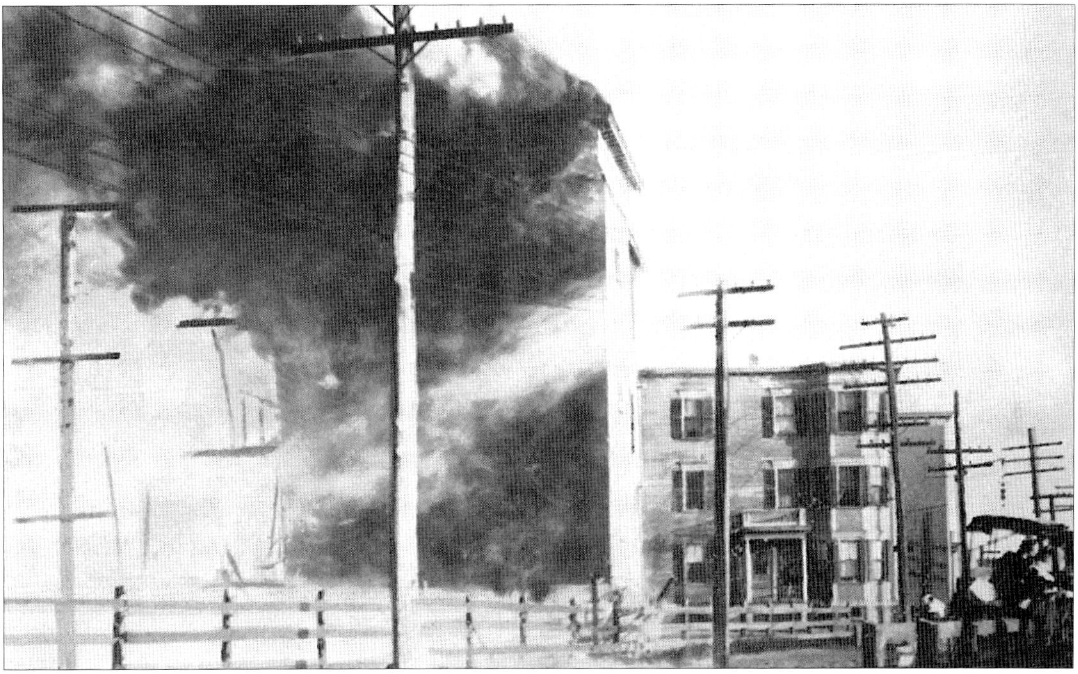

These three images were photographed by Malcolm E. Robb of Salem. The image on the top of the facing page (*Neg #15081*) depicts the blaze on Pope Street, 250 feet away from the Korn Leather Company, where the conflagration started. The two images of the three-decker homes on Canal Street show the progression of the fire after it crossed the Boston and Maine Railroad tracks. (*Neg #15154, previous page; Neg #32860, this page*)

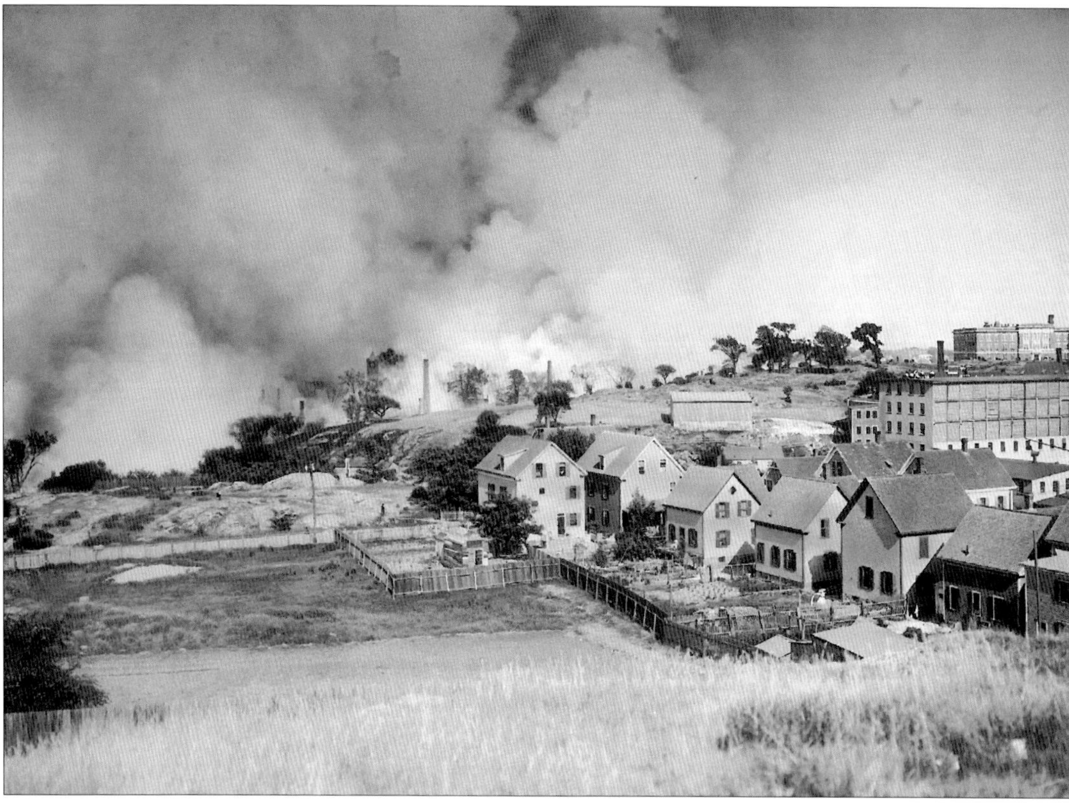

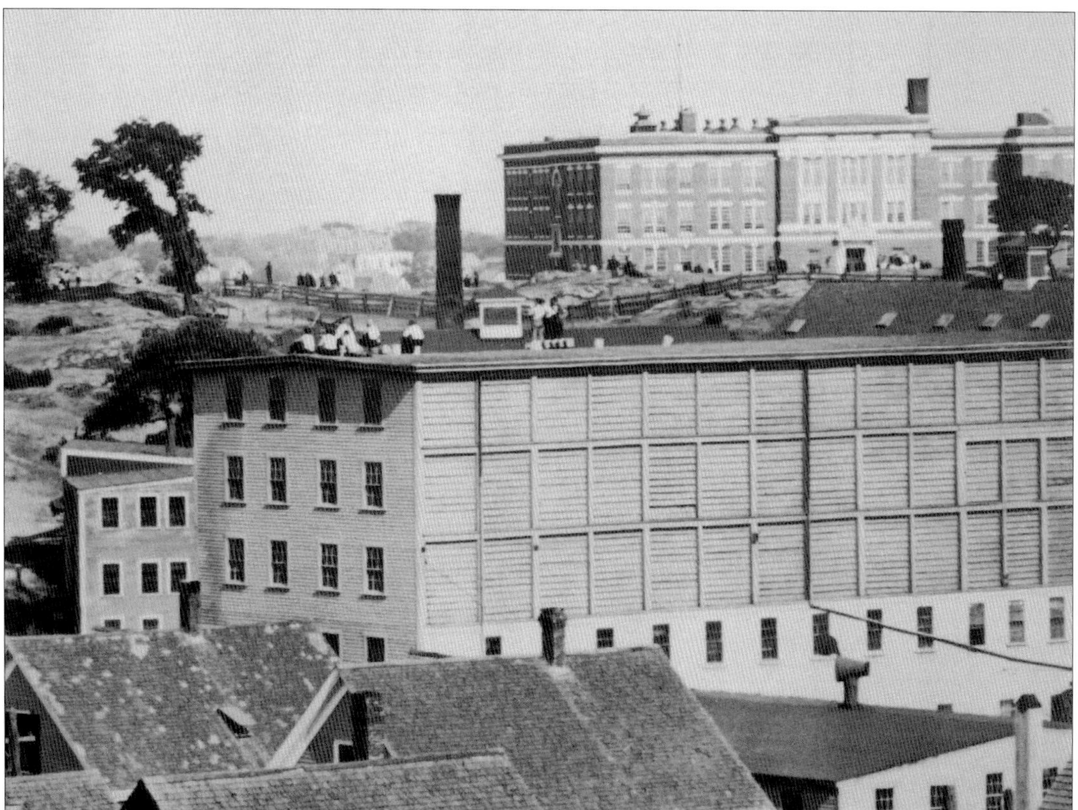

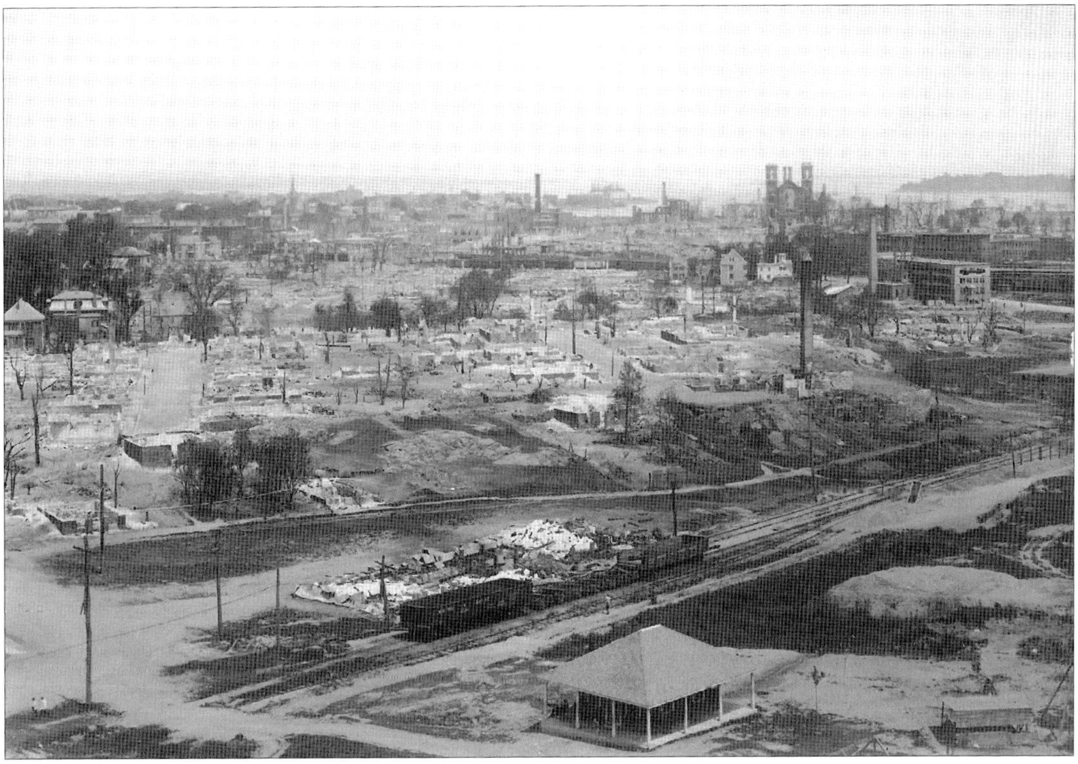

The top image on the facing page shows the view of the fire from Looney Hill near the newly built high school, which was built in 1909 on Highland Avenue; the high school is in the upper right corner of the image. The open landmass of the Broad Street Cemetery prevented the fire from reaching the high school, as did the ravine at the bottom of the hill, which provided a natural barrier to the fire on its western front. (*Neg #4019*) Close inspection of the Looney Hill image reveals Salem residents sitting on the roof of a manufacturing building near the high school. They were safe from the fire and could watch its spread as the afternoon and the destruction of the fire progressed. (*Neg #4019Detail*) The image on this page looks toward Lafayette Street across the Boston and Maine Railroad tracks and reveals the devastation created by the fire. The towers of St. Joseph's Church are in the distance in the upper left portion of the image. The P.A. Field Shoe Factory ruins are to the left of the church, and Jackson Street is in the foreground. (*Neg #4675*)

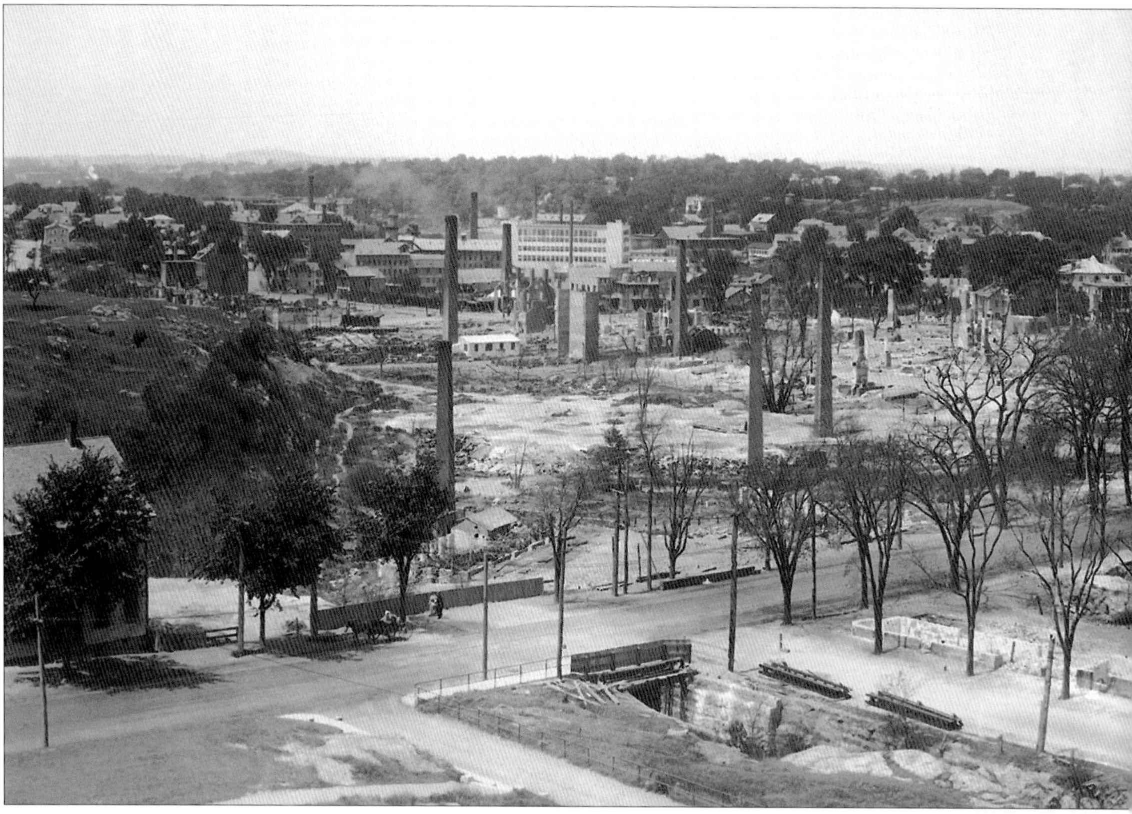

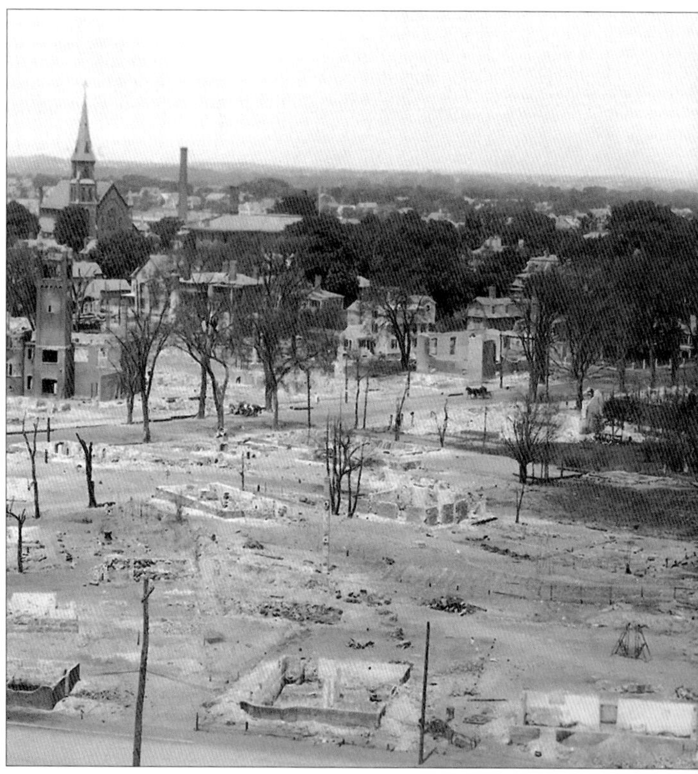

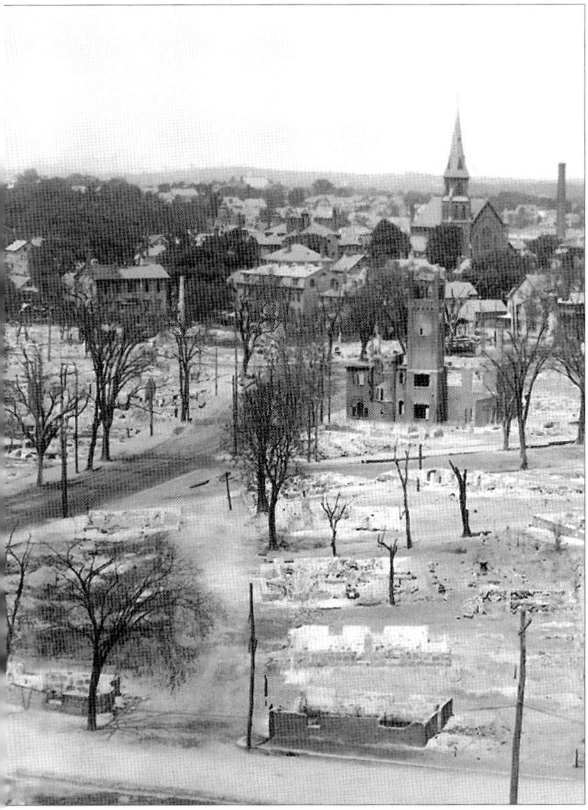

The split panoramic image shows the destruction to the city as seen from the high school after the fire had been contained. Reading left to right, Highland Avenue is in the foreground of the picture, with the Immaculate Conception Church and the ruins of the Essex Street Firehouse depicted in the right portion of the upper split of the panoramic view. The lower portion of the split panoramic view starts from the church and firehouse and reveals the scene looking north from the high school. (*Neg #32865a and Neg #32865b*)

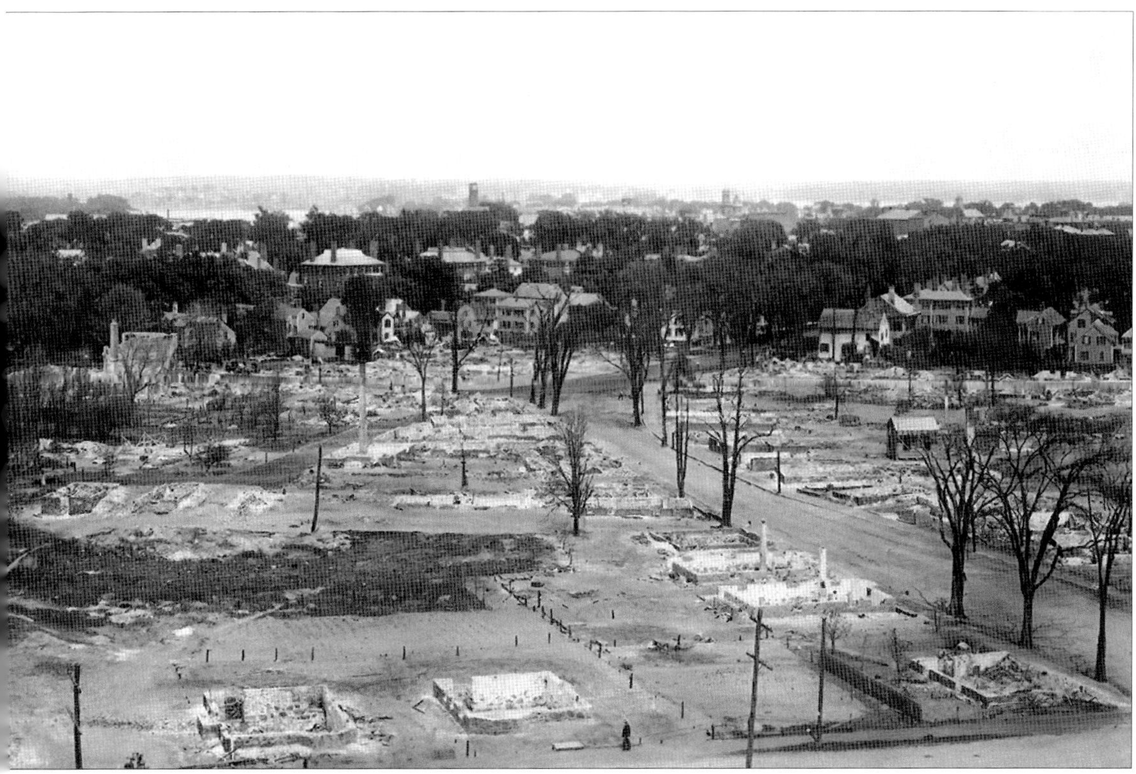

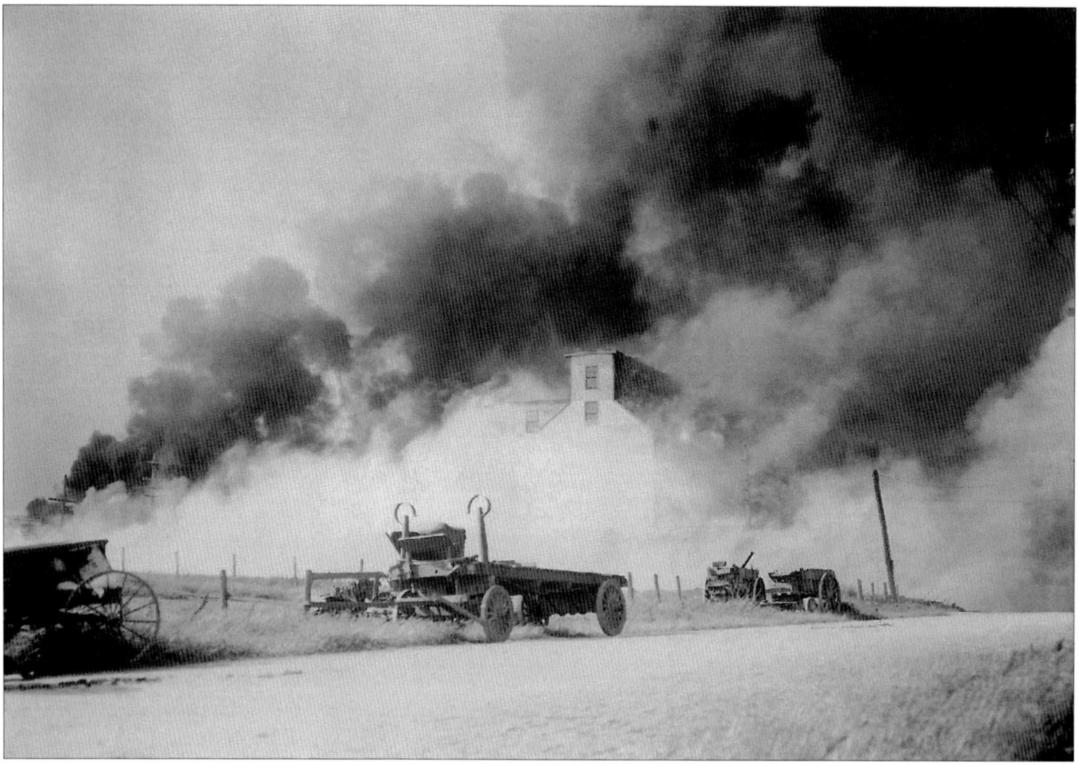

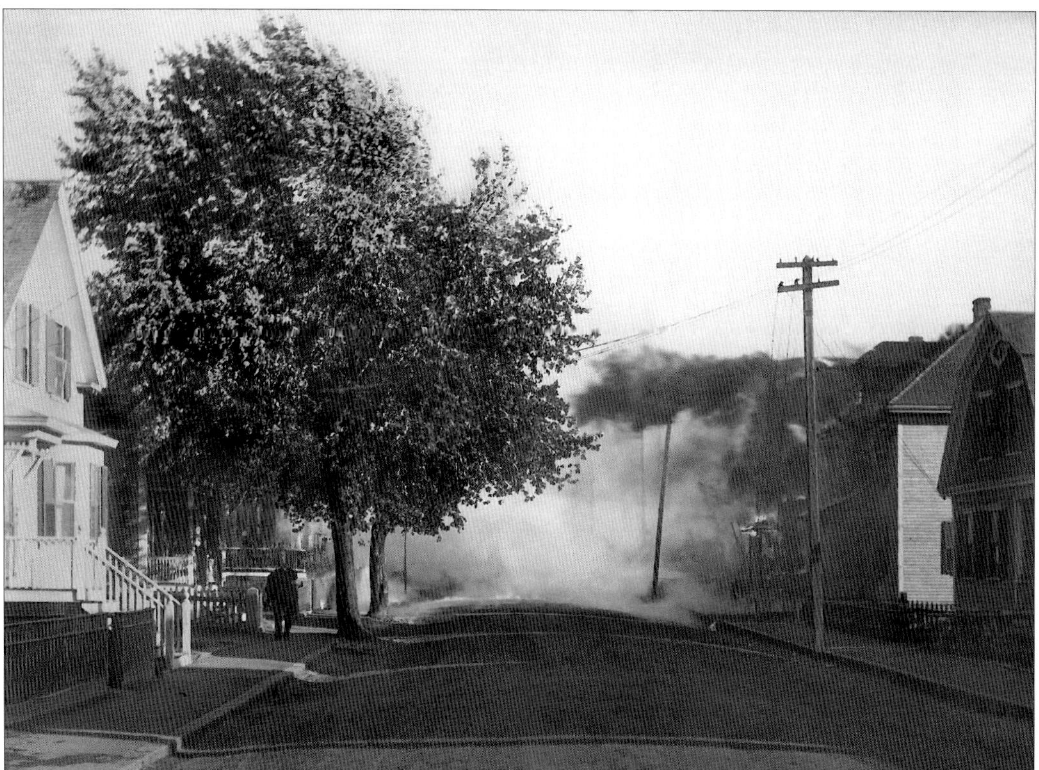

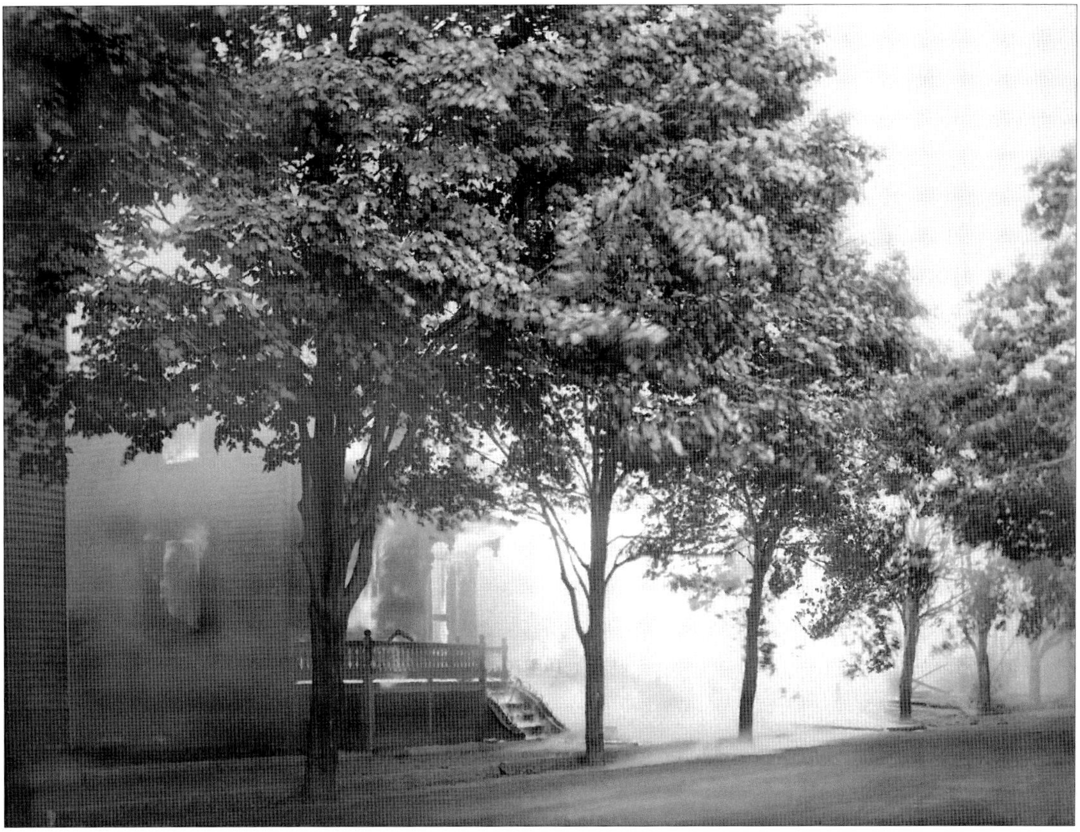

At the height of the fire, teamsters cut their horses loose to escape the impending flames and abandoned their vehicles to reach safety. (*Neg #32814*) Most descriptions of the fire discuss the destruction to the factories and homes in the path of the inferno. However, hundreds of trees were also destroyed. Warren J. Hale, superintendent of Parks and Shade Trees, reported that over 1,000 trees were replanted in the burned area; over 100 of these trees did not survive and more than 350 trees were planted than were burned. Including the cost of loam and tree guards, the cost of replacement averaged $6.53 per tree. Trees planted included maples, elms, pin oaks, sycamores, poplars and lindens. The image on the previous page shows Cabot Street as the fire progressed toward it. (*Neg #32796*) The image above shows the trees on Gardner Street before the fire. (*Neg #32813*)

Following pages: The Phillips Library collection has several photograph albums containing images of the fire. The illustrations on the next two pages are from the album compiled by William F. Abbott (1892–1980), a building contractor in Salem. Each of these images has been labeled by the photographer, identifying location and time of the actual photograph. (*Neg #32870 and Neg #32871*)

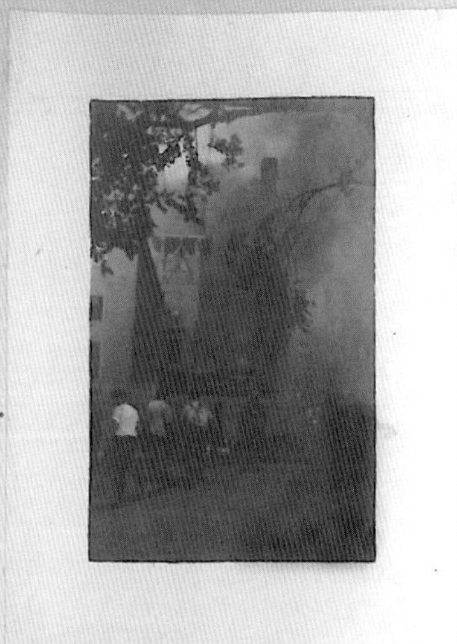

WILKINSON COUNTER CO
JEFFERSON AVE 4.15 PM

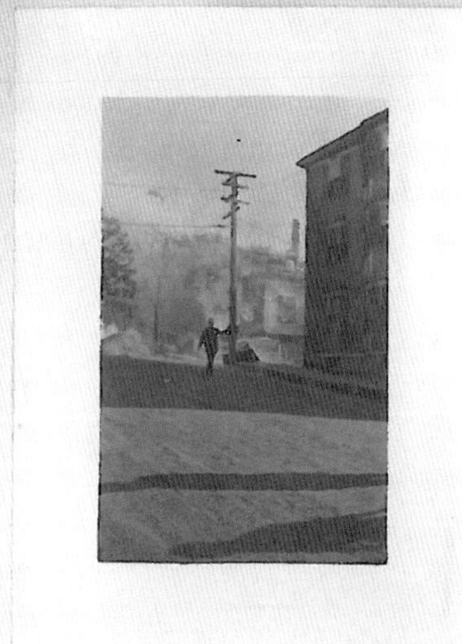

CABOT COR. ROSLYN 4.30 PM

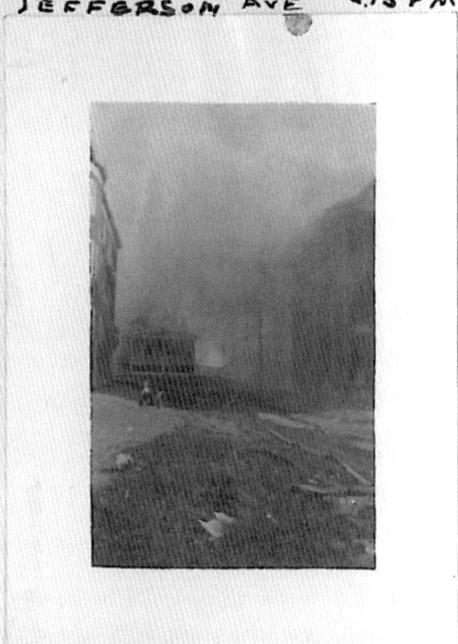

GENEVA ST. NEAR ROSLYN
4.30 PM

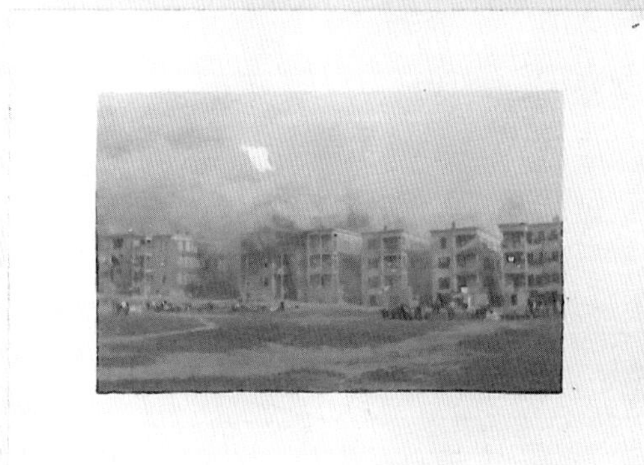

ROSLYN ST. TENEMENTS
4.30 P.M.

UPPER BROAD ST.
5 P.M.

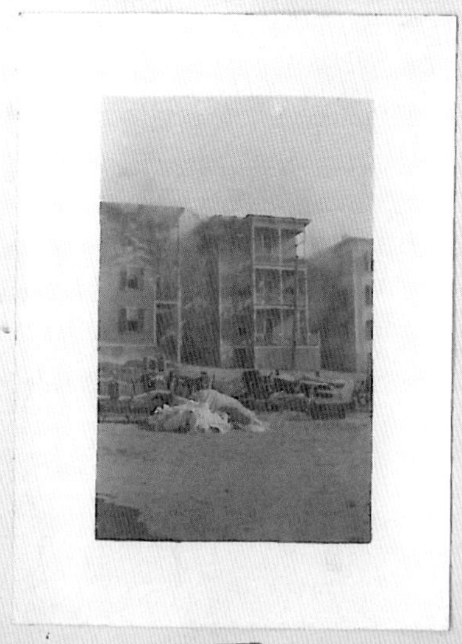

ROSLYN ST. TENEMENTS
4.30 PM

JUNE 25
1914

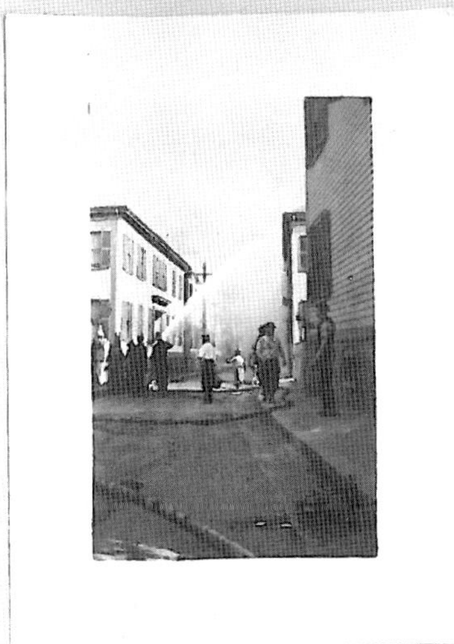

ORNE SQUARE
3.45 P.M.

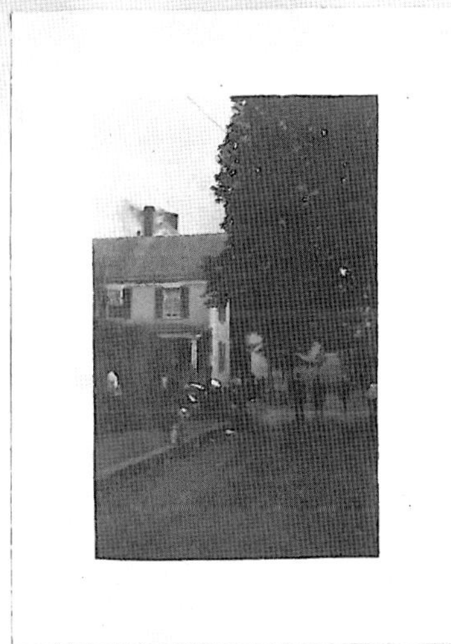

ENDICOTT NEAR SUMMER
4.20 PM

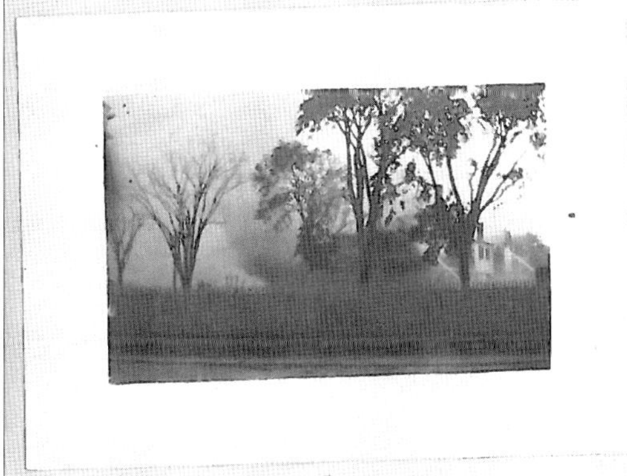

DOWNING ST
4.10 P.M.

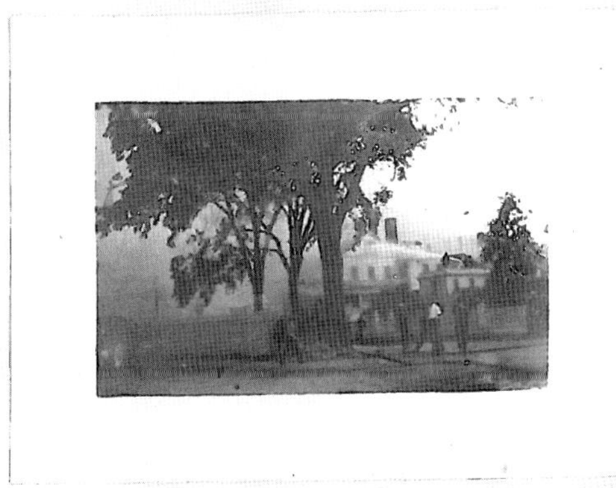

MARGIN
4.15 P.M.

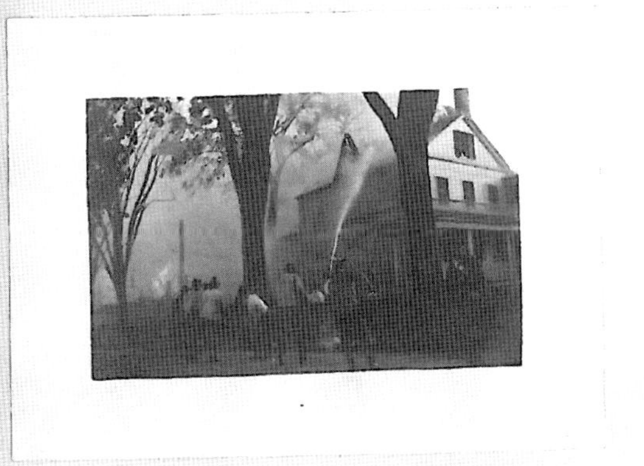

MARGIN
4.15 PM

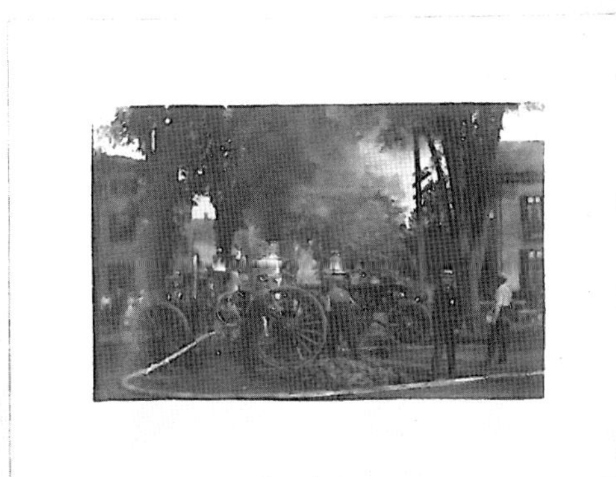

BROAD AND HATHORN
ENGINE 2 PEABODY
" 3 SALEM
5 PM.

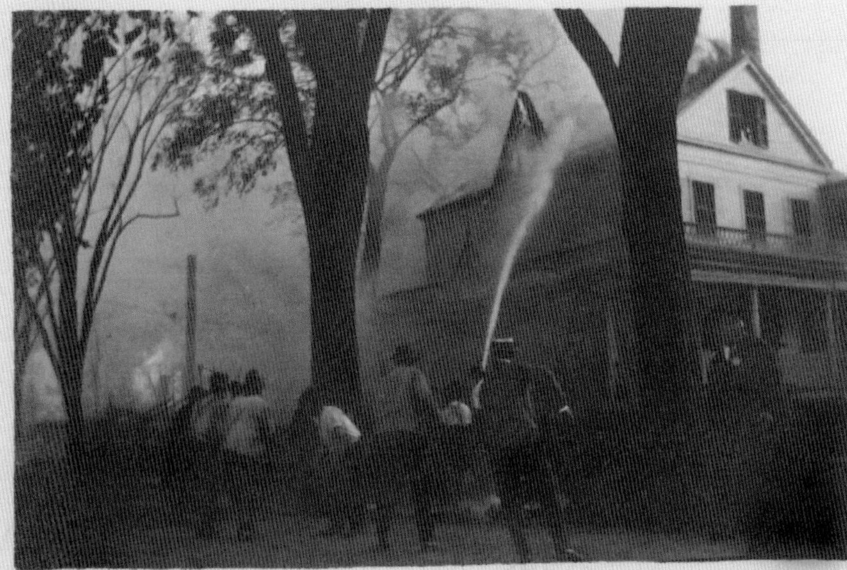

MARGIN
4.15 PM

X

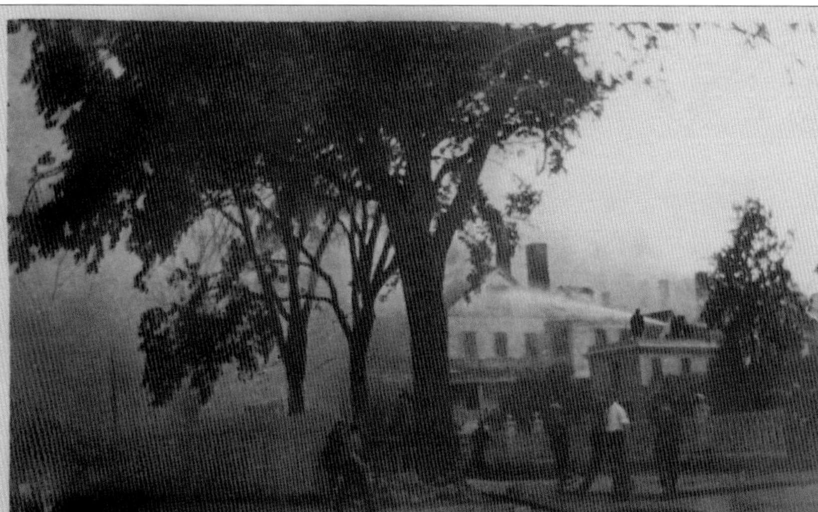

MARGIN
4.15 P.M.

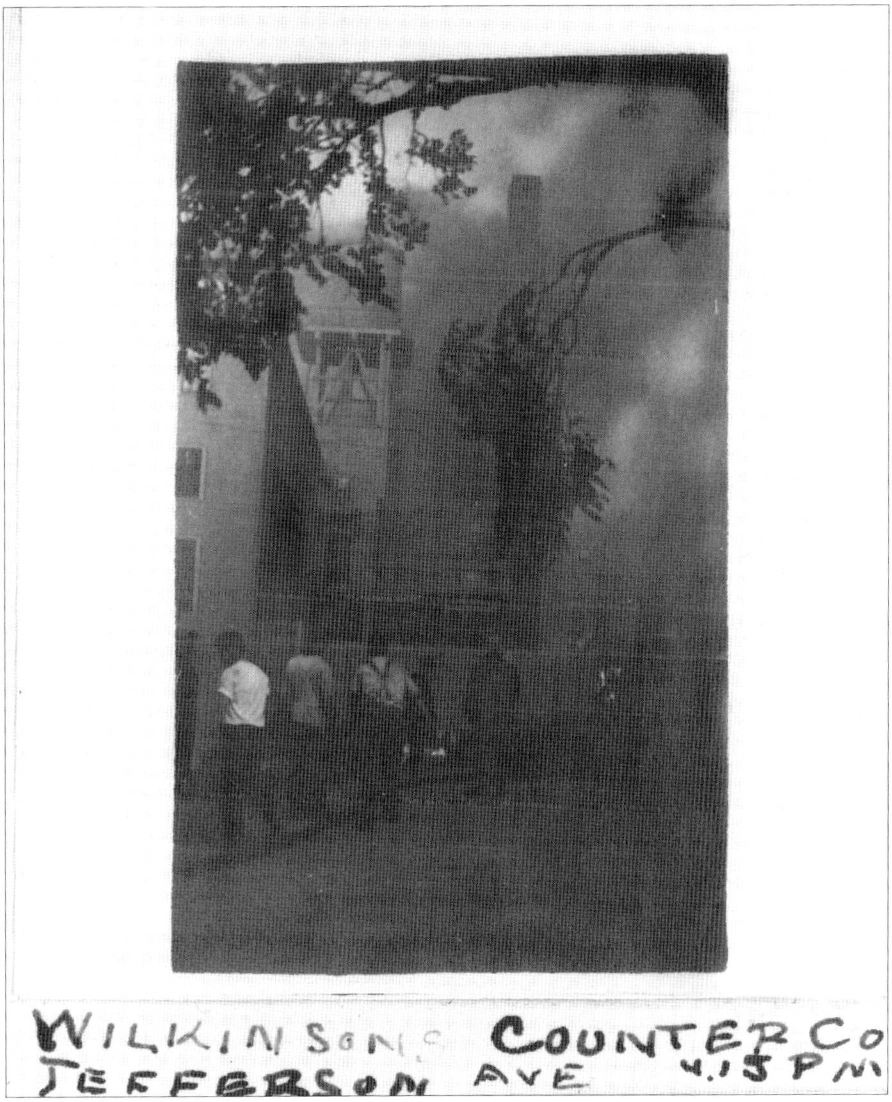

Enlarged images from William Abbott's album show the efforts of firemen attempting to stop the fire. The Margin Street photographs show how the fire spread—sparks from one burning building were caught by the wind and landed on neighboring buildings. The firemen fought from all angles—on the street and on the roofs of buildings—attempting to save the wooden structures. (*Neg #32887 and Neg #32886*) The Wilkinson Counter Company on Jefferson Avenue was Salem's most modern factory. The building was mill constructed and sprinklered and had slate-shingled walls with a twenty-thousand-gallon tank of water on the roof. With the help of Engine 6 from Boston, the building was saved because of the efforts of the firemen and the constantly running sprinklers. Only the top floor of the building was damaged; once the water supply in the tank was gone, the low water pressure from the city water supply was insufficient to reach the top floor. The factory was restored immediately after the fire and business resumed July 15. (*Neg #32905*)

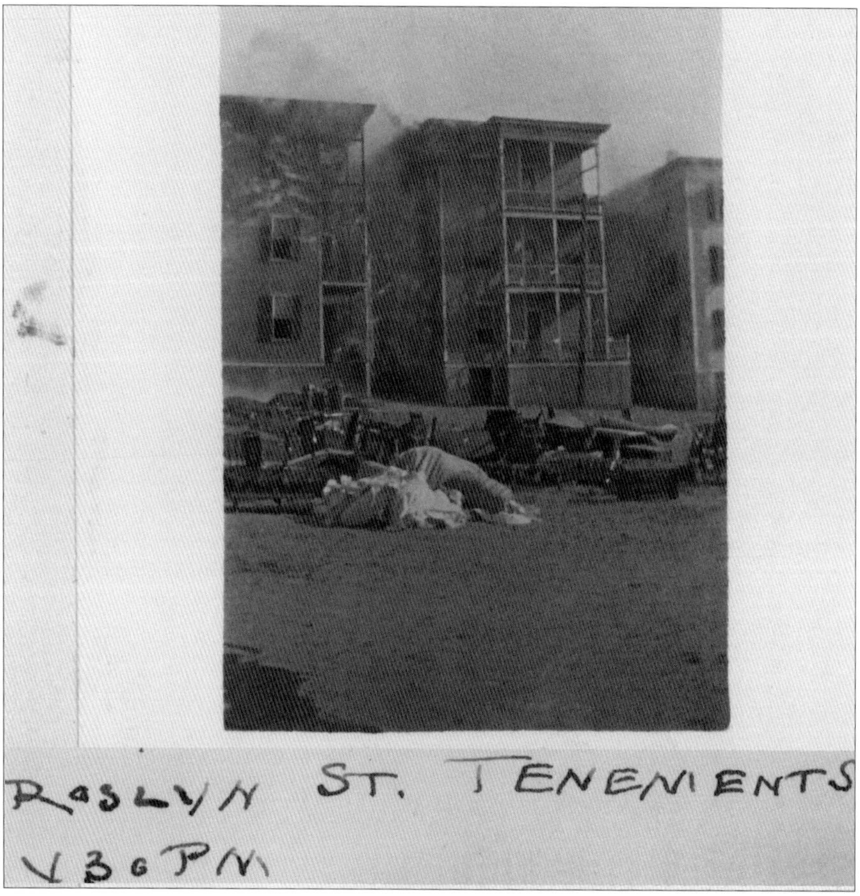

The times listed in the captions of these enlarged images from the Abbott album show that although the fire had jumped across the Boston and Maine Railroad tracks, firemen continued to fight the fire on several fronts. The Roslyn Street tenements were between Canal and Lafayette Streets, beyond the tracks. (*Neg #32890*) The image of the steamer engine fighting the fire on Upper Broad Street is northwest of the tracks, as is the image on Broad and Hathorn Streets. In this image, the Peabody Fire Department is assisting the Salem Fire Department; the firemen were successful in preventing the fire from spreading into an area of Salem filled with historic architecture. (*Neg #32904, top; Neg #32888, bottom*)

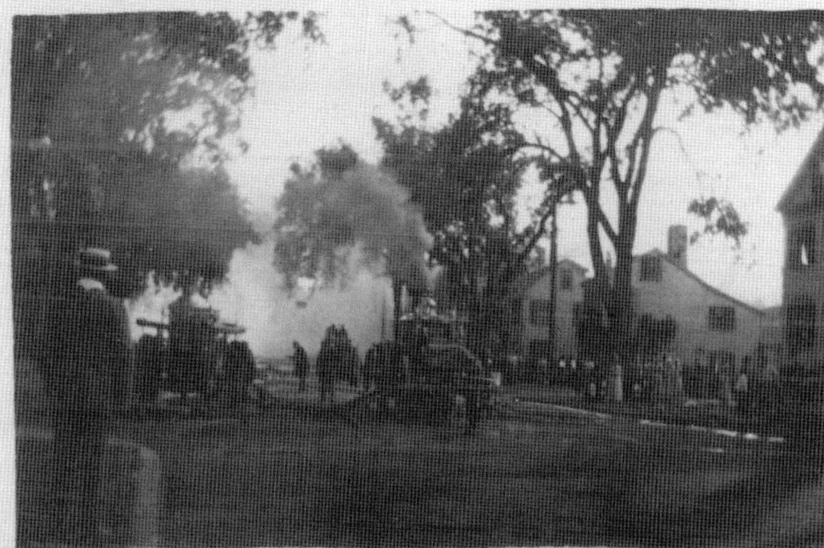

UPPER BROAD ST.
5 P.M.

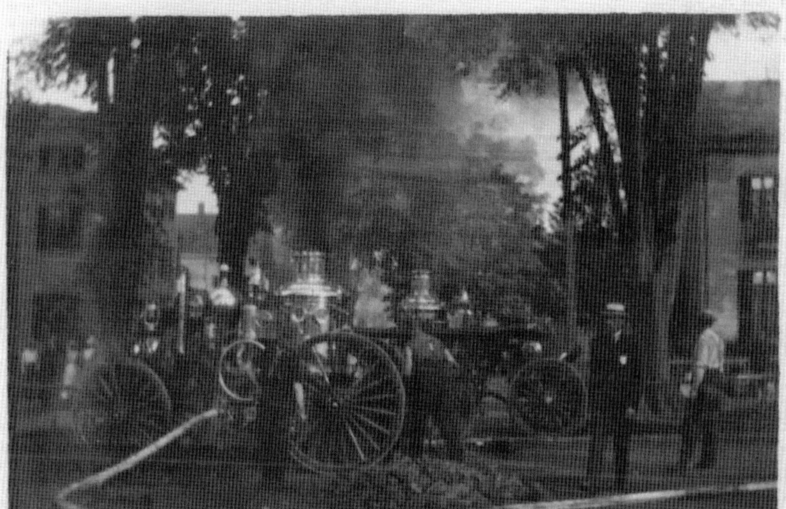

BROAD AND HATHORN
ENGINE 2 PEABODY
3 SALEM
5 PM.

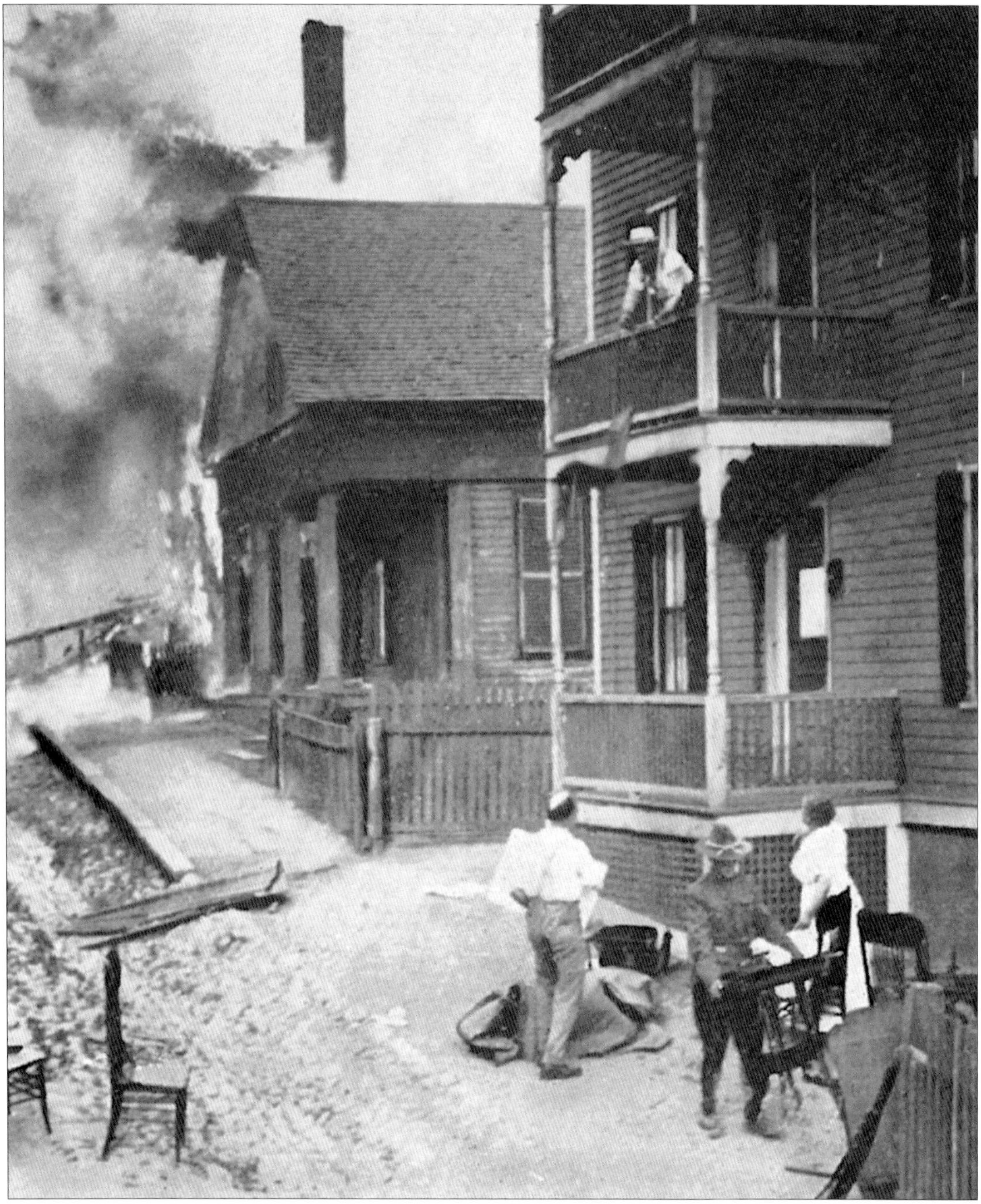

The Salem Fire, by Arthur Barnett Jones, states that the militia was called in at 2:31 p.m. on June 25; this placed the city under semi-martial law. Seventeen hundred militiamen from neighboring towns in Massachusetts served the residents of Salem during the fire and assisted them in the relief efforts afterward. The militia remained in Salem until the evening of July 7. In this image, a militiaman helps to remove possessions from a tenement on Roslyn Street, which was in the direct path of the fire. (*Neg #32830*)

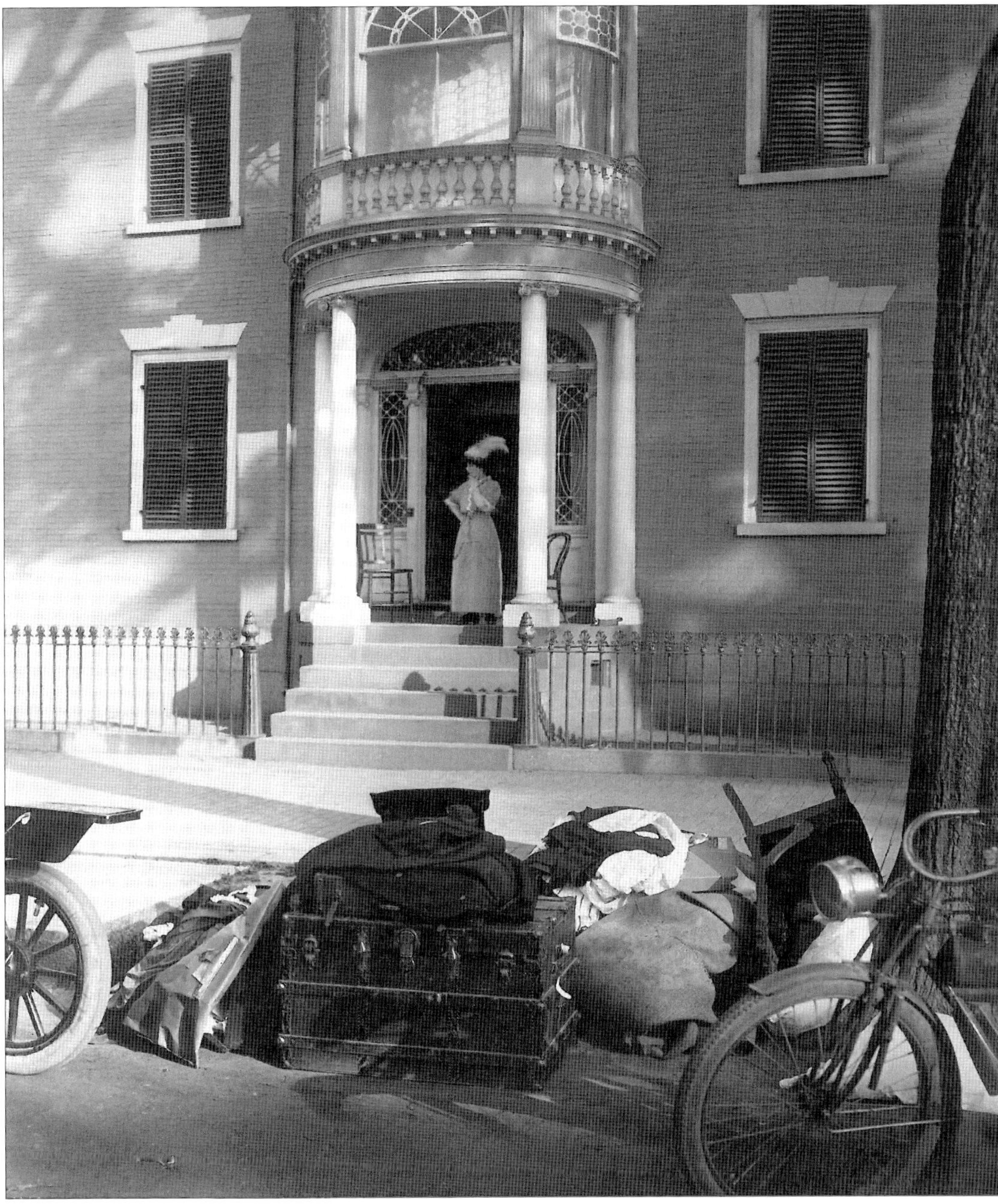

The west to northwest wind caused the fire to spread east and southeast through the city. By 3:00 p.m., the fire approached the historic section of Salem known for its colonial architecture, homes of the nineteenth-century Salem merchants and sea captains. A concerned resident on Chestnut Street prepared to evacuate as she watched the fire from the steps of her home, which was originally built for a Salem sea captain. Because of a change in the wind, the historic district was spared. Had the fire damaged the historic district of the city as it did the immigrant section, the impact of the fire would have been quite different. (*Neg #16752*)

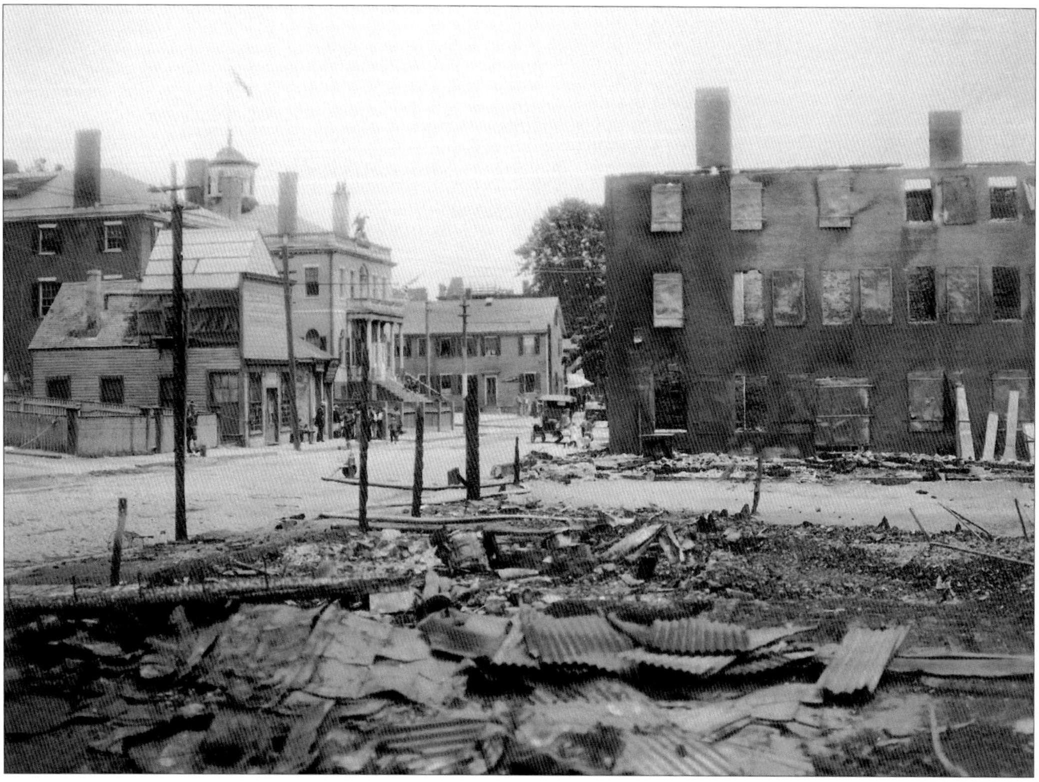

In *The Salem Fire*, Jones identifies this photo as being taken by M.E. Robb of Salem. The fire department took its last stand along Front and Derby Streets, finally stopping the fire about 3:00 a.m. Friday, June 26, within one block of the mercantile district. According to Jones, a three-story brick building at Lane's Wharf with a blind wall on the east gave firemen the opportunity to control and stop the fire. This image also depicts the Custom House, the building with the eagle on the roof, which was built in 1818. In the early nineteenth century, at the height of the Salem shipping industry, the Custom House collected 8 to 12 percent of the nation's revenues. The introduction to *The Scarlet Letter* by Nathaniel Hawthorne has a sketch about the Custom House, in which Hawthorne claims to have found the records that inspired the novel. (*Neg #4261*)

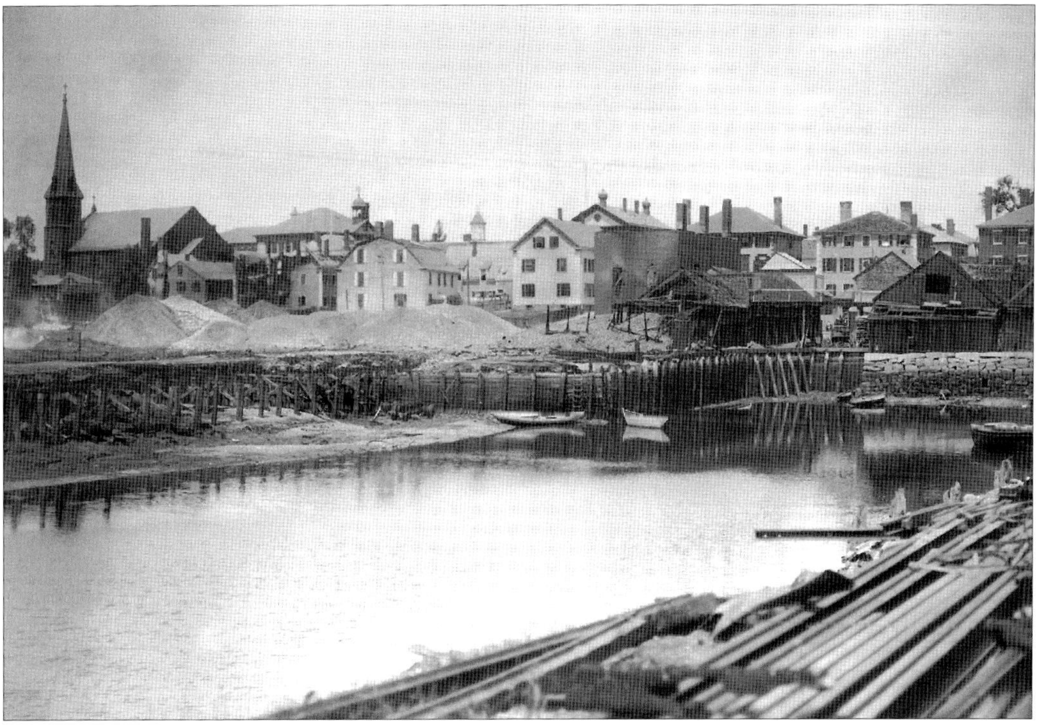

This picture shows another view of Derby Street where the fire was stopped. Derby Wharf was one of the oldest landing sites in the country during the height of the maritime industry in Salem. The Immaculate Conception Church is in the upper left portion of the image. Piles of burning coals can be seen on Union Wharf, which was destroyed during the fire. (*Neg #4436*)

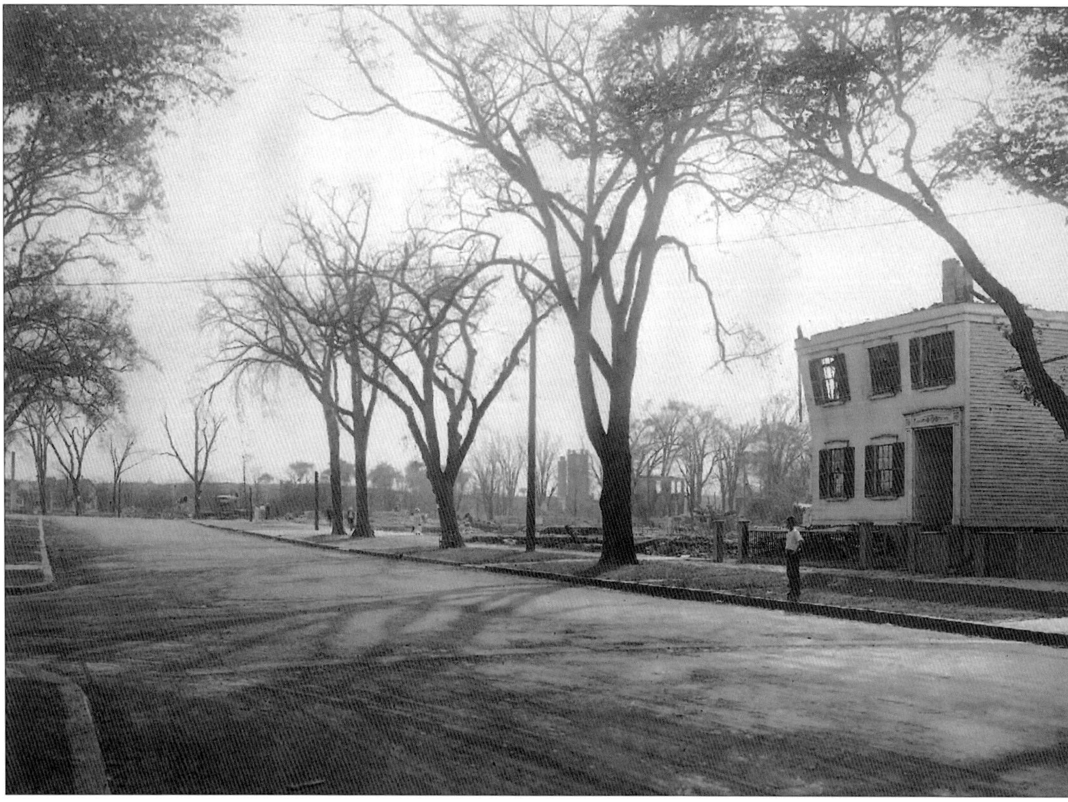

This image shows a Salem resident standing in utter dismay as he views the destruction of the fire on Upper Broad Street. Before long, the streets of Salem would be thronged with tourists who came to view the city after the fire. (*Neg #4207*)

Chapter 3

Salem After the Fire

Immediately after the fire, many contradictory statistics were provided about the amount of damage, the number of people made homeless, the number of deaths caused by the fire, etc. On the one-year anniversary of the fire, June 25, 1915, *The Salem Evening News* ran a series of commemorative articles and an itemization of fire statistics, which is shown below:

First alarm sent from Box 48 at 1:37 p.m. on June 25, 1914.
General alarm sounded four minutes later.
Number of alarms during course of fire, 21, including military and water call.
Fire under control in 13 hours, although it raged for hours afterward.
Area devastated by flames, 251¼ acres.
Streets wiped out entirely, 51.
Streets partially wiped out, 48.
Families rendered homeless, about 3500.
Number persons burned out, estimated at 15,000.
Number persons other than above, whose employment was affected by fire, 3000.
Lives lost in conflagration, three; many other deaths, however, due to the effects of the fire.
Buildings of all kinds destroyed in the fire, rising 1800, including 41 factories of different kinds.
Number of places of business, including variety stores, approaching 400.
Total loss caused by fire, estimated at $15,000,000.
Insurance carried $14,661,652, of which $11,737,639 was paid.
Taxable property loss to the city, $5,300,000.
Trees destroyed in fire, 594.
Miles of curbing put hors de combat, nine.
Number of cities and towns which sent apparatus, 22, with a total of 67 pieces.
Military companies on duty in city after fire, 26.
Police from all over Essex County, Boston and other points to number few hundred on duty here.
Relief fund contributed $635,810.01 from all over country, $100,000 from state, and national appropriation of $200,000 partially expended. Amount of general and state funds expended up to present, $705,724.62.

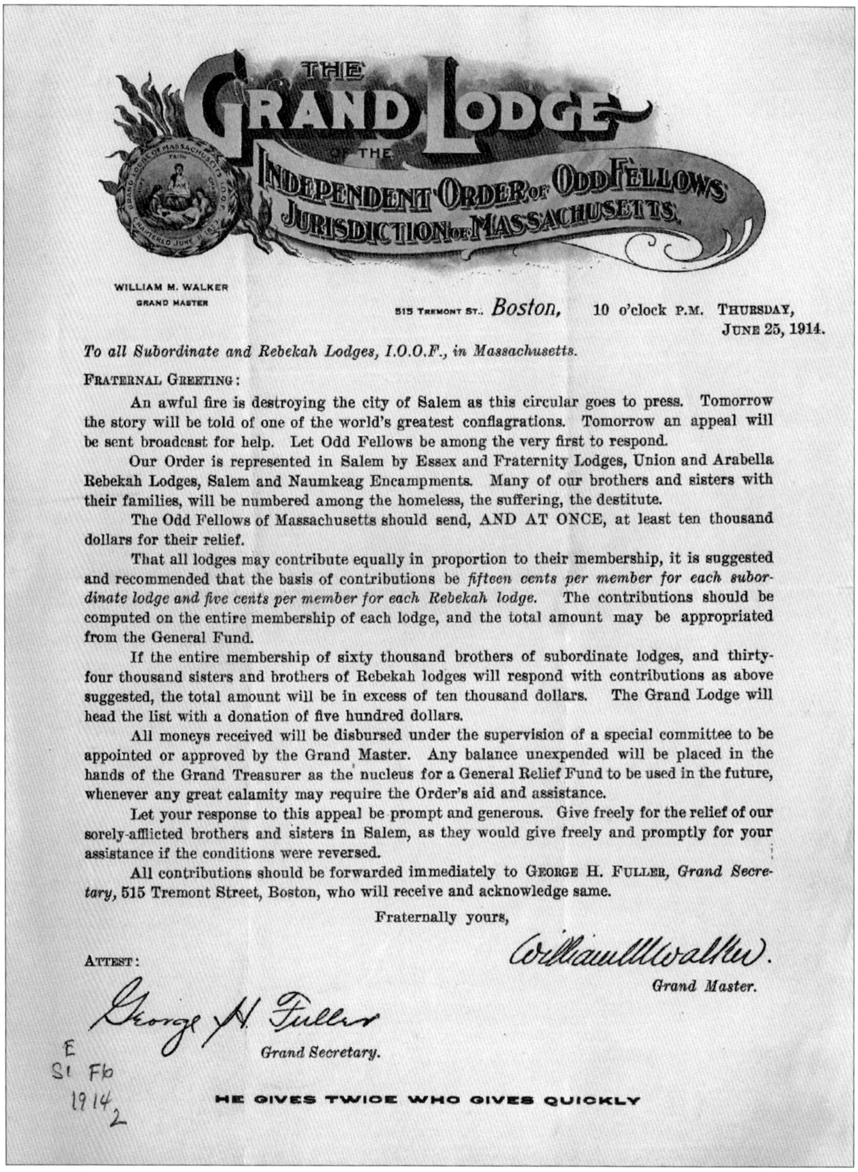

WILLIAM M. WALKER
GRAND MASTER

515 TREMONT ST., *Boston,* 10 o'clock P.M. THURSDAY,
JUNE 25, 1914.

To all Subordinate and Rebekah Lodges, I.O.O.F., in Massachusetts.

FRATERNAL GREETING:

An awful fire is destroying the city of Salem as this circular goes to press. Tomorrow the story will be told of one of the world's greatest conflagrations. Tomorrow an appeal will be sent broadcast for help. Let Odd Fellows be among the very first to respond.

Our Order is represented in Salem by Essex and Fraternity Lodges, Union and Arabella Rebekah Lodges, Salem and Naumkeag Encampments. Many of our brothers and sisters with their families, will be numbered among the homeless, the suffering, the destitute.

The Odd Fellows of Massachusetts should send, AND AT ONCE, at least ten thousand dollars for their relief.

That all lodges may contribute equally in proportion to their membership, it is suggested and recommended that the basis of contributions be *fifteen cents per member for each subordinate lodge and five cents per member for each Rebekah lodge.* The contributions should be computed on the entire membership of each lodge, and the total amount may be appropriated from the General Fund.

If the entire membership of sixty thousand brothers of subordinate lodges, and thirty-four thousand sisters and brothers of Rebekah lodges will respond with contributions as above suggested, the total amount will be in excess of ten thousand dollars. The Grand Lodge will head the list with a donation of five hundred dollars.

All moneys received will be disbursed under the supervision of a special committee to be appointed or approved by the Grand Master. Any balance unexpended will be placed in the hands of the Grand Treasurer as the nucleus for a General Relief Fund to be used in the future, whenever any great calamity may require the Order's aid and assistance.

Let your response to this appeal be prompt and generous. Give freely for the relief of our sorely-afflicted brothers and sisters in Salem, as they would give freely and promptly for your assistance if the conditions were reversed.

All contributions should be forwarded immediately to GEORGE H. FULLER, *Grand Secretary,* 515 Tremont Street, Boston, who will receive and acknowledge same.

Fraternally yours,

William M. Walker.
Grand Master.

ATTEST:

George H. Fuller
Grand Secretary.

HE GIVES TWICE WHO GIVES QUICKLY

This circular, dated Thursday, June 25, 1914, was printed at 10:00 p.m. before the Salem Fire had been completely controlled. It was signed by the grand master of the Grand Lodge of the Independent Order of Odd Fellows in Boston, Massachusetts, William M. Walker, and attested to by Grand Secretary George H. Fuller. The Odd Fellows felt it was their duty to set an example and immediately called for funds to establish a relief effort for the victims of the fire. They pledged $10,000 to a fund, suggesting that all lodges in Massachusetts contribute in proportion to their membership, using a basis of 15 cents per person for each lodge and 5 cents per person for each Rebekah lodge. (*Neg #32832*)

While doing research for this book, I found additional statistics about the fire:

3,000 change of address cards processed at the Post Office; "burned-outs" would receive their mail via General Delivery until a new address was provided.

Henry C. Frick, the Pittsburgh millionaire who had a summer house in the Salem Willows, donated $25,000 to relief fund.

John D. Rockefeller donated $10,000 to relief fund.

Monies received by the Relief Committee on the first day after the fire, $97,000; $325,000 received by the beginning of July.

Red Cross donation for relief efforts, $5,000.

Estimated 200,000–300,000 visitors to Salem to view the ruins.

Cost to the state for 13-day duty of the militia, $26,000.

Number of sprinklered buildings destroyed, 11.

Number of unsprinklered factories destroyed, 7.

Additional acres burned in North Salem, 3.

Population in Salem before fire, 47,000.

Population in Salem immediately after fire due to relocation, 37,000.

Loss of leather factory jobs due to plant relocation after fire, 1,450.

Stephen Phillips, Salem resident, lent his car and chauffeur to Salem police to assist town in any way possible, car drove 1,500 miles in 10 days within Salem city limits.

Boy Scouts assisting with relief work, 300–400.

Naumkeag Directory of Salem, 1914 *changes in number of names listed in directory from previous year, 5,708.*

By July 3, Massachusetts House of Representatives introduced bill to permit city to abate taxes in the burned area and to borrow $100,000 in anticipation of taxes lost.

Firemen whose homes were destroyed during the fire, 30.

Firemen's Relief Fund sent to Chief Arnold after the fire to assist the burned-out firefighters, $3853.80, providing $126.46 for each man.

Examples of fire equipment lost in the fire: 650 feet hose, 2 fire axes, 12 rubber coats, 25 feet garden hose, 3 street blankets, 27-foot ladder.

Following pages: This aerial shot was photographed by George T. Murray, staff photographer of the *Boston Journal*, on June 26, 1914. It depicts the ruin wrought by the fire on the previous day. The ruins of the Naumkeag Cotton Steam Company can be seen at the top of the image. This is the first known instance of aircraft being used to take photographs for newspaper illustrations. The airplane, probably a Burgess-Dunne seaplane, provided by the Burgess Company of Marblehead, Massachusetts, was piloted by Clifford L. Webster, a member of the U.S. Naval Reserve. According to the souvenir of this photograph created by the *Boston Journal*, Murray was in the air for more than one hour and "crept from one place to another on the air machine, hanging perilously by one hand while he directed his camera with the other. The success of his daring is shown in this wonderful photograph." The circulation department of the *Boston Journal* indicated that never before had an image created such a response as the aerial shot did; anyone searching for copies offered fifty cents a copy for the paper, which sold for one cent an issue. (*Neg #17033*)

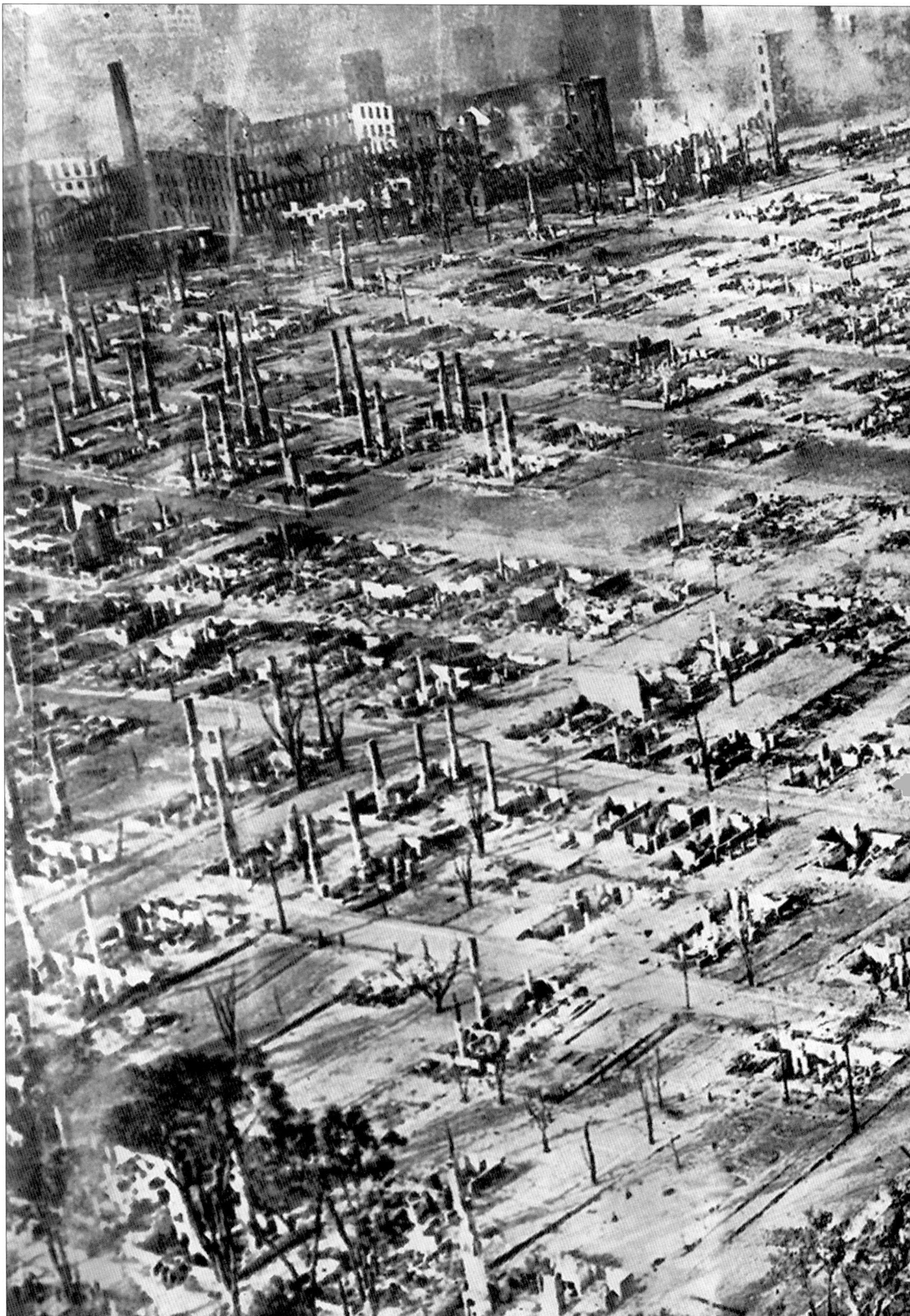

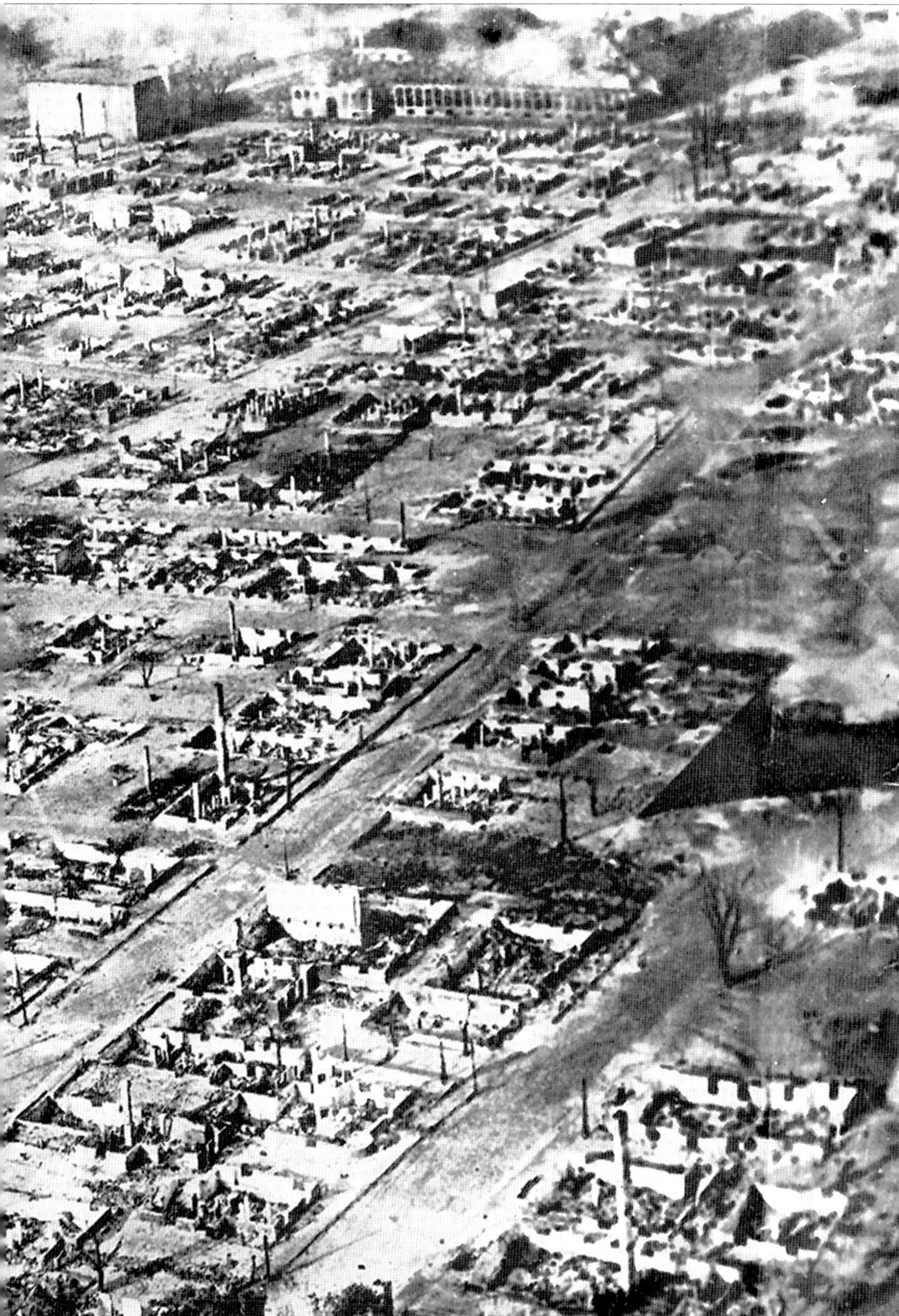

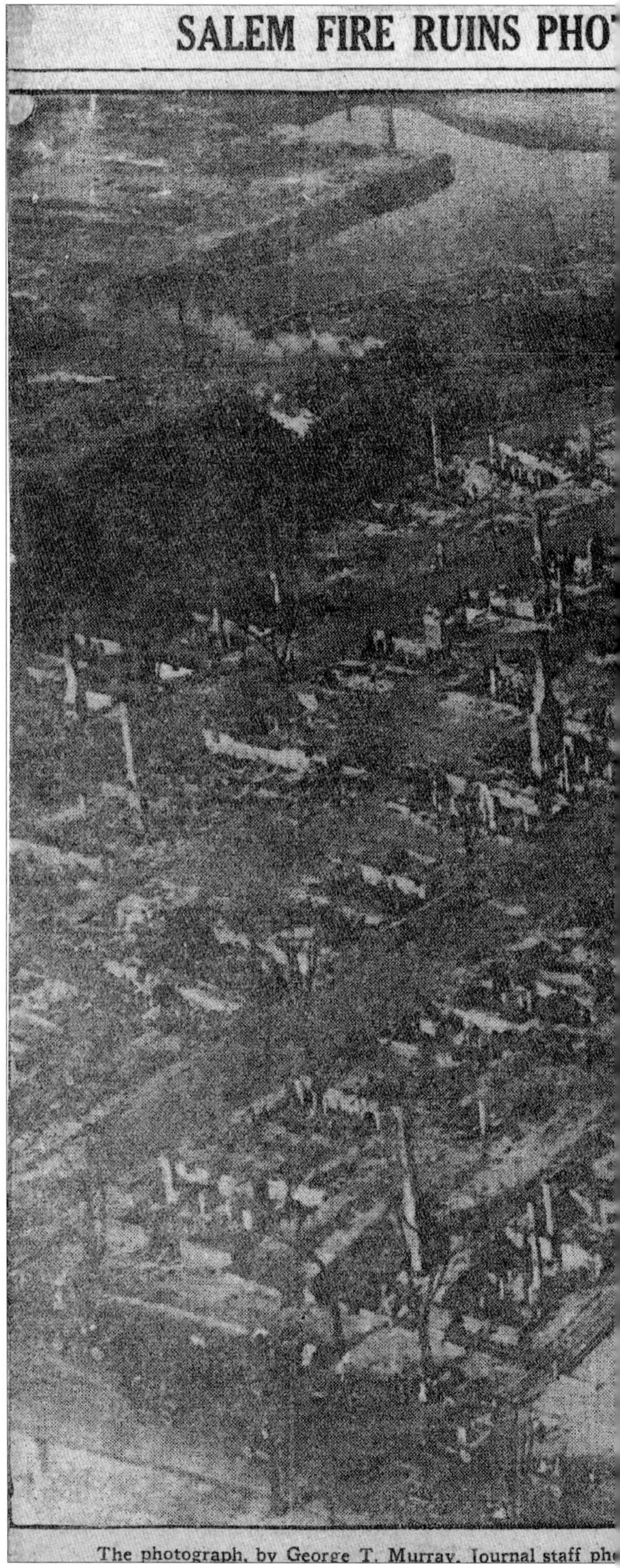

George T. Murray took several shots for the *Boston Journal* while flying above Salem. This view looks toward the harbor and the ruins of the Naumkeag Steam Cotton Company. St. Joseph's Church, in the foreground, was gutted by the fire; an aerial shot depicts this destruction more readily than would images taken from the ground. An article in the *Boston Journal* dated June 29 quotes Mayor Hurley of Salem as saying that the aerial images of "the devastated landscape, with its bare streets and gaunt chimneys, in the midst of the flame-swept area, will call to the attention of everyone the magnitude of the damage and our immediate need for ready money." (*Neg #32945*)

RAPHED FROM THE JOURNAL HYDRO-AEROPLANE

Journal
June 27-1914

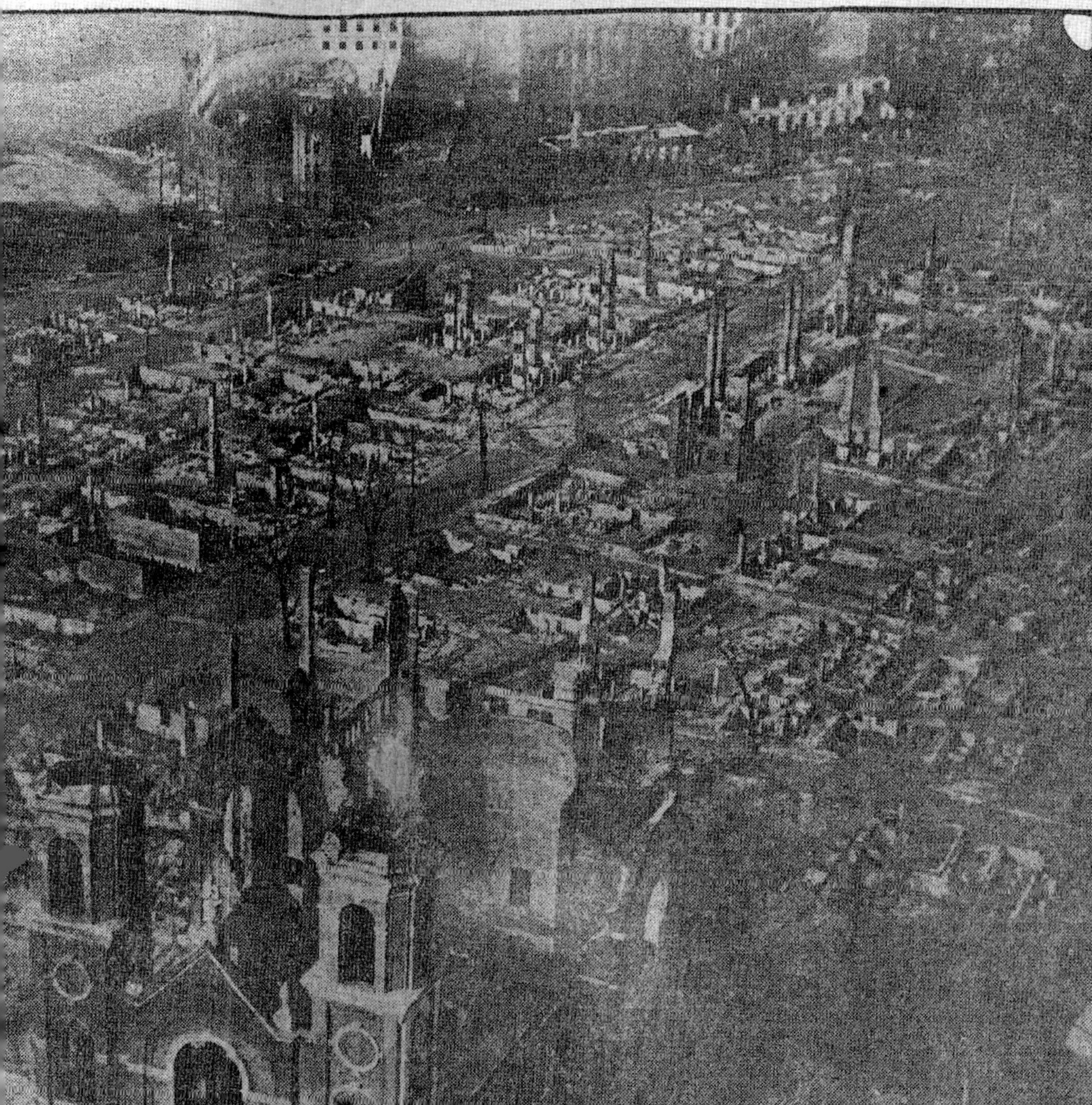

her, shows St. Joseph's Church in the foreground. The photograph is copyrighted, 1914, by The Boston Journal

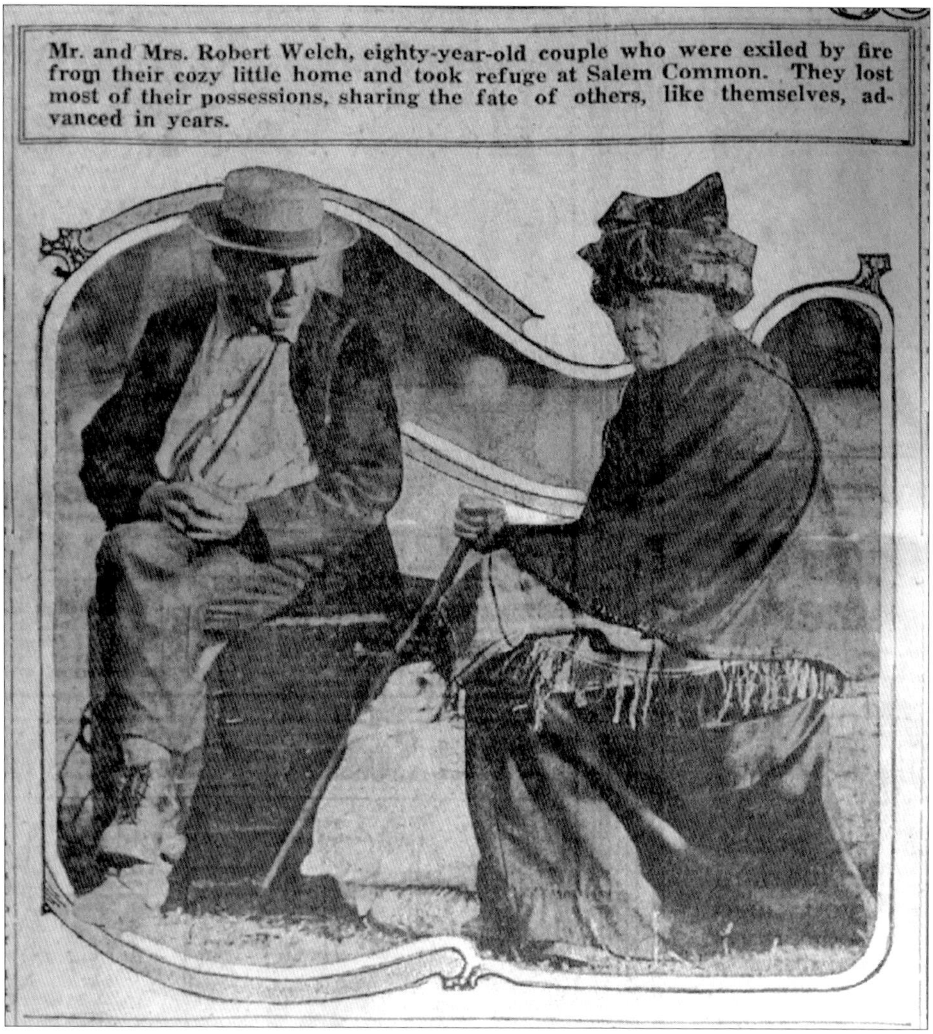

Mr. and Mrs. Robert Welch, eighty-year-old couple who were exiled by fire from their cozy little home and took refuge at Salem Common. They lost most of their possessions, sharing the fate of others, like themselves, advanced in years.

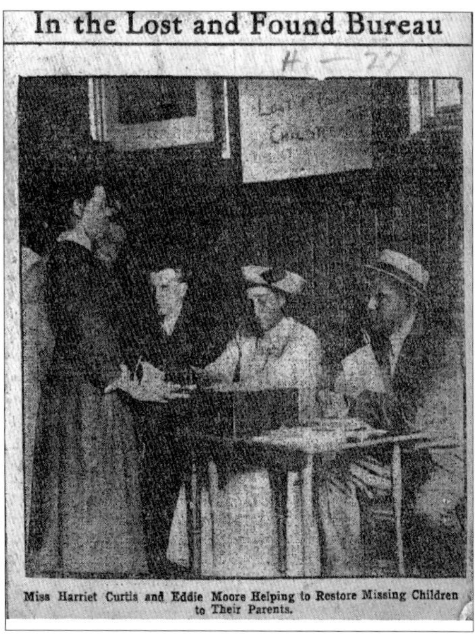

In the Lost and Found Bureau

Miss Harriet Curtis and Eddie Moore Helping to Restore Missing Children to Their Parents.

Mr. and Mrs. Robert Welch, an eighty-year-old couple, took refuge on the Salem Common after being forced out of their home because of the fire. Like many elders in their situation, they lost most of their possessions. According to the 1914 Naumkeag Directory of Salem, the Welch home was at 22 Ward Street. (*Neg #32916*) Identified only with the letter "H" and the date "27th," the newspaper clipping on the bottom of this page deals with missing children after the fire. The caption states: "Miss Harriet Curtis and Eddie Moore help to restore missing parents to their children." (*Neg #32915*)

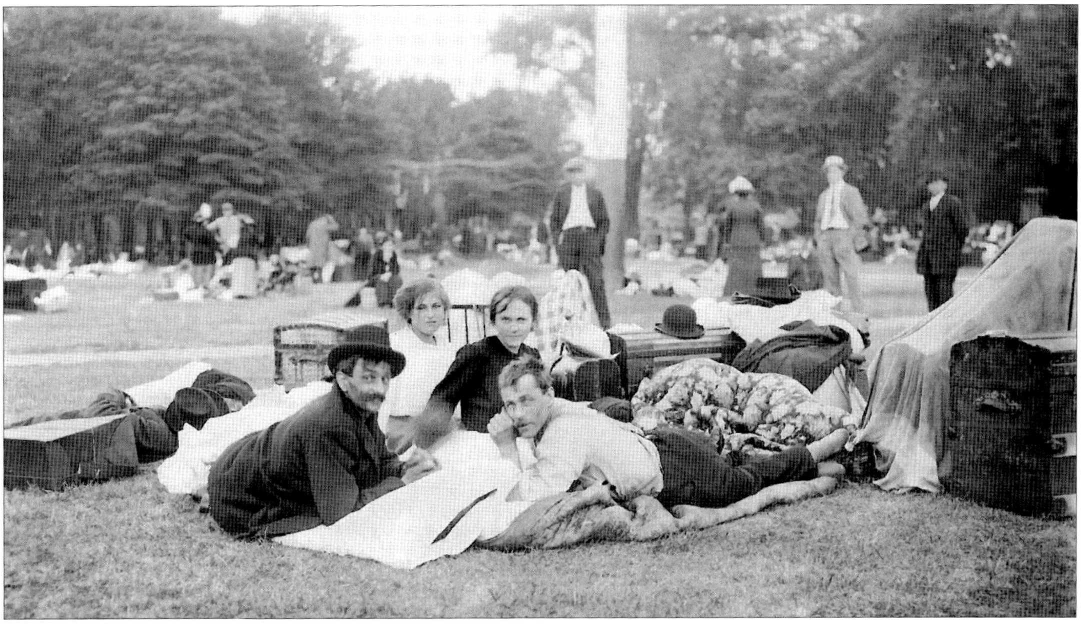

People made homeless by the fire flocked to the eight acres of the Salem Common, some with the possessions they had saved from their homes and others with nothing but what they wore. Several nationalities were represented by the families on the common. The refugees in the top image (*Neg #32788*) are probably wondering where they will live in the coming days and how their families will be clothed and fed. According to news reports, the lower picture was taken at dawn, 4:30 a.m. on June 26, 1914. (*Neg #32820*) Homeless families slept outside near the high school, at Salem Willows, at Forest River Park and among the graves in the Broad Street Cemetery. Babies were born in the cemetery and on the common on the night of the fire.

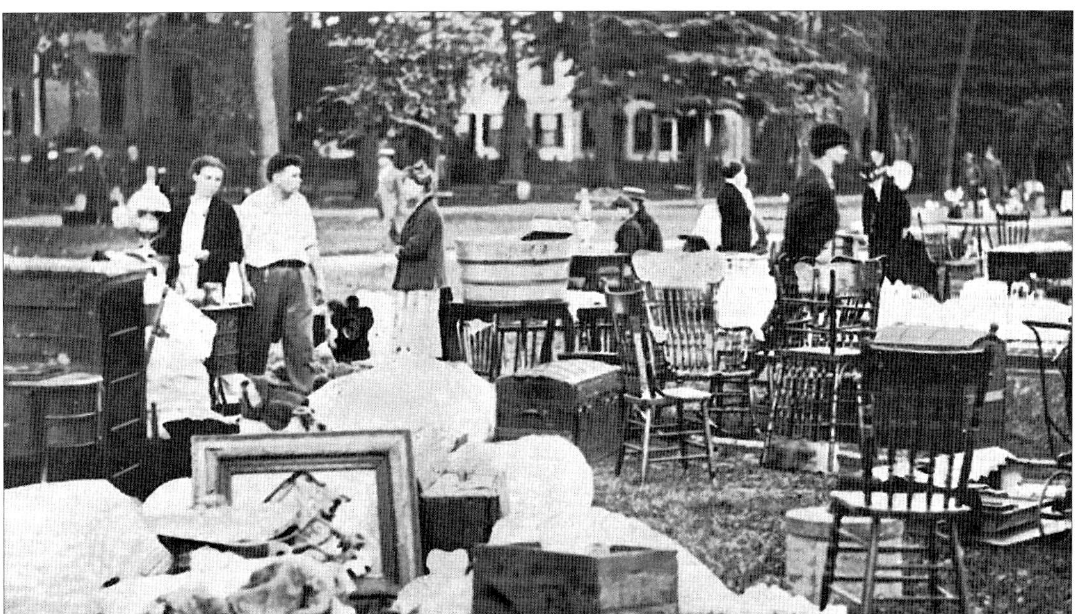

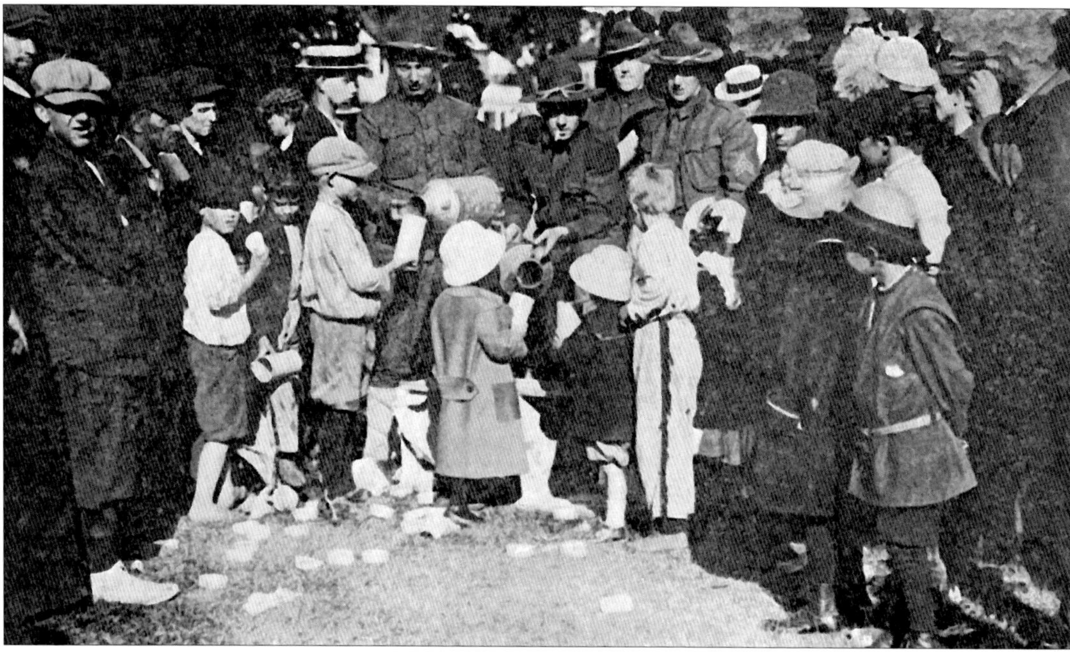

The day after the fire over 2,500 people were given food. Distribution of rations began at six o'clock that morning. The top image (*Neg #32821*) depicts the militia giving bread and milk to homeless children. The bread line shown at the bottom of the page may have been taken at Highland Park or at Forest River Park, locations for refugees that were set up shortly after the fire was contained. (*Neg #32819*)

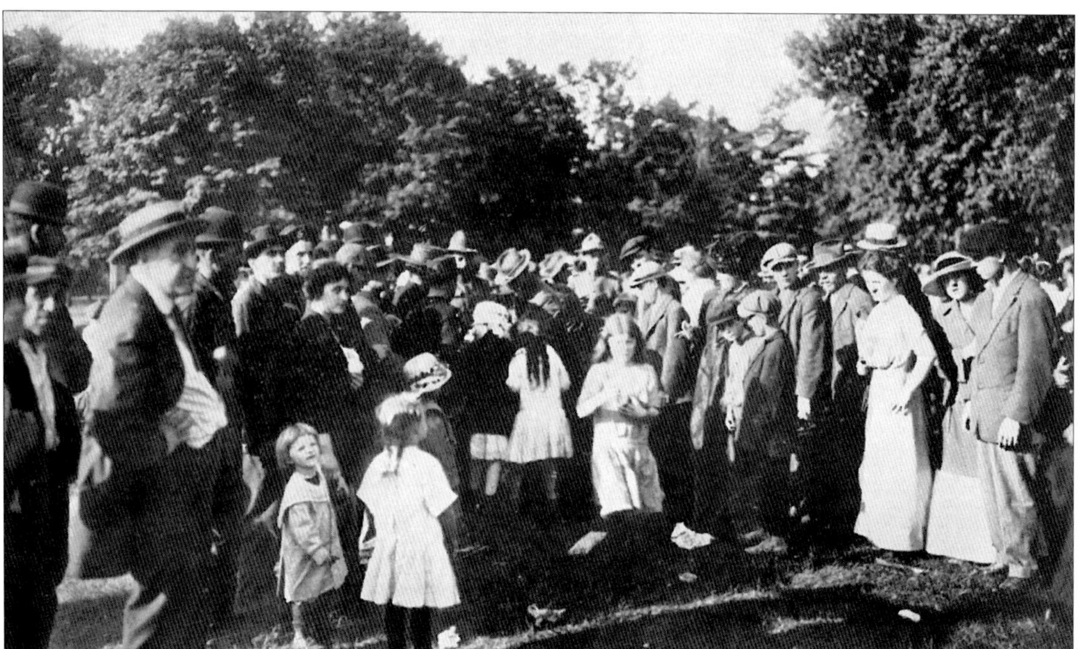

The Story About the Salem Electric Lighting Company

The power plant at the Salem Electric Lighting Company was one of the three fireproof buildings to survive the fire. Five men stayed in the building as the fire roared past the property. Their story is told in the *Tenney Service*, dated September 1914, published by Charles H. Tenney & Company, the parent company to the Salem Electric Lighting Company.

The five men in the building were E.A. Harris of Malden; William O. Regan, first assistant engineer; Joseph Veil, a fireman for the company; an unnamed young man who came over with Harris (who had brought a boat to escape the fire); and George F. Heustis, superintendent of the Malden Electric Company. Harris was Malden's office manager who accompanied Heustis to Salem to help. The flames swept down on the building and beat against its brick walls and wired-glass windows with a fury that only those who were there could appreciate. As the fire passed, it burned every structure in the vicinity of the power plant. The employees who remained in the plant continually threw water on the wooden doors that were not fireproof and kept up the steam in the boilers. Although the turbines had been shut down, an exciter operated at normal voltage the entire time. At 1:00 a.m., Heustis opened the door to see how things were progressing:

> *I went to the boiler room door facing Lafayette street and threw the latch. I had no sooner done this, than I found myself lying flat in the middle of the floor. The wind pressure against the door was so great that when the doors were freed, they opened with such force as to throw me, but the breath of good fresh air more than paid me for the bump. I called to the other fellows and we all stood in the door-way enjoying the cool wind which came from the unburned part of Lafayette street down the river. At frequent intervals we would get a blast of hot air that would drive us back.*

When they looked out, Heustis wrote:

> *Where the stable had been, there was an empty space. The office building was nothing but bare brick walls. The huge piles of lumber which were across the river were nothing but smoldering piles of ashes…Our coal crusher and the conveyer into the bunker were all burned and a slight fire had started in the bunker. The great mills of the Naumkeag Steam Cotton Company were just starting to burn around the wooden coping. This was the only building that I saw burn. All other buildings were standing when we shut ourselves in, and burned when we could not see out.*

Every electric pole in stock was consumed by the fire. Wire, cross arms, tools and other stock necessary to install power were also lost. During the fire, arrangements were made to obtain necessary supplies from other branches of the company and by June 26, poles, arms, tools, wire and workmen arrived in the station yard to begin the repair to the electrical system in the city. Power was completely restored to the city by June 27.

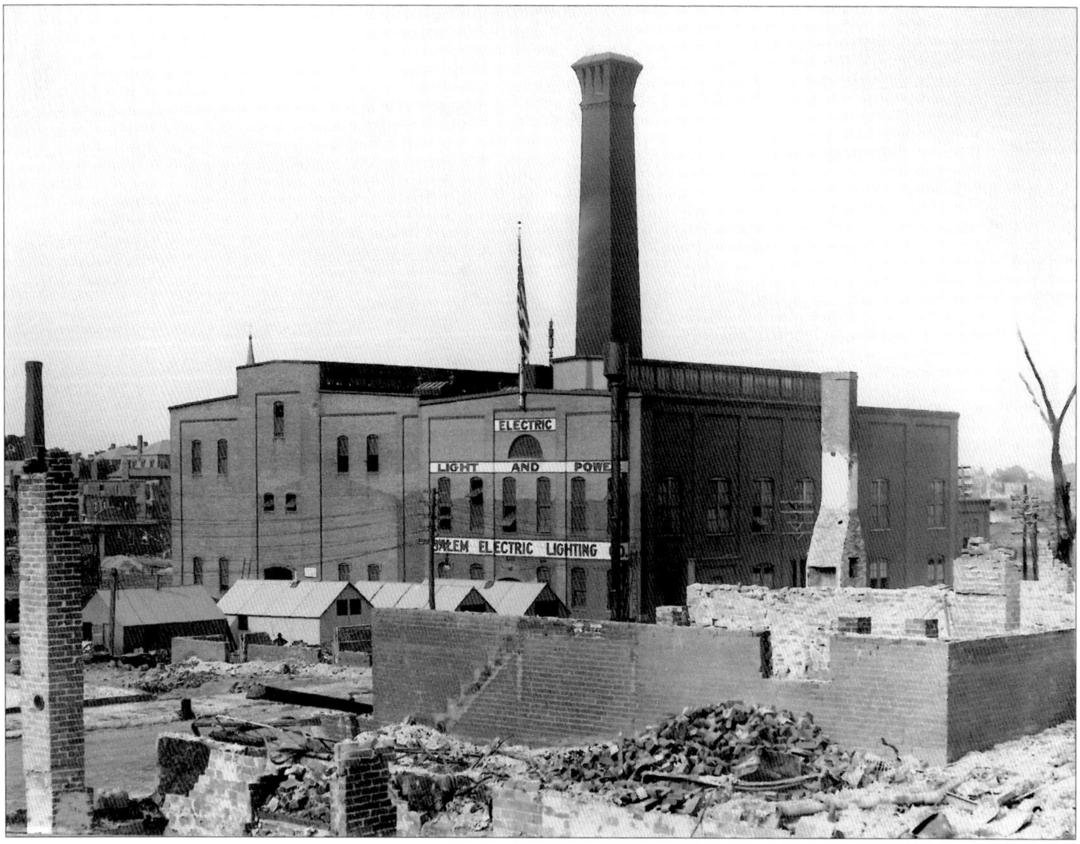

The three buildings depicted were lone survivors of the fire. The Salem Electric Light Company's Power House, shown on this page, was on Peabody Street and South River, east of Lafayette Street. The one-story building was protected by concrete walls sixteen inches thick with a tile roof supported by unprotected steel trusses and a floor of heavy concrete. Windows on three sides of the building were wired glass, either set in metal or protected by fireproof shutters; windows on the north side facing the South River were unprotected. Portable voting booths outside the building were set up immediately to house homeless workers so the plant could reopen to furnish power on Friday, June 26. At the time of the fire, the Electric Light Company supplied power to 455,000 incandescent lights, 400 of which were streetlights, as well as 977 nernst lights and 725 arc lights, of which 350 were streetlights. (*Neg #32794*) The two buildings shown on the next page were part of the Naumkeag Steam Cotton Company. The upper image shows Storehouse No. 10, formerly a gas house known as the Gasometer. It stood on the northeast corner of the property; it was circular and one story tall, with a joisted roof, asbestos-covered shingles, metal cornices and brick walls, and had two double fire doors. It was not sprinklered and contained 1,500 bales of cotton, all of which escaped damage from the fire. (*Neg #32868*) The lower image is Storehouse No. 1, also known as the Finished-goods Storehouse. Located on the corner of East Gardner and Lynch Streets on the western side of the mill property, the storehouse was within five feet of fire exposure. None of the finished product stored in the four-story building was damaged because of concrete construction and wired-glass windows protected by tin-clad fire shutters on the inside, which closed when the heat melted a fusible link. Protection was so effective that the sprinklers never opened during the fire. (*Neg #32777*)

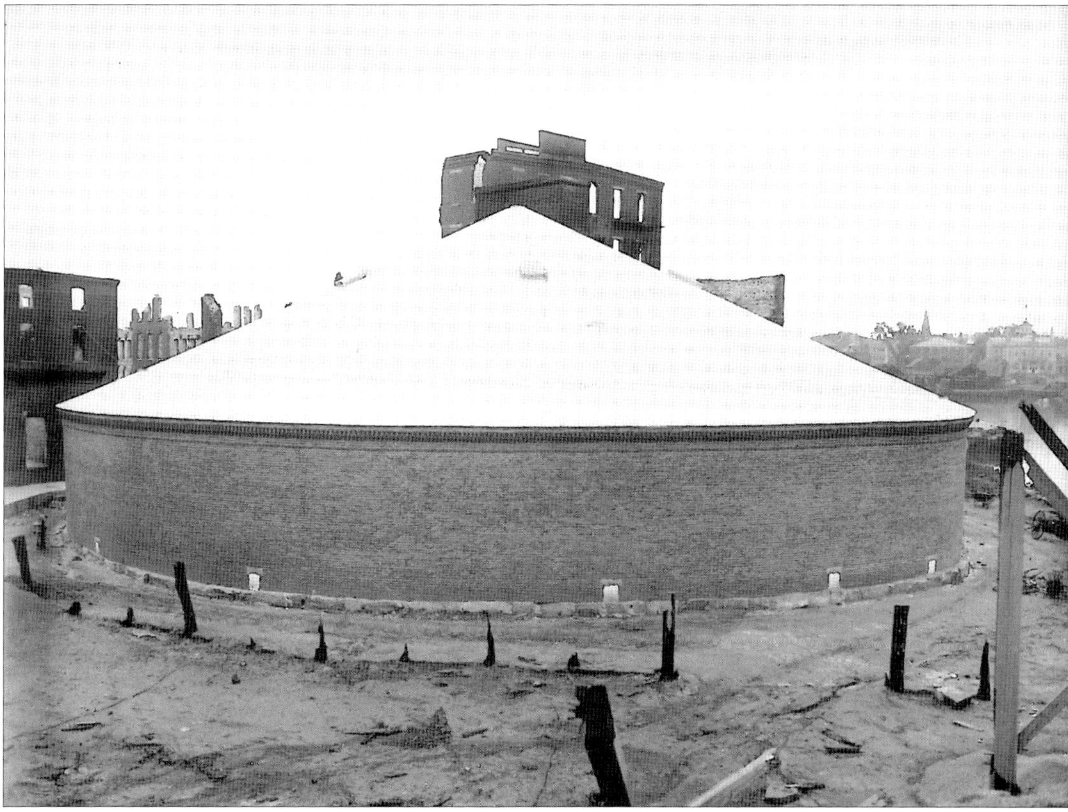

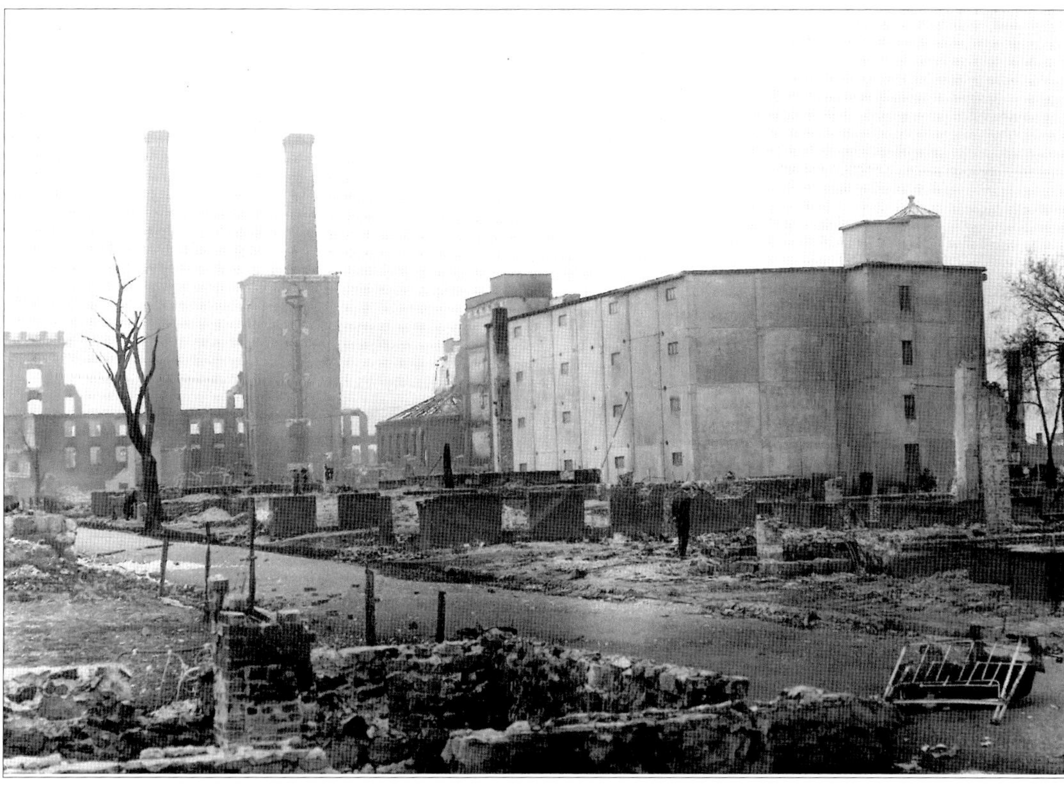

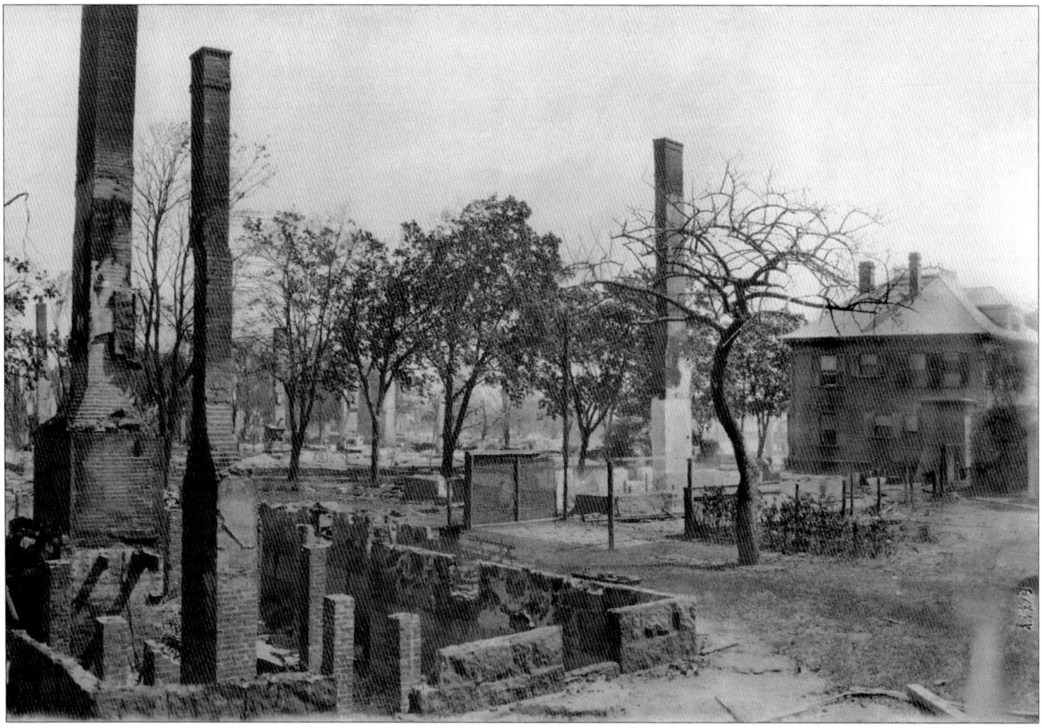

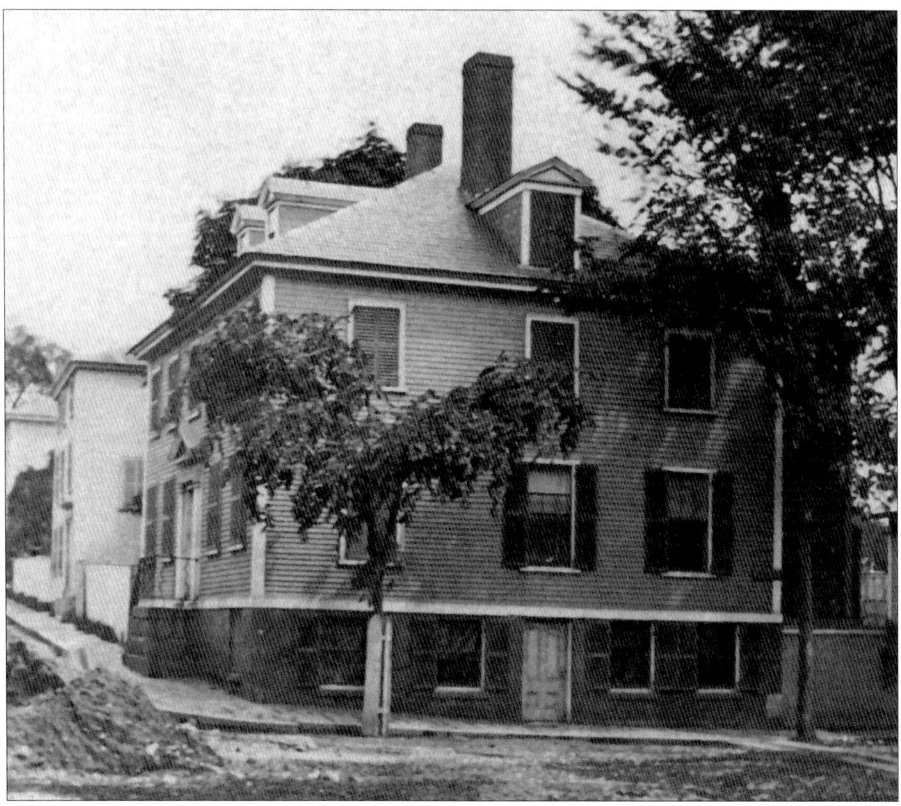

According to the 1908 *Visitor's Guide to Salem*, Lafayette Street was one of the "longest and finest streets in Salem." It was lined on both sides with great elms, most of which were planted in 1808 by Ezekial Hersey Derby. The fire destroyed the street from Derby Street to Holly and Leach Streets. The image at the bottom of this page looks down Lafayette Street from the vantage point of Leach and Holly Streets. Notice that many of the great elms have been destroyed by the fire. (*Neg #4617, a Robb photo*) The bottom image on the previous page shows the only wooden home to survive the fire on the portion of Lafayette Street that was damaged. (*Neg #32827*) Although not specifically identified as such, the top image depicts a home among the ruins surrounding it, which may be the house on Lafayette Street that survived. (*Neg #32942, a Rankins & Ward photo*) An unidentified newspaper clipping found in a scrapbook, dated June 25, indicates that Lieutenant Governor Barry told the Boston firemen fighting flames on Lafayette Street that the only way to prevent the flames from destroying the entire street was to save the home of William L. McMorrow at the corner of Lafayette and Holly Streets. Water had run out at this point, but Barry convinced the firemen to obtain chemical tanks, which they did, and the house was saved. In 1985, the Lafayette Street Historic District was created, which contains Salem's collection of late nineteenth-century Victorian architecture that survived the fire.

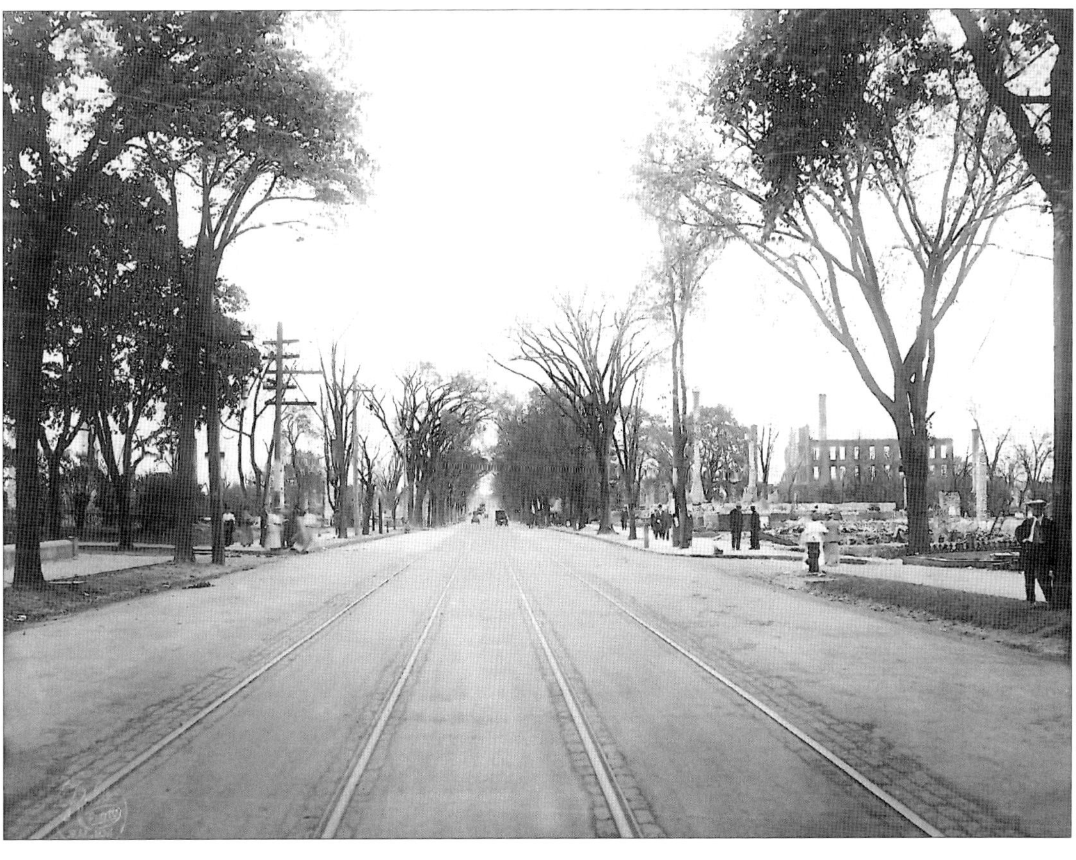

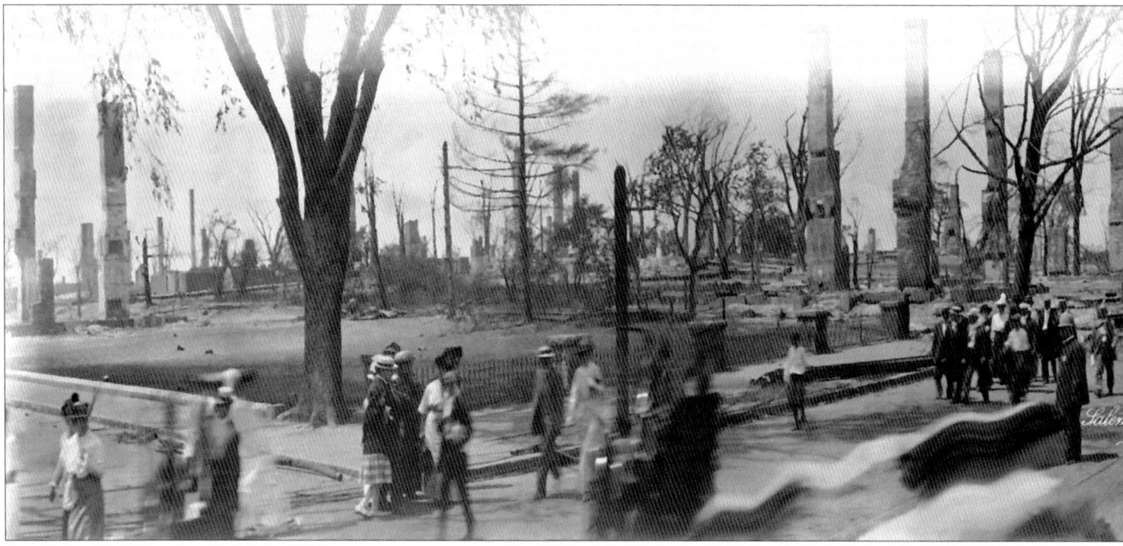

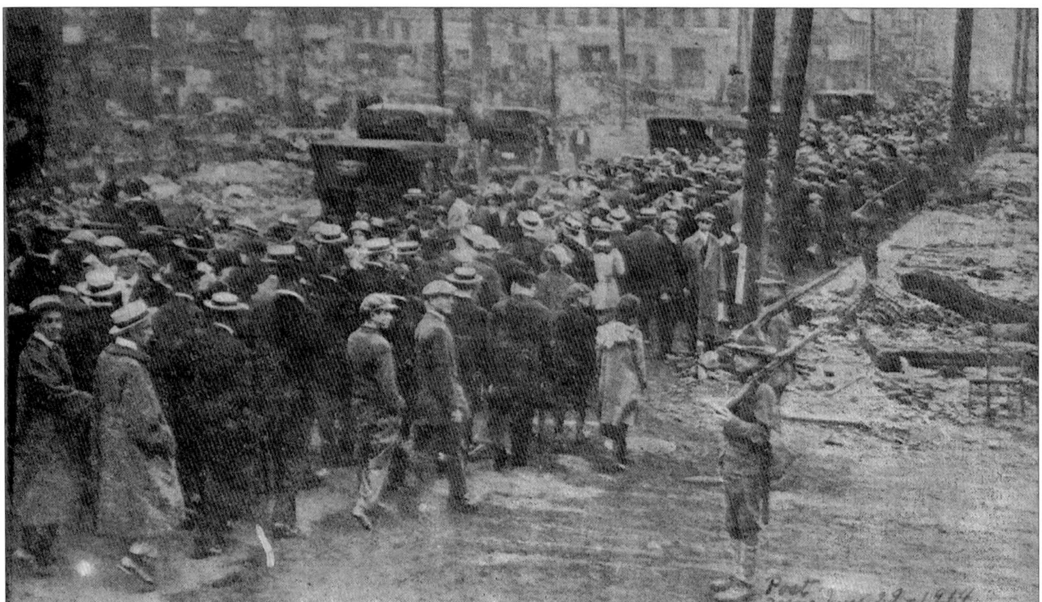

THOUSANDS OF VISITORS WENT TO SALEM YESTERDAY FROM BOSTON AND NORTH SHORE CITIES AND TOWNS. HERE IS A GLIMPSE OF SOME OF THEM. THEY ARE IN GOOD ORDER BUT ARE EAGERLY SCRUTINIZING THE WORK OF THE FLAMES.

Hundreds of thousands of visitors came to Salem to view the ruins. On June 28, *The Salem Evening News* wrote that Salem had entertained the largest number of visitors in its history on June 27. Estimates ranged from 200,000 to 500,000 visitors; license plates were from all the states in New England and the middle states, along with Ohio, Indiana and Illinois. The following week a license plate was seen from California as large numbers of visitors continued to flock to the city. At one point it was estimated there were over thirty thousand cars in the city. Relief efforts took advantage

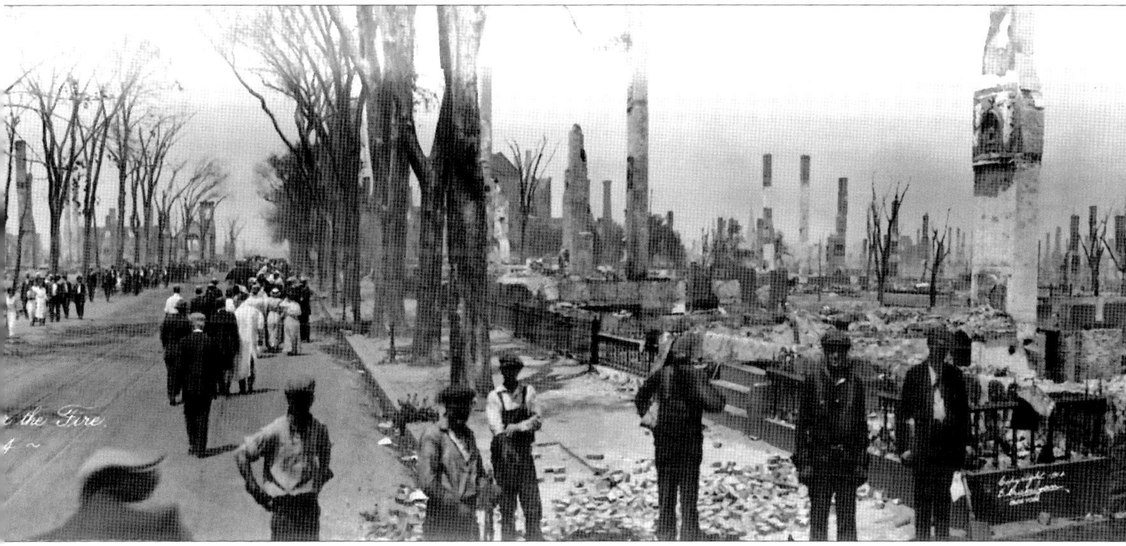

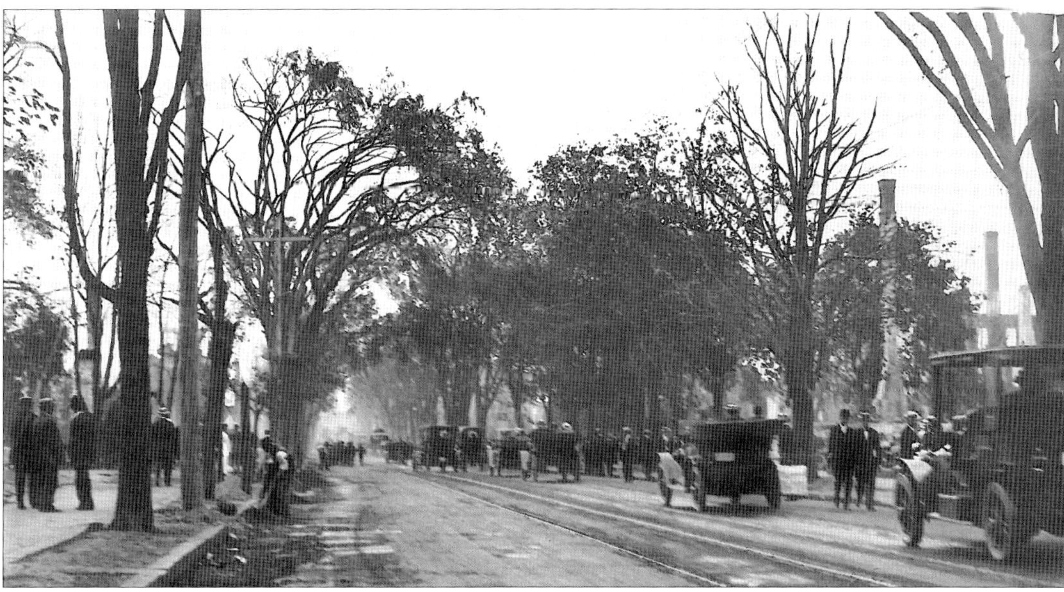

of the visitors to raise funds for the homeless by selling tags, which the donator then wore to indicate they had helped the relief effort. (*panoramic view, Neg #32880; Neg #32817*) The newspaper clipping is from the *Boston Post*, dated June 29. The caption states, "Thousands of visitors went to Salem yesterday from Boston and North Shore cities and towns. Here is a glimpse of some of them. They are in good order but eagerly scrutinizing the work of the flames." Militiamen were available to manage the crowds. (*Neg #32908*)

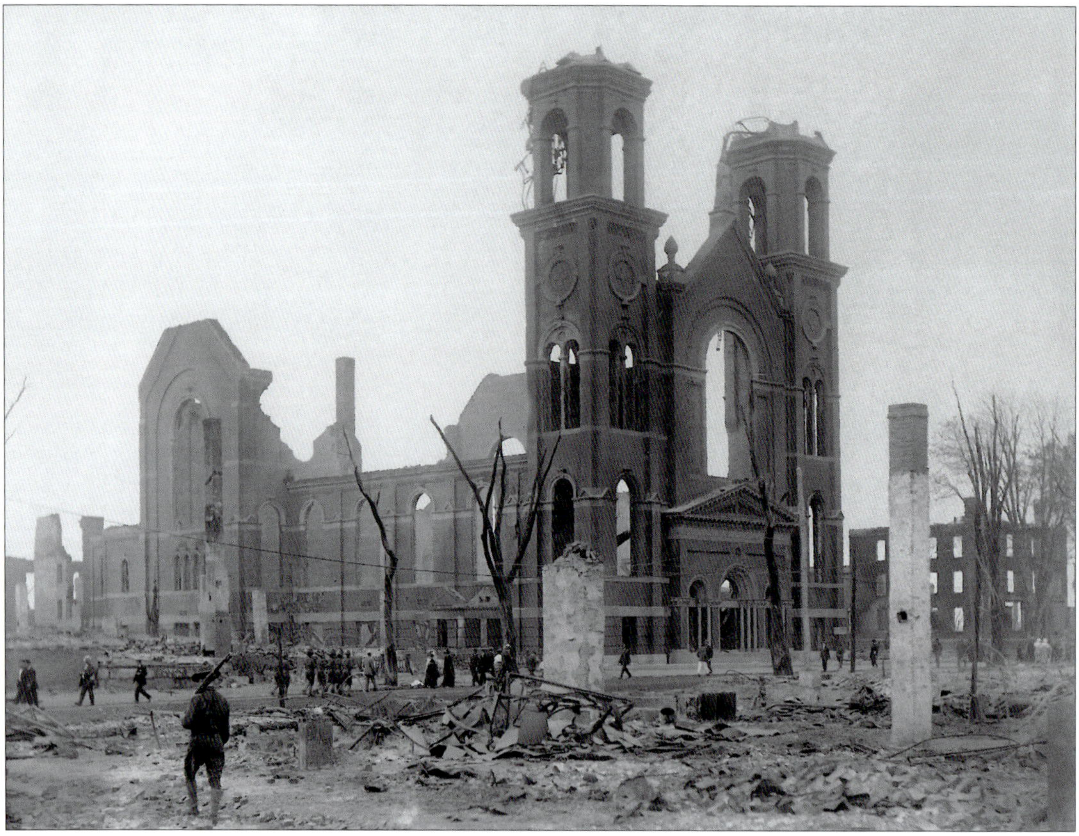

The image on this page shows the destruction to St. Joseph's Church. The transept walls preserved the outline of the church but the interior was completely consumed by the blaze. The towers, still standing, continued to protect the statue of St. Joseph. (*Neg #13638*) The two images on the next page, shot by M.E. Robb, depict ruins on Boston Street. In *The Salem Fire*, Jones identifies the house in the middle of the top picture as belonging to Mrs. Mark Smith. The tent in the far right of the picture was one of the first businesses to reopen after the fire. (*Neg #4469*) As identified by a newspaper article in the *Lynn Item* dated July 1, A.D. Fraser ran a barber shop at 38 Boston Street. He was one of the first Salem residents to lose his property. When the ruins cooled down the next day, he erected a tent a few feet away, borrowed a chair and opened the tent for business. His old customers returned. "He enjoys the distinction of being the first business man to start up after the fire." *The Salem Evening News* identified the barber as Emile D. Fraser; Jones's book identified the barber as John Frazier. Regardless of his name, he was enterprising. (*Neg #16788*)

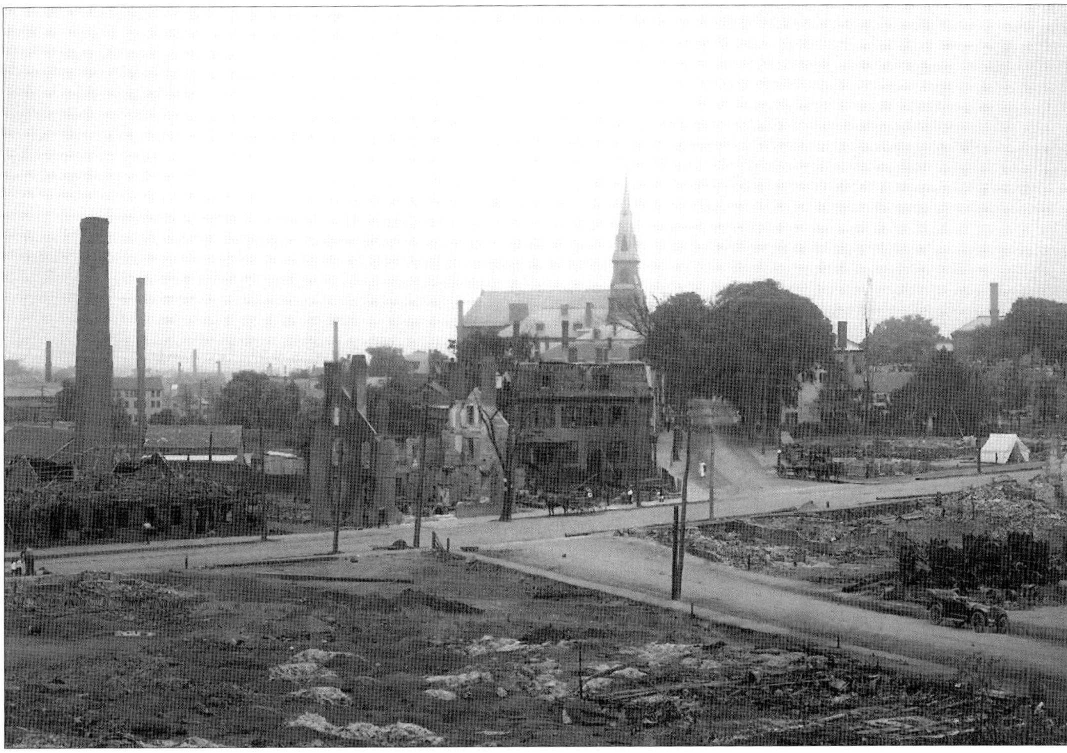

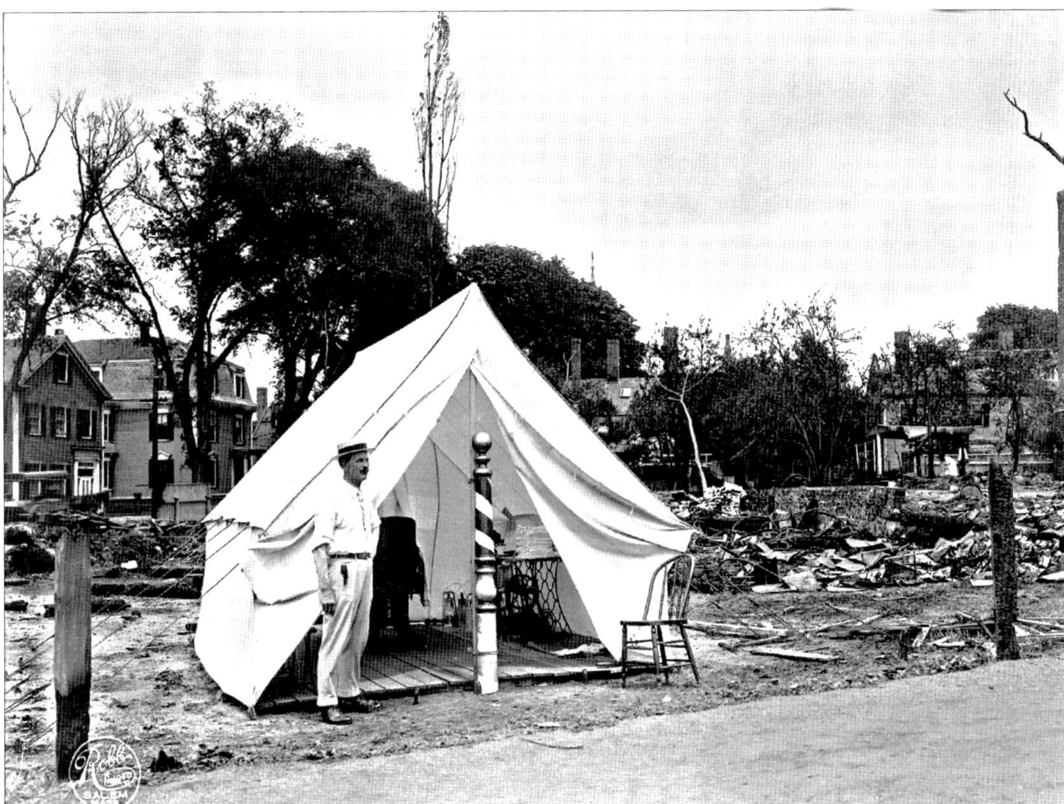

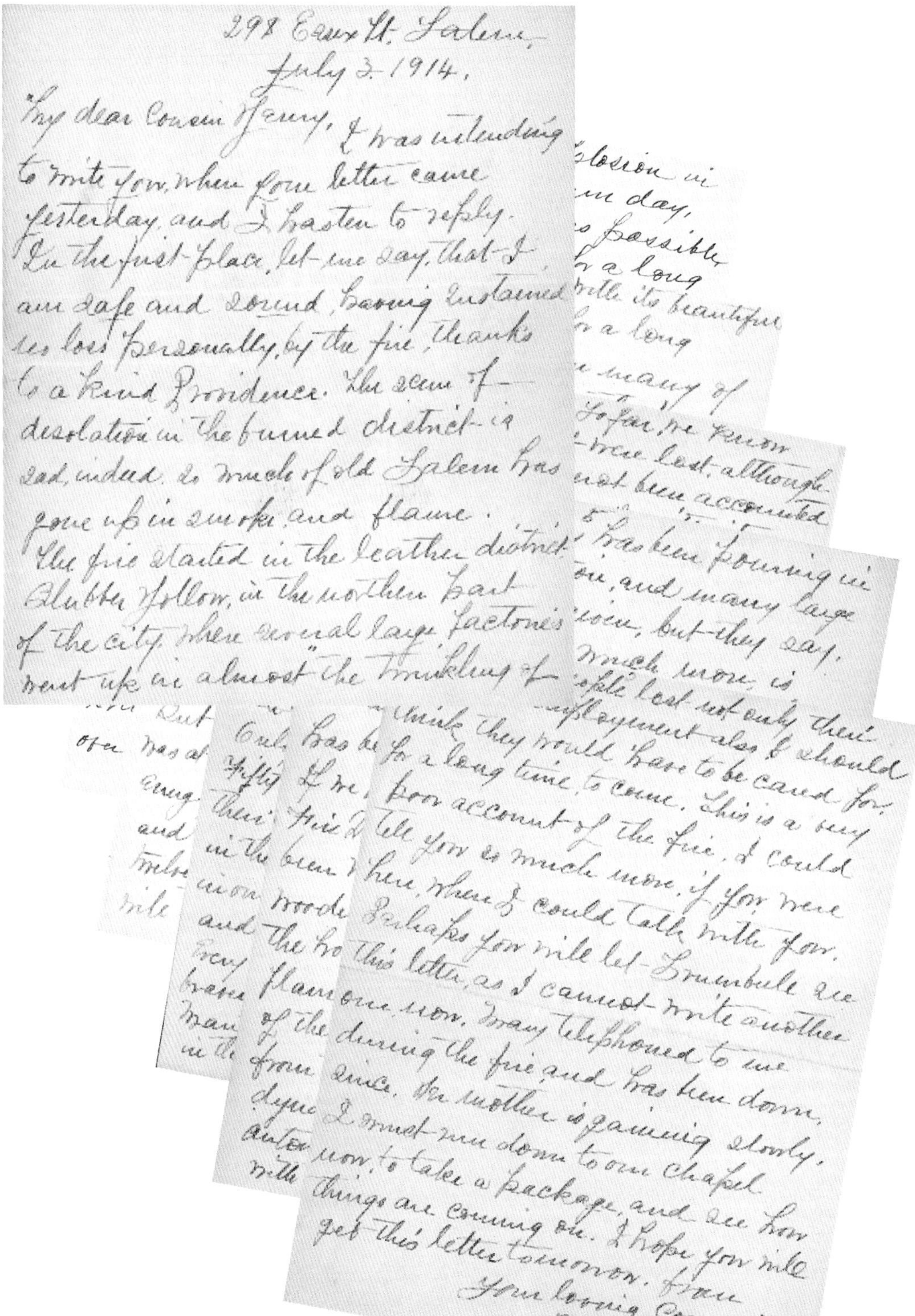
298 Essex St. Salem,
July 3. 1914.

My dear Cousin Henry, I was intending to write you, when your letter came yesterday, and I hasten to reply. In the first place, let me say, that I am safe and sound, having sustained no loss personally, by the fire, thanks to a kind Providence. The scene of desolation in the burned district is sad, indeed. So much of old Salem has gone up in smoke and flame. The fire started in the leather district, Blubber Hollow, in the northern part of the city, where several large factories went up in almost "the twinkling of

think they would have to be cared for for a long time, to come. This is a very poor account of the fire. I could tell you so much more, if you were here, when I could talk with you. Perhaps you will let Grumbelle see this letter, as I cannot write another one, now. Mary telephoned to me during the fire, and has been down since. Her mother is gaining slowly. I must run down to our chapel now, to take a package, and see how things are coming on. I hope you will get this letter tomorrow. From

Your loving Cousin,
Mary H. Cleaveland

Transcript of letter from Mary Cleaveland
To her Cousin Henry Concerning the Salem, Massachusetts Fire of 1914

298 Essex Street, Salem
July 2, 1914

My dear Cousin Henry,
I was intending to write you when your letter came yesterday and I hasten to reply. In the first place, let me say that I am safe and sound, having sustained no loss personally by the fire, thanks to a kind Providence. The scene of desolation in the burned district is sad, indeed, so much of old Salem has gone up in smoke and flame. The fire started in the leather district, Blubber Hollow, in the northern part of the city, where several large factories went up in almost "the twinkling of an Eye," as there was an explosion in one of them. It was a very warm day, and everything was as dry as possible. We had had almost no rain, for a long time. Then, the houses on Boston St., which crosses the upper part of Essex Street, were the next to go. The wind changed several times, during the fire, and the flames burst out in unexpected places. The upper part of Broad St. is entirely consumed, and the house which you may remember, where dear Father and I lived so happily for many years, was burned. There the houses on the streets leading out of Broad, were swept by the flames. Then the burning brands and sparks, carried the fire over to South Salem, and a large part of Lafayette, the main st., with its beautiful residences, was destroyed for a long distance, and the houses on many of the side sts. also went up in the flames. Then, to our horror, as night was coming on, the fire burst out in North Salem, and sparks were flying thick and fast in our direction, and seemed very near us. Every one thought the whole city would go then. We, in this house, packed up our valuables, and were ready to leave on a moment's notice. But the Chief of the Boston Fire Department was at the head of things there, and Every energy was directed to staying the flames, and at last they were quelled, after twelve houses had been burned. Now, I will tell you of course of the bright things, for which to be thankful. So far, we know of only four lives that were lost, although all the missing have not been accounted for. Beautiful Chestnut St., with its fine colonial houses, was untouched by the flames, and <u>every</u> <u>historic</u> <u>landmark</u> was saved, the Essex Institute, the Museum, The House of Seven Gables, etc. etc. still remain. The center of the city, with all the stores, was spared. Only two churches were burned. Fifty families in our church, the Tabernacle, lost their homes and nearly everything in them. A Relief Station

Opposite: Mary W. Cleaveland was a resident of Salem during the time of the fire. In its collection, the Phillips Library holds an original letter she wrote to her cousin Henry, dated July 3, 1914. Mary wrote to her cousin to tell him that she was safe and unharmed; she described the fire and how she felt about the destruction she viewed afterward. The letter was donated to the library by Margaret E. Moore from Madison, Wisconsin; Ms. Moore's mother, Ruth Hungerford Cleaveland, was the daughter of Trumball Williams Cleaveland, referred to in the letter, and a cousin of Mary W. Cleaveland. Shown below is the full transcript of the letter. Underlined words and discrepancies in spelling and grammar appear as written by Mary Cleaveland. (*Neg #32940*)

was opened in our Chapel, on Monday morning, and we are trying first to help our own. Every one remarks on the patience and bravery, shown so largely, by the sufferers. Many of the refugees are living in tents, in the various camps, and are being fed by the city. Money has been pouring in from every direction, and many large sums have been given, but they say, that much, very much more, is needed. Salem has learned a lesson in this disaster, and a Commission has been formed to direct the rebuilding. If we had not had a very efficient Fire Department, the city would have been wiped out long ago, with so many wooden buildings. I shall never forget the horrors of June 25, the smoke, and flames, the almost constant sound of fire alarm whistles, the explosions from the houses which were being dynamited, and the Fire Engines and automobiles filling our streets, the latter loaded with furniture from the burning houses. So many people lost not only their houses, but employment also. I should think they would have to be cared for for a long time to come. This is a very poor account of the fire, I could tell you so much more, if you were here where I could talk with you. Perhaps you will let Trumbull [T. Williams Cleaveland of New York City] *see this letter, as I cannot write another one now. Mary telephoned to me during the fire and has been down since. Her mother is gaining slowly. I must run down to our Chapel now, to take a package, and see how things are coming on. I hope you will get this letter tomorrow, from*

Your loving cousin,
Mary W. Cleaveland

Chapter 4

Relief Efforts

At dawn the day after the fire, relief efforts for the homeless began with the distribution of food to the victims. Food supplies were brought into the armory, and tent camps were set up for the homeless at Forest River, Bertram Field at the high school and at the Willows, near the Tuberculosis Camp. As the *Boston Globe* reported on June 26, "Humanity, not charity, was the dominating feature of the relief work." The same article indicated that by 9:00 a.m. wagons, motorcars, trucks, trolley cars and trains brought supplies to the city. At the end of the day, enough supplies had been brought to the armory to feed the homeless for a week: 20,000 pounds of beef, 5,000 pounds of ham, 2,600 pounds of coffee, 5,000 loaves of bread, 5,000 pounds of canned fruit, 1,600 pounds of butter, 500 bags of potatoes and 500 dozen eggs. Hundreds of cans of milk were available and a supply of 7,500 tents arrived to house the homeless and the militia. Montayne Perry states in *The Salem Fire Relief* that "helpfulness and hopefulness were the dominant notes of that day. In the face of utter desolation and disaster, Salemites of every age and condition lifted their heads valiantly and faced the future with a courage that was nothing less than marvelous."

Two main relief camps existed: Forest River near the ocean and Bertram Field at the high school on Highland Avenue. All tents had floors and cots, and many housed up to 11 people. Bertram Field had 152 tents with 470 registered occupants. Forest River Camp accommodated 1,500 people in 400 tents. Each tent was numbered and registered at militia headquarters; registration identified the family and occupants living in the tent. The parks had electricity and flush closets were set up at both facilities a short time after the camps opened; prior to that, all public buildings in Salem were open for sanitation facilities.

An unidentified newspaper clipping, dated July 3, in one of the scrapbooks in the library collection wrote about a group of men sitting in a smoking club:

> *Say, Jack, do you realize that we have passed through one of the greatest disasters in this country and that at no time was there any panic? That within 24 hours the people were being systematically taken care of and that supper and breakfast were served so that many of the people burned out never lost a meal? Why say, Jack, to my mind the people of the state of Massachusetts should be proud of what was done in Salem by the rank and file…So I say, Jack, when you start finding fault with every one, put yourself in their place and see how much better you would do. When you realize that one-third of Salem was wiped out by fire and that in less than 48 hours comfortable quarters and plenty of food were to be had and that there was no epidemic of sickness, very little if any suffering, and practically no looting…Conservative we may be, bitter in our politics, strong in our opinions, but when the test comes we all belong to that brotherhood of man whose first thought is not of himself but of his less fortunate brother.*

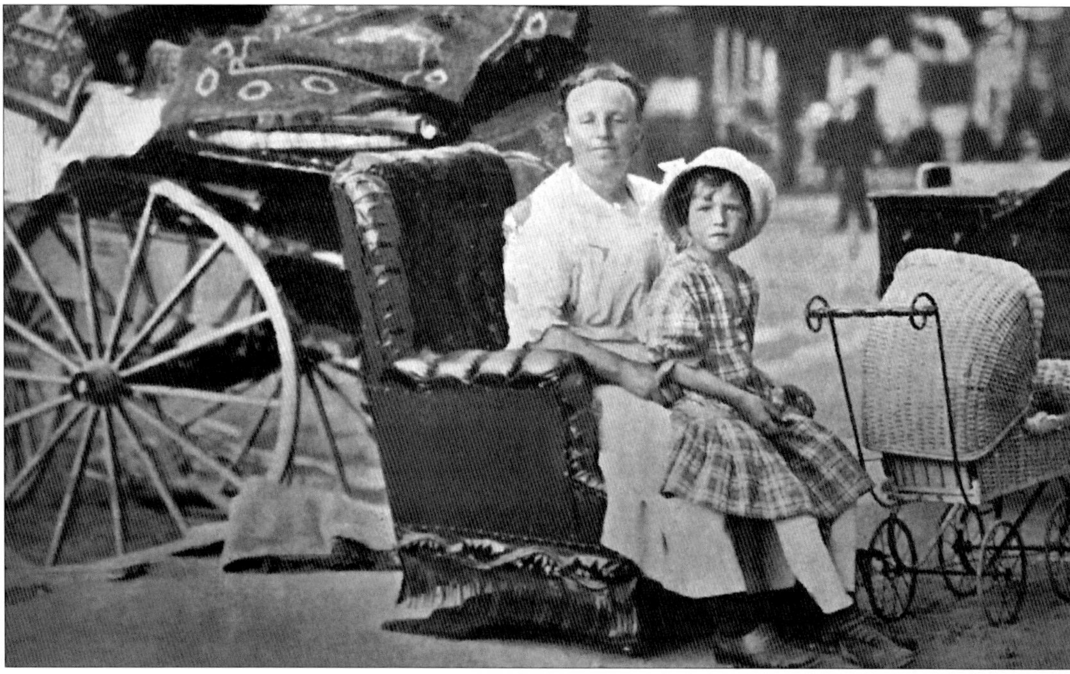

Both of these images are in one of the scrapbooks in the Phillips Library collection. The newspaper sources were not identified. The caption for the top image reads: "Mrs. Henry Reed and her daughter Helen are guarding their possessions on Harvard Street." (*Neg#32818*) The newspaper caption for the bottom image states: "Charlie Grey's family is ready to leave their home as the fire rages near Orne Street." (*Neg #32840*) Each of these images deftly portrays the pain and confusion of homelessness brought on by the fire.

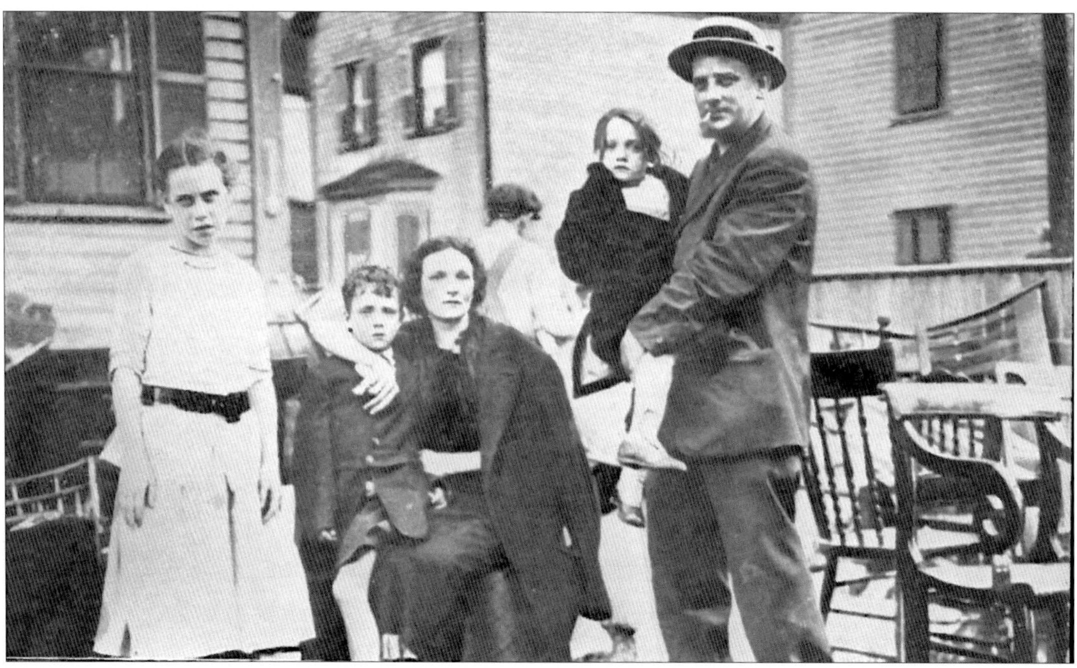

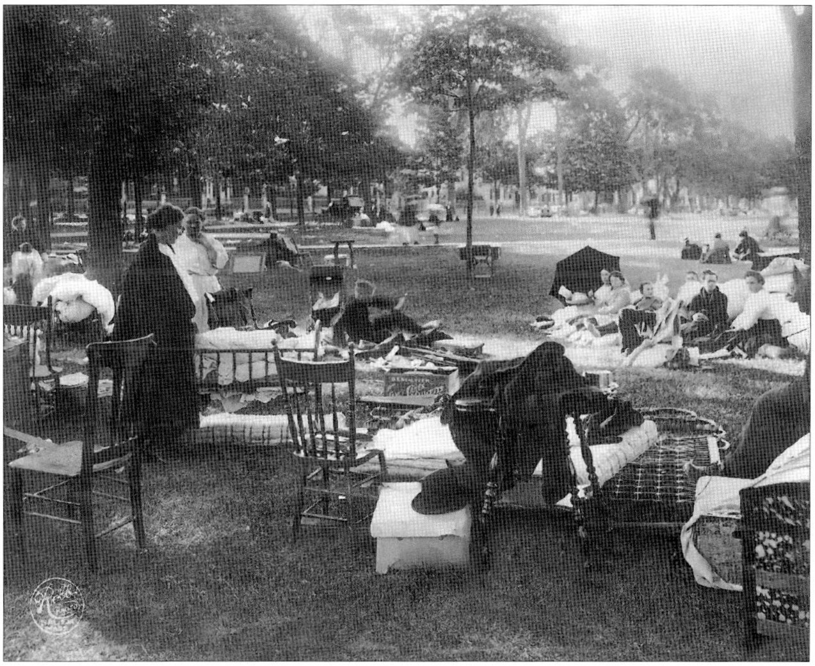

Families saved whatever they could from burning homes: furniture, carpenter tools, cook stoves, baby carriages and headboards and footboards of iron beds, which when draped with blankets could provide shelter. Some possessions were abandoned on the streets as the residents fled or were brought with them to Salem Common. In his *Reminiscences of the Salem Fire of June 25, 1914*, John Porter Sumner talks about carrying a tea table up the steep grade of Wallis Street toward Main Street. "I felt I couldn't go any farther, carrying that tea-table, which became heavier at every step. I wanted to leave it right there. But Mother insisted it was a family heirloom and that some day it would be mine…I acquired it after Mother's death in 1957 and own it still. It is one of my most prized possessions." (*Neg #4228, above, a Robb photo; Neg #32786, below*)

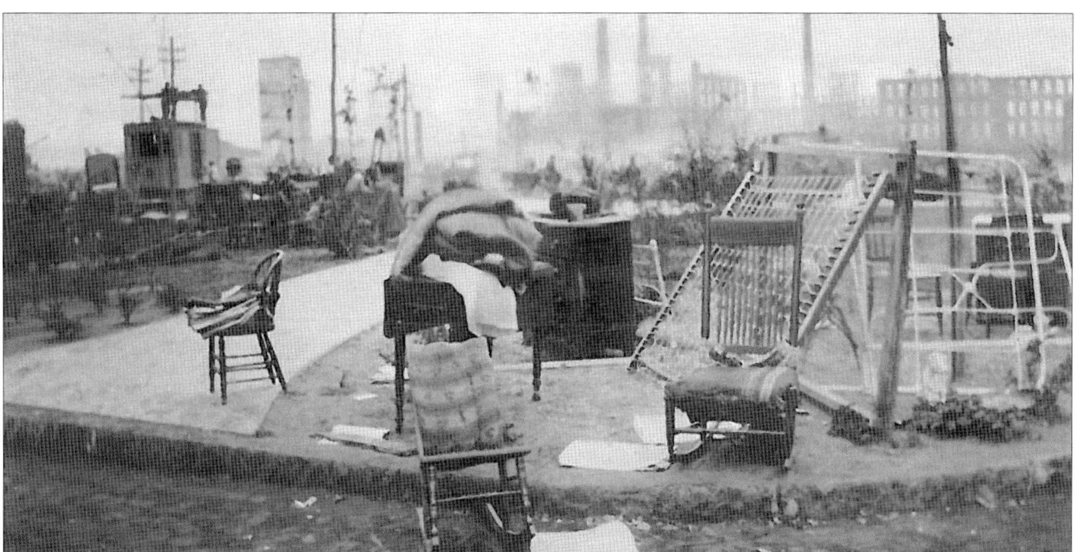

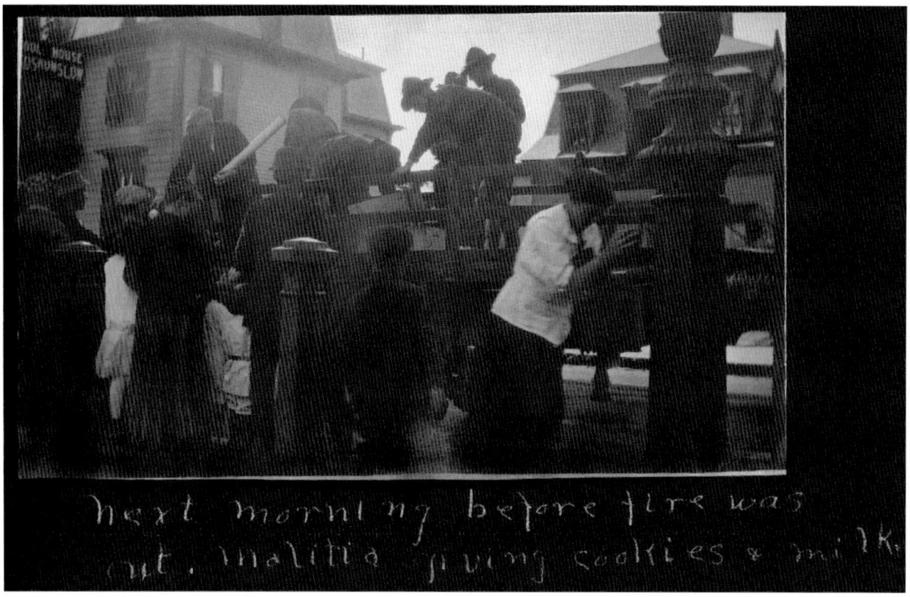

These photographs were in a family album compiled by Florence Foss Swan from 1908 to 1915. Each of the images is captioned. The upper photo caption reads, "Next morning before fire was out. Malitia [*sic*] giving cookies & milk." (*Neg #32892*) The lower photo caption reads, "Tents for Refugees on High School grounds." (*Neg #32893*)

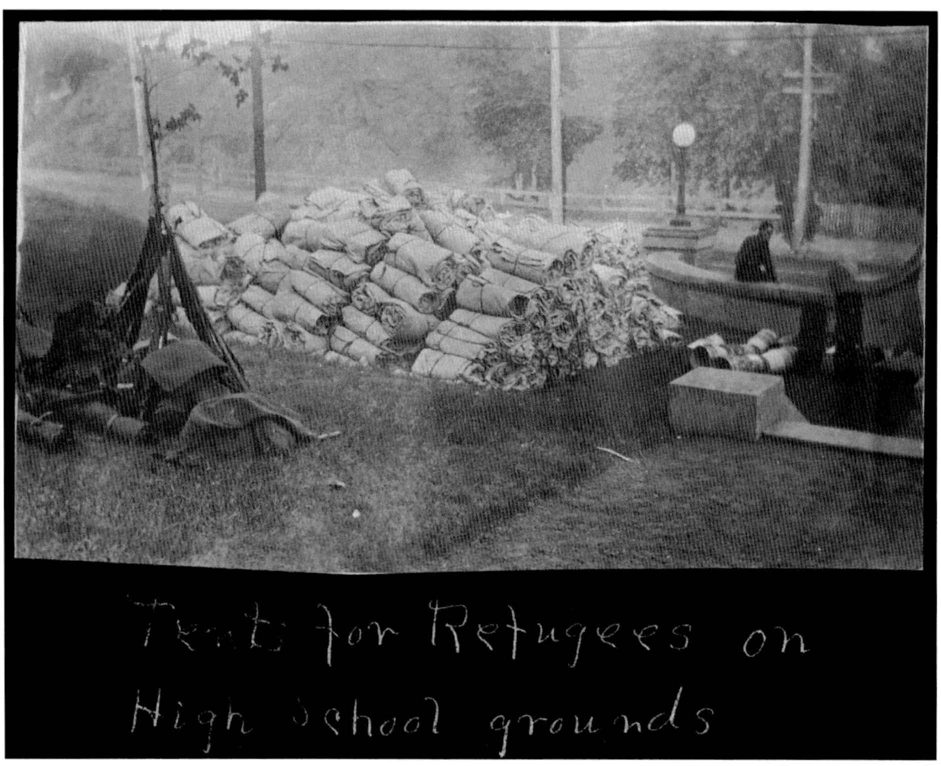

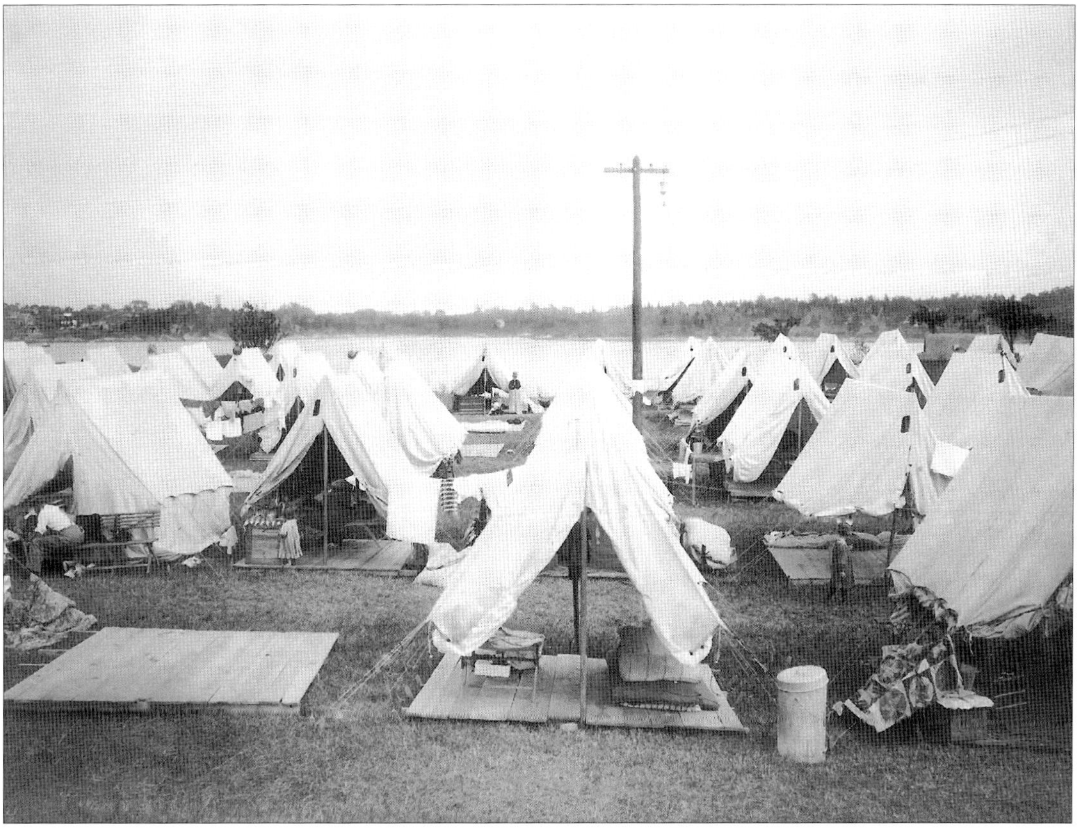

The tent city depicted below is the Forest River Camp along the banks of the Salem Harbor. (*Neg #4610*) On June 28, *The Salem Evening News* reported that Governor Walsh of Massachusetts spoke to the refugees at Forest River Park on June 27. "I first want to express the great sympathy that Massachusetts has for you all, and I promise you that this Commonwealth is really interested in you and concerned about your welfare, and nothing will be left undone to provide you with homes and employment." He asked refugees to be patient and "to obey regulations given you by militiamen and nurses, particularly in the matter of health in the camps."

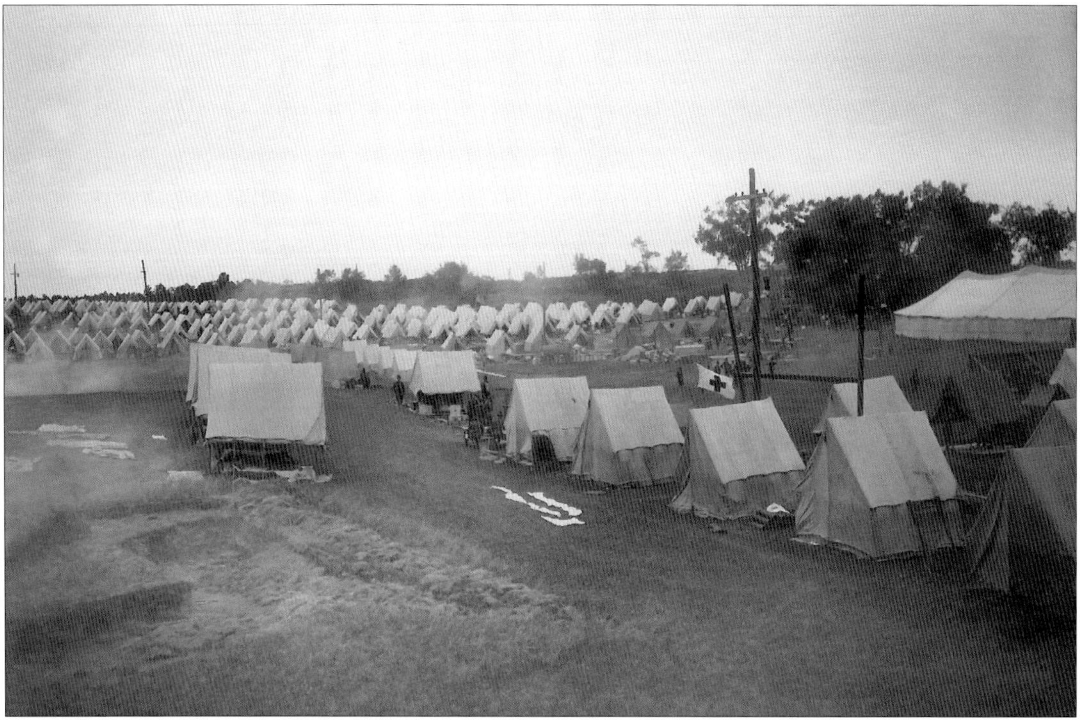

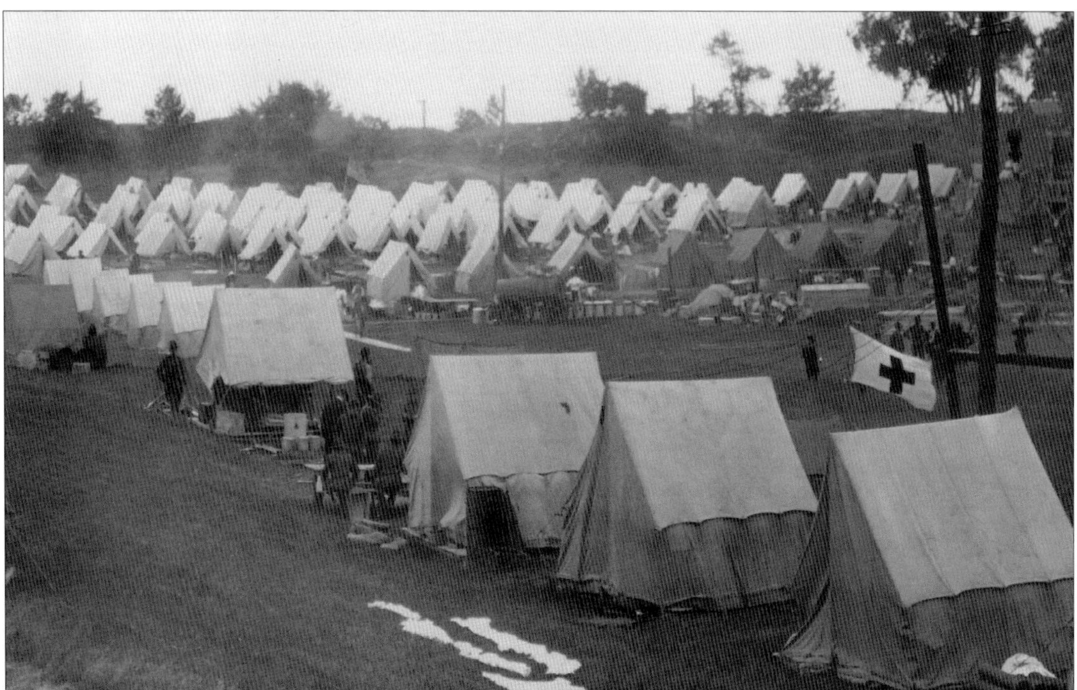

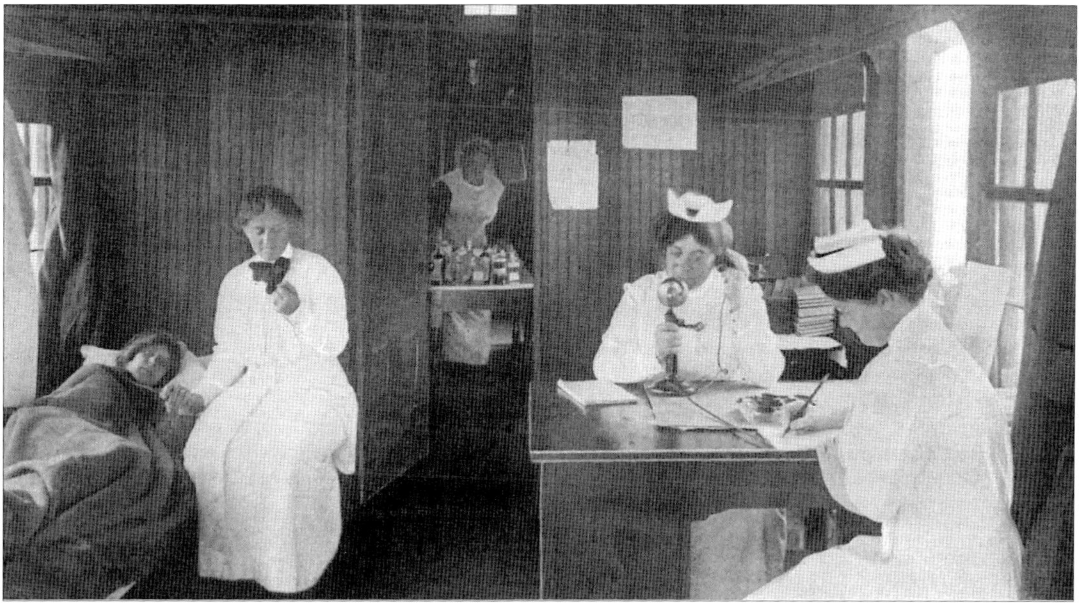

Forest River Camp, near the ocean (*Neg #4110*), accommodated over fifteen hundred people. Of the four hundred tents erected, three hundred were occupied by refugees; militiamen, the Red Cross and caretakers occupied the others. The large tent in the upper right portion of the complete image was the dining hall for the camp. The detail image displays the militia presence at the camp as well as the daily work of the volunteers managing the camp. (*Neg #4110Detail*) The Red Cross personnel at this camp maintained the health of the refugees. By July 13, the Red Cross had provided nine nurses, six of whom were regular Red Cross nurses and three of whom were volunteers. Immediately after the fire, physicians assisted the homeless free of charge. Records of the Salem Fire Relief Committee include a letter, dated August 17, 1914, asking that physicians receive some remuneration for their efforts; the committee agreed to pay for physician services. Hospital workers from Boston also came to assist the refugees; Margaret J. Fallon, the assistant superintendent of the Boston City Hospital, was in charge of the hospital at the Forest River Camp. John J. Lydon of Boston was the chief medical officer at Bertram Camp. The nurses in the photograph served the refugees at the Salem Willows Camp Infirmary. (*Neg #32835*)

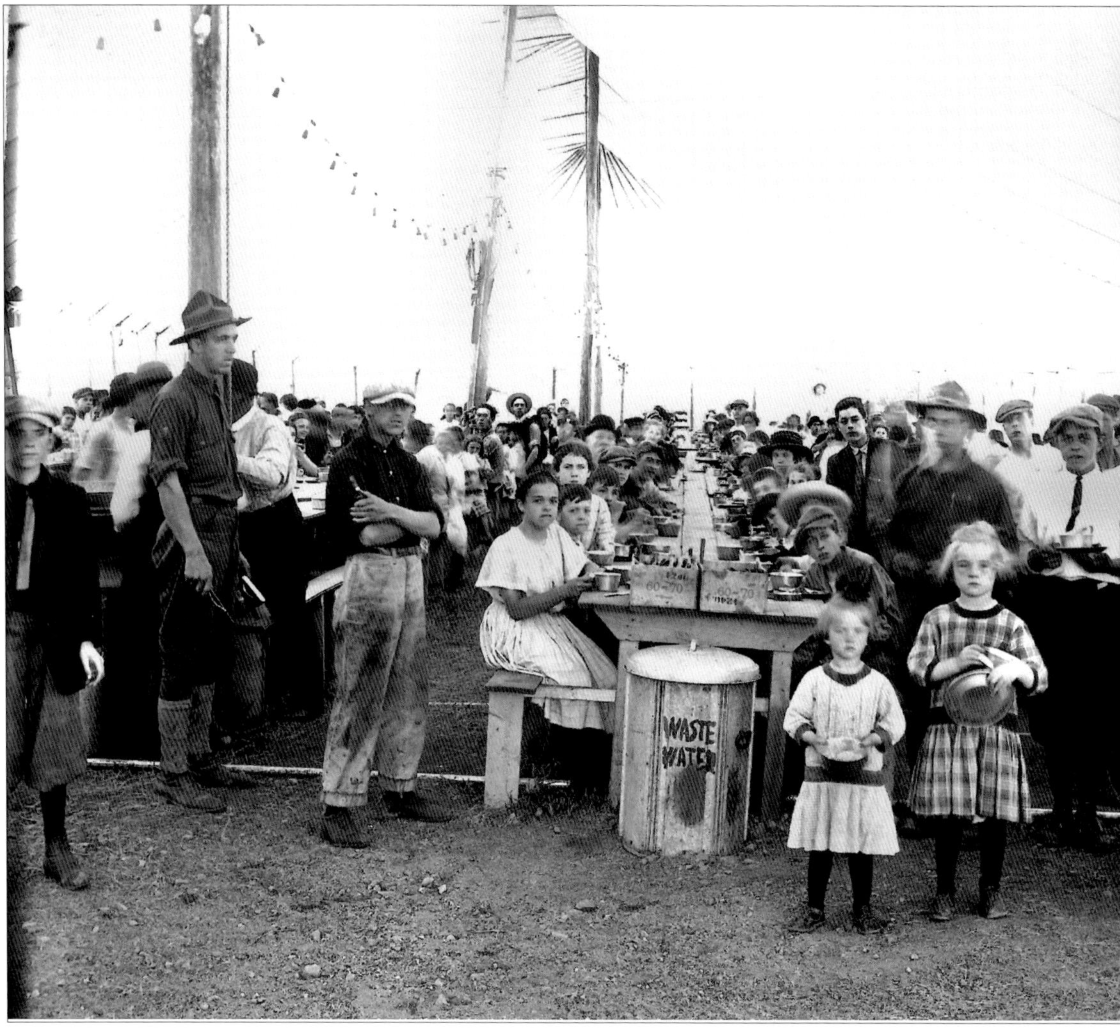
WASTE
WATER

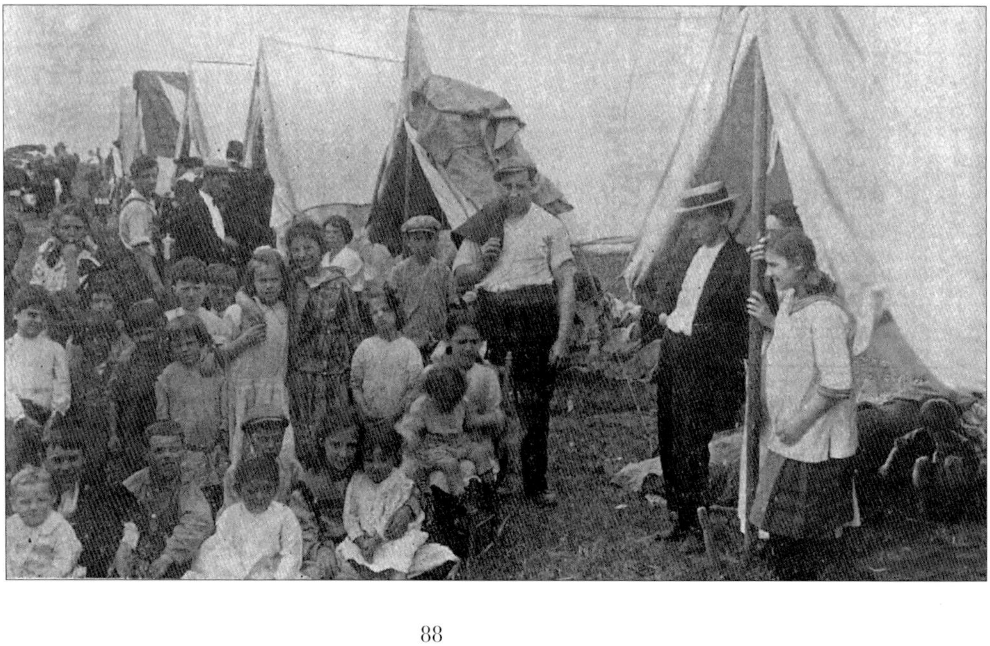

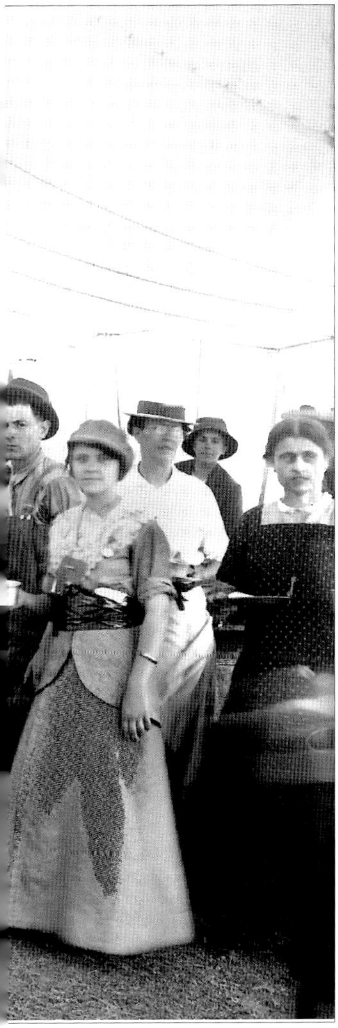

The Mess Hall at Forest River Camp was housed in a large tent erected near the road that entered the park. The tent was the dining hall and another source of shelter during the day. It sat hundreds of refugees at one time; obtaining a meal could take almost an hour. (*Neg #16789, a Robb photo*) *The Salem Evening News* carried many articles about the refugee tent camps. An article dated June 30 wrote about the Forest River Camp, in which a reporter stated that one is "reminded very forcefully of Longfellow's tale, 'Evangeline,' by a sight of this camp of exiled people." Although the surroundings may have been different, as much as possible the daily aspects of life continued at the tent camps. (*Neg #32808*) An undated image from the *Utica Journal* in a scrapbook depicts a group of children living in one of the tent camps. The caption to the image states: "Many of the children gleefully hailed the adventure as a lark." It is to the parents' credit that these children could respond so positively to their new environment. (*Neg #32944*)

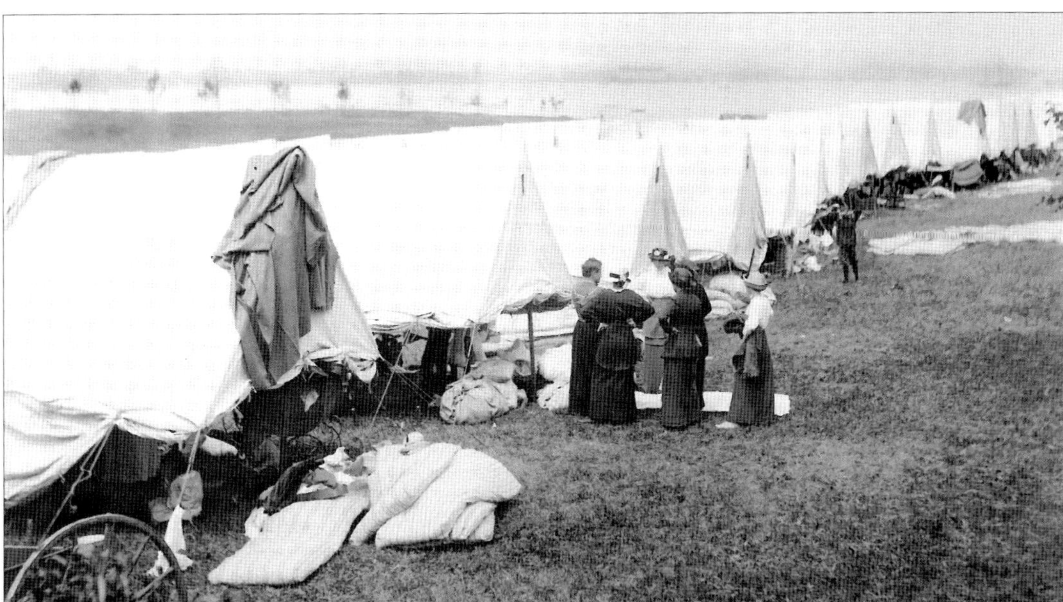

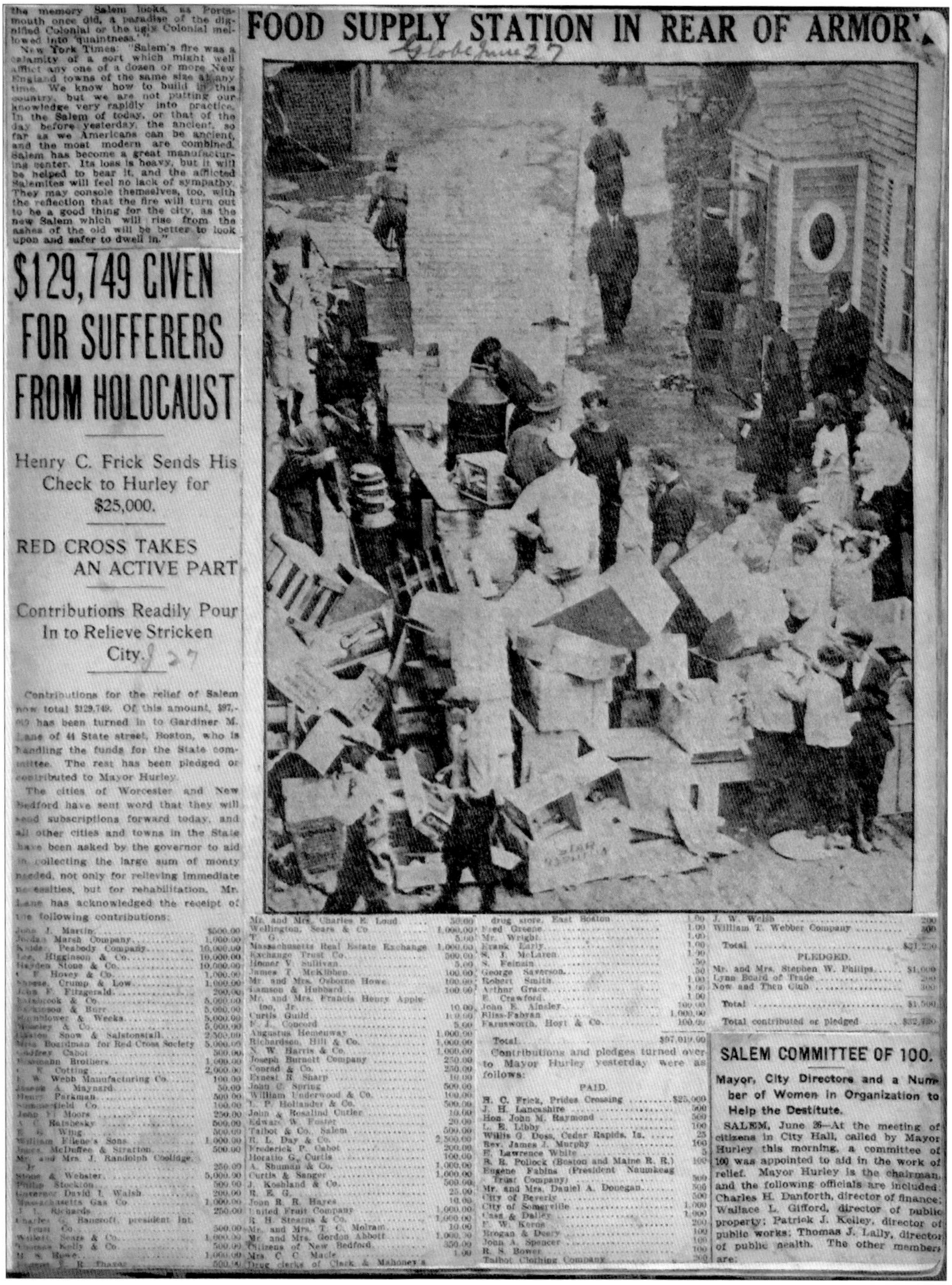

the memory Salem looks, as Portsmouth once did, a paradise of the dignified Colonial or the ugly Colonial mellowed into 'quaintness.'"

New York Times: "Salem's fire was a calamity of a sort which might well afflict any one of a dozen or more New England towns of the same size at any time. We know how to build in this country, but we are not putting our knowledge very rapidly into practice. In the Salem of today, or that of the day before yesterday, the ancient, so far as we Americans can be ancient, and the most modern are combined. Salem has become a great manufacturing center. Its loss is heavy, but it will be helped to bear it, and the afflicted Salemites will feel no lack of sympathy. They may console themselves, too, with the reflection that the fire will turn out to be a good thing for the city, as the new Salem which will rise from the ashes of the old will be better to look upon and safer to dwell in."

$129,749 GIVEN FOR SUFFERERS FROM HOLOCAUST

Henry C. Frick Sends His Check to Hurley for $25,000.

RED CROSS TAKES AN ACTIVE PART

Contributions Readily Pour In to Relieve Stricken City.

Contributions for the relief of Salem now total $129,749. Of this amount, $97,019 has been turned in to Gardiner M. Lane of 44 State street, Boston, who is handling the funds for the State committee. The rest has been pledged or contributed to Mayor Hurley.

The cities of Worcester and New Bedford have sent word that they will send subscriptions forward today, and all other cities and towns in the State have been asked by the governor to aid in collecting the large sum of money needed, not only for relieving immediate necessities, but for rehabilitation. Mr. Lane has acknowledged the receipt of the following contributions:

Total ... $97,019.00

Contributions and pledges turned over to Mayor Hurley yesterday were as follows:

FOOD SUPPLY STATION IN REAR OF ARMORY

SALEM COMMITTEE OF 100.

Mayor, City Directors and a Number of Women in Organization to Help the Destitute.

SALEM, June 26—At the meeting of citizens in City Hall, called by Mayor Hurley this morning, a committee of 100 was appointed to aid in the work of relief. Mayor Hurley is the chairman, and the following officials are included: Charles H. Danforth, director of finance; Wallace L. Gifford, director of public property; Patrick J. Kelley, director of public works; Thomas J. Lally, director of public health. The other members are:

On June 27, the *Boston Globe* printed a photograph of the Food Supply Station at the rear of the State Armory on Essex Street. The number of strewn boxes emphasizes the large number of refugees requiring provisions after the fire. Found in a scrapbook, the image depicts the state of the city in the aftermath of the fire, labeled as a "holocaust" in an accompanying article on the same scrapbook page. Lieutenant John E. Osborn, in charge of the armory, was a member of the First Corps of Cadets under General White and was assisted by Lieutenant Eugene H. Clapp. (*Neg #32907*)

LEAVING ARMORY WITH THEIR RATIONS.

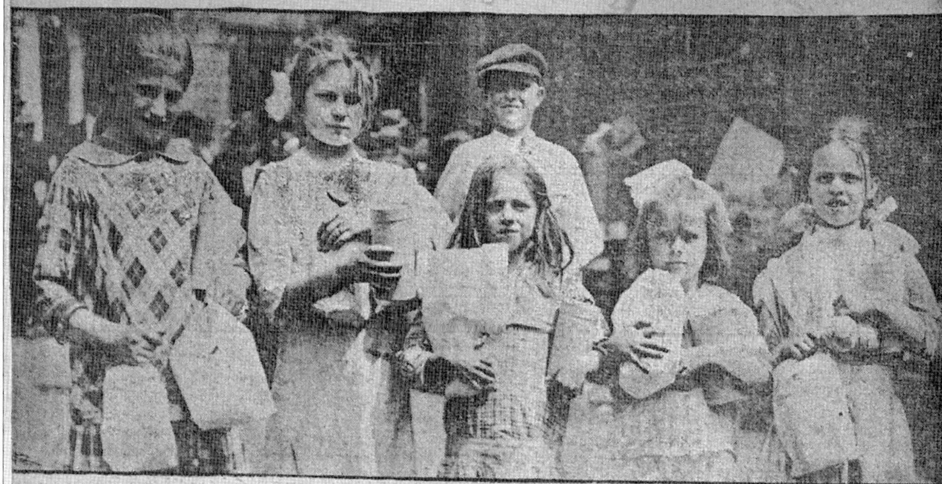

LITTLE SALEM FIRE SUFFERERS.
Leaving Armory of 2d Corps of Cadets With Their Food Allotments.

Grand Army hall at the Plains, and the vestry of the Baptist church at Danversport, and to have some one there every day for a week at least to take charge of clothing and bedding that is to go to Salem. The committee in charge of the relief work in Salem have asked for clothing of all kinds, especially simple, plain clothes that may be put right on, and underclothes and stockings, even those which are mended; for household and kitchen furniture, pans and dishes and most especially for bedding. Bedding and money in the form of cash they are most in need of, but those who wish to give money will please give it to the treasurer for the fund from Danvers to be appointed later by the town.

M. C. Pettingell has kindly offered to carry to Salem all the things that are brought to the three above-named placed.

Men and women who can volunteer to give their services to help the relief work in Salem will please report to Miss Kimball at Franklin hall, Essex street, near Washington square, Salem. Both trained workers and volunteers are needed.

"Riverbank," the William Penn Hussey estate, was turned into a hospital during Thursday night. Scores of refugees camped out in the vicinity of Clinton avenue and other places at Danversport during the fire. The selectmen and other citizens took steps yesterday to have them supplied with food through the Salem relief committee.

Dr. Curnick's subject at the Methodist church tomorrow morning will be: "The Salem fire." At the 7 o'clock service Mrs. O. A. Hartman and Mrs. Bertha Hood will sing a duet. Dr. Curnick will preach on: "The broad and the narrow way."

Miss Marjorie Lynch of the Specialty shop has provided a box for contributions of the Salem relief fund and suggests that the storekeepers generally follow out this plan.

Rev. Edward H. Cotton has spent several days at Brewster, arranging for the Boy Scout camp.

Out of respect and sympathy for the city of Salem, believing that her overwhelming problem subordinates all else just now, the Danvers store-

Principal H. E. Wentworth of the Saltonstall school of Salem, formerly master of the Tapley school, was among the Summer street residents burned out in the Salem fire, while several other former Danvers people were victims.

FLYING BRICKS ENDANGER LIFE

Hundreds of Pieces of Shattered Brick Fall in Crowded Streets; Several Injured This Morning.

Unless special precautions are taken there will be some casualties from the dynamiting of the standing walls of the burned buildings.

About 10.45 this morning a blast was fired at the P. A. Field factory, Canal and Mill streets, and a shower of shattered bricks fell all the way up Washington street as far as Town House square, several pieces as large as hen's eggs being being up in front of the *News* office.

Several citizens who did not even know blasting was going on were struck by flying missiles.

One man sustained asevere cut on the back of the head from which the blood flowed freely.

The most serious result of this expolsion was an injury sustained by oJhn Collier of 27 Andrew street. He was standing just outside of the railroad station, facing away from the fire district. When the explosion came a piece of brick, nearly half the original size, struck him in the back, knocking him down. He was rendered helpless by the shock and was picked up and carried into the station agent's office in the railroad station, and was later taken home.

Another blast at 11 o'clock sent up another shower of shattered brick, but no injuries were reported from that.

MANUFACTURING INTERESTS SUFFERED HEAVILY IN FIRE

Half the Industries of Salem Were Completely Wiped Out; Invested Capital Sustained Loss of Fully $5,000,000 While Over 3000 Persons Are Thrown Out of Employment; This Last Means Loss of $30,000 Weekly in Wages, Consequent Hardship and Much Suffering; Suggestion That Building Corporation Be Formed Giving Employment to the Needy and Ensuring Modernly Constructed Factories.

About one-half of the manufacturing industry of Salem was wiped out by the fire. That means a loss of $5,000,000 in invested capital, the throwing out of employment of more than 3000 persons and the cutting off of wages at the rate of $30,000 weekly. The last loss is the greatest loss of all. It means cutting off the income of bread winners of hundreds of families and consequent hardship and suffering. Thousands of persons, who never knew what it was to ask for aid, will have to be helped until they can find new means to earn their livelihood.

The fire came just as the manufacturing industries were about to emerge from a dull period and to increase their business and to give em-

received $50,000 worth of the latest textile machinery from the Draper works and was about to increase its output. It will take two or three years to rebuild and re-equip the mills and get them running as smoothly as before the fire.

The shoe and leather manufacturers suffered much. The Wilkinson Counter Co.

Had Just Moved In

to its new factory on Margin street. It had not had time to get it thoroughly insured. It had it covered with a blanket policy, a temporary arrangement. It was getting ready to make 60,000 pairs of counters a day, and was anticipating the best year of business in its history. When

An unidentified newspaper clipping with only the letter "G" and "27" written at the top of the image shows children with food rations. The caption at the bottom of the image states: "Little Salem Sufferers, leaving Armory of 2d Corps of Cadets with their food allotments." The armory was the commissary for food supplies, which included loaves of bread, beef in cans and in bulk, biscuits, crackers, potatoes, butter, evaporated and fresh milk, ready-to-drink and raw coffee and all kinds of canned goods. (*Neg #32914*)

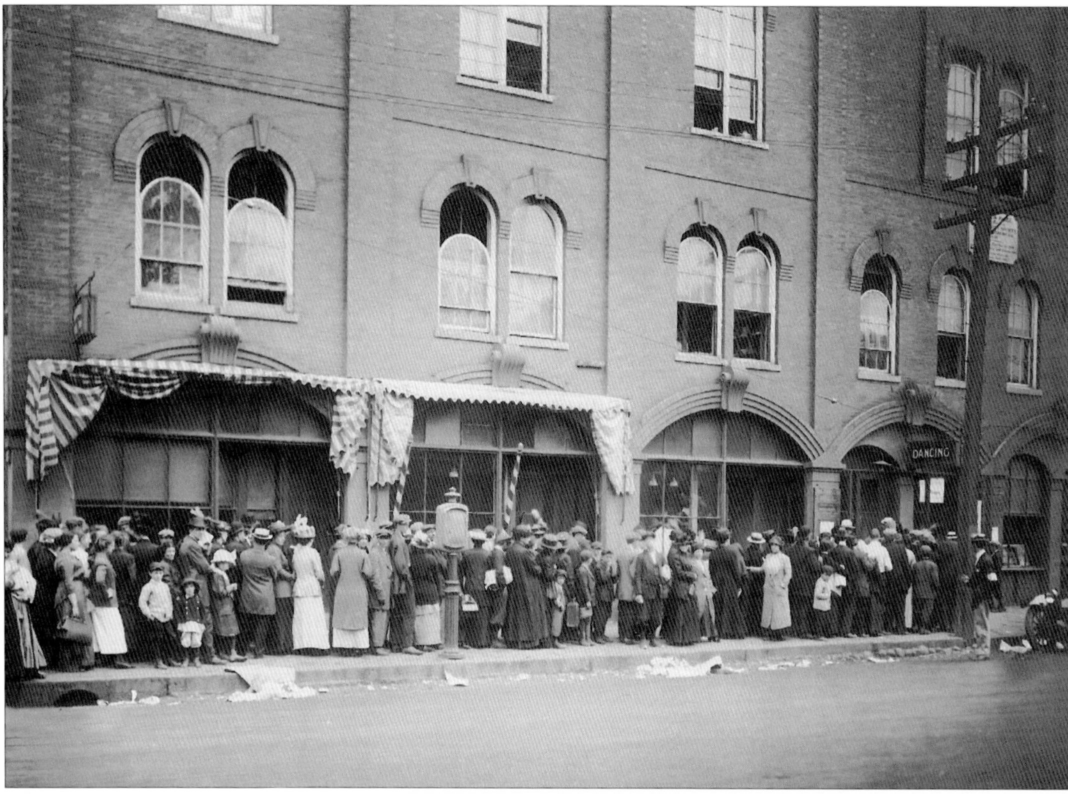
DANCING

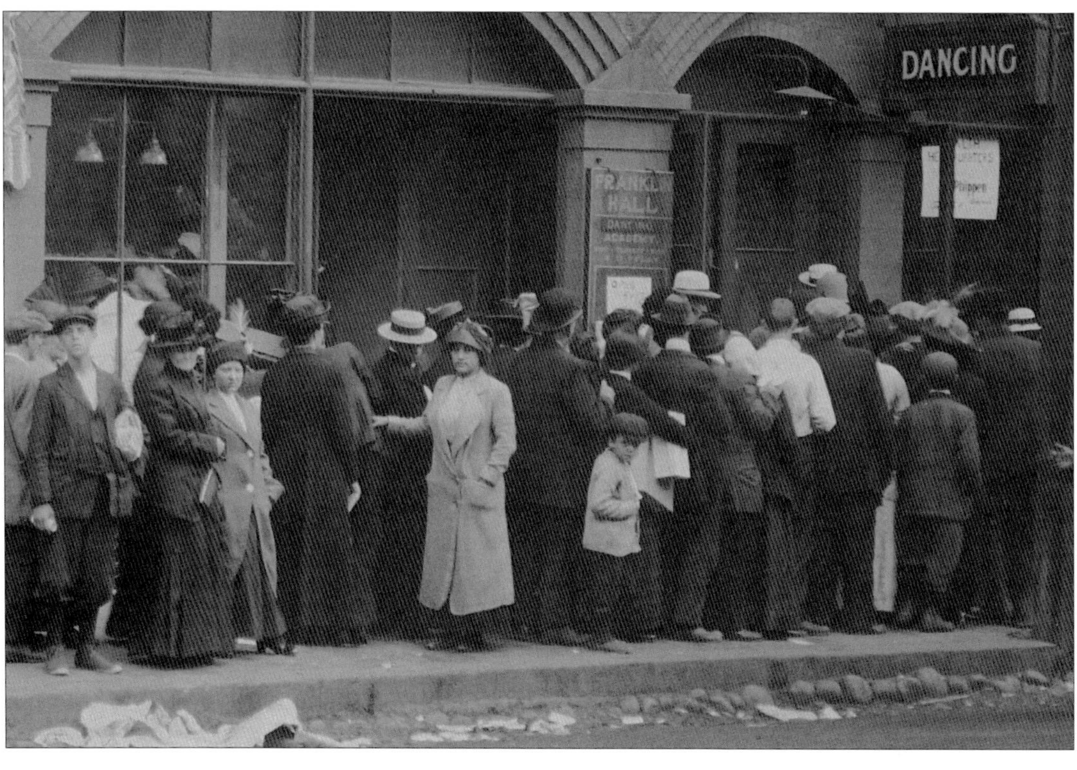
DANCING

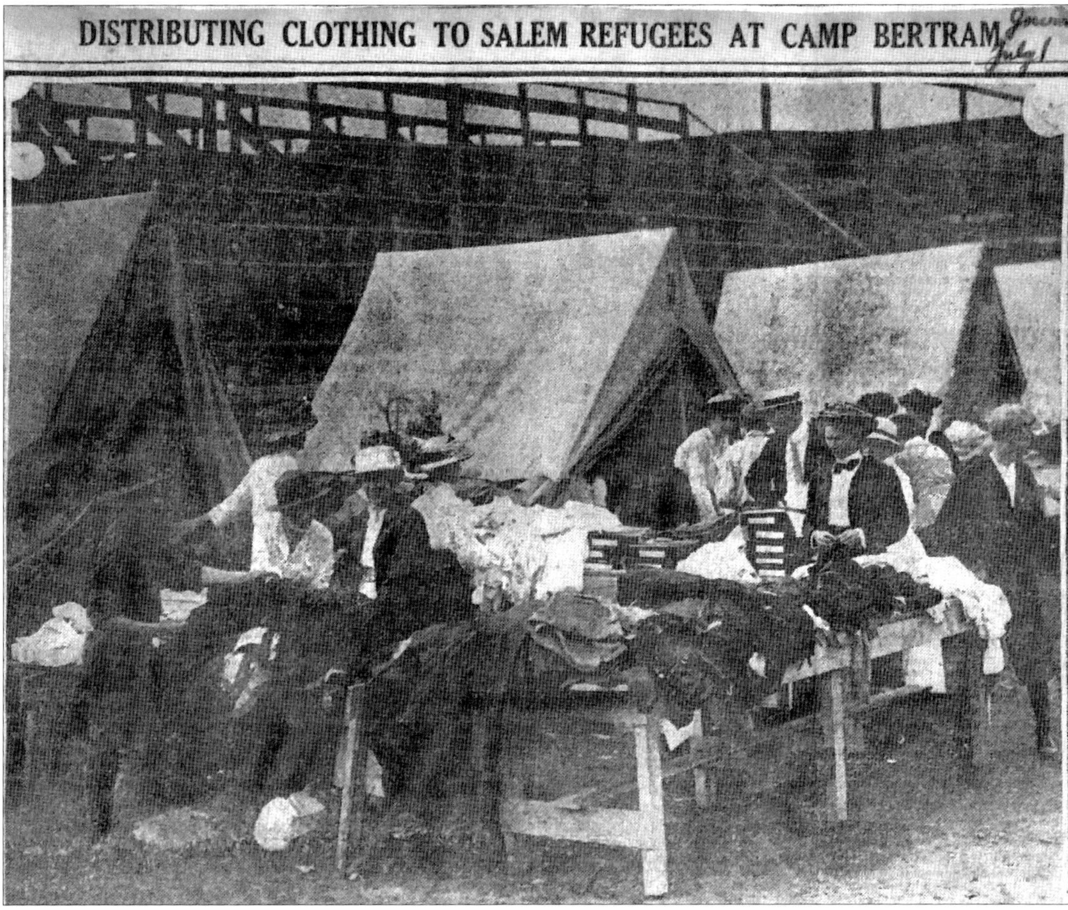

A storehouse was opened in Franklin Hall to distribute clothing to those who had suffered in the fire. A steady stream of refugees, about 250 per hour, came to receive clothing, many of whom were parents looking for children's clothing. Distribution of the clothing was under the direction of Mrs. J. Willard Helburn, who had experience as a social worker and had also suffered during the fire. Salem Fire Relief Records state that donated clothing was fumigated at a central receiving station before being passed out to the refugees. (*Neg #4536, a Robb photo; Neg #4536Detail*) The newspaper image, dated July 1, with *Journal* listed as its source, shows that clothing was also distributed at the refugee camps. Camp Bertram identified in the caption was actually Highland Park, located in Bertram Field, on Highland Avenue, the athletic field of Salem High School. (*Neg #32911*)

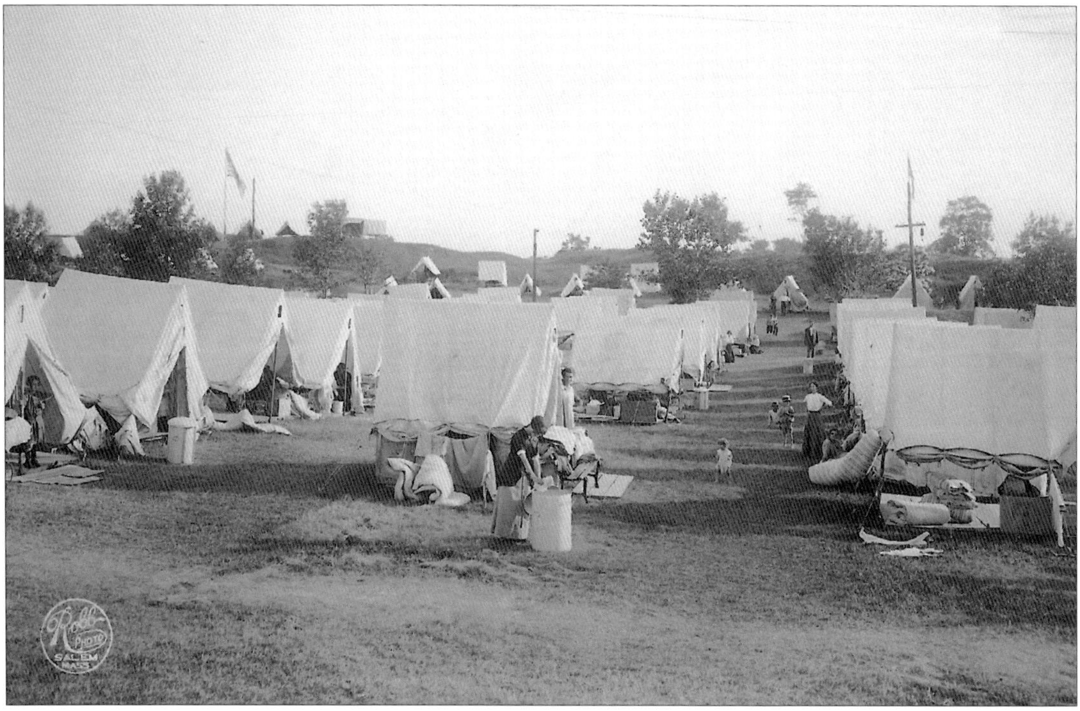

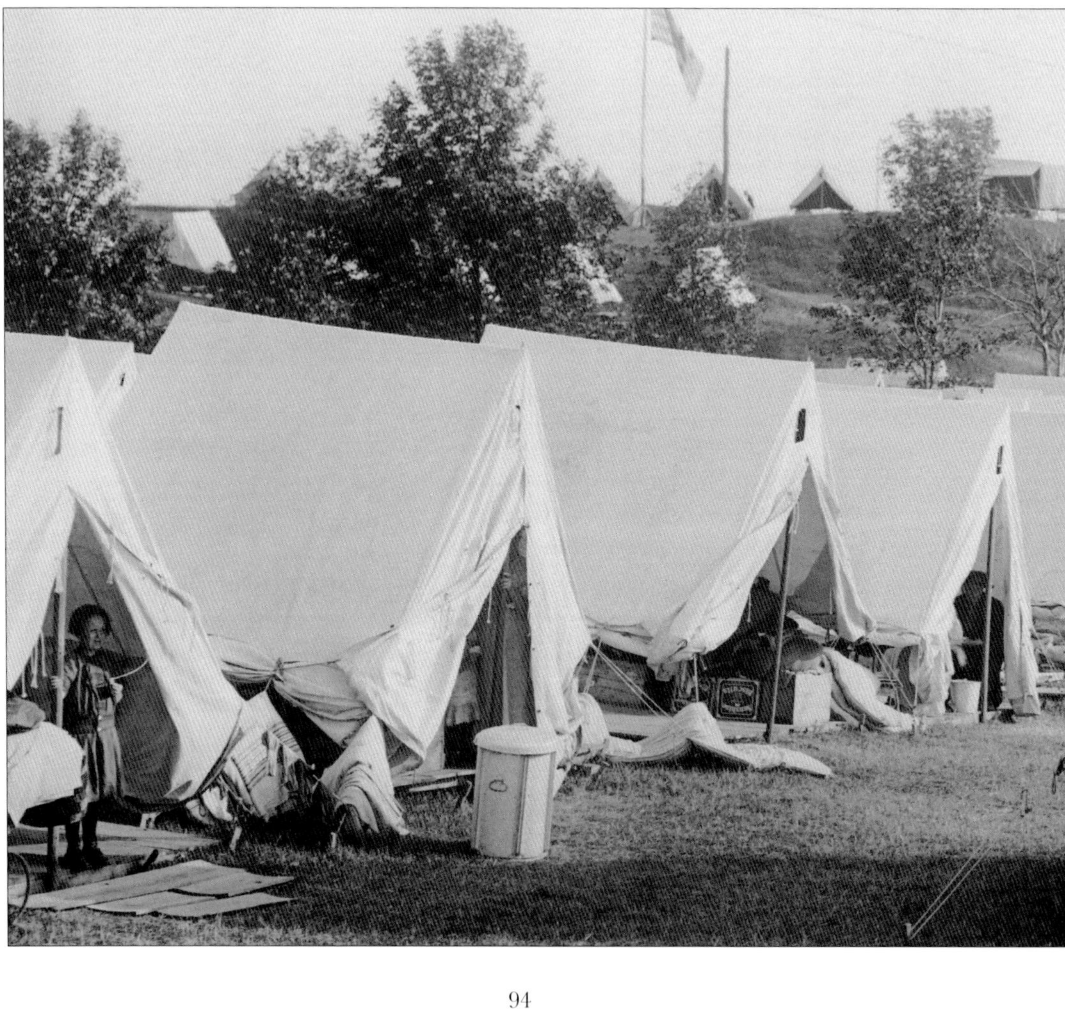

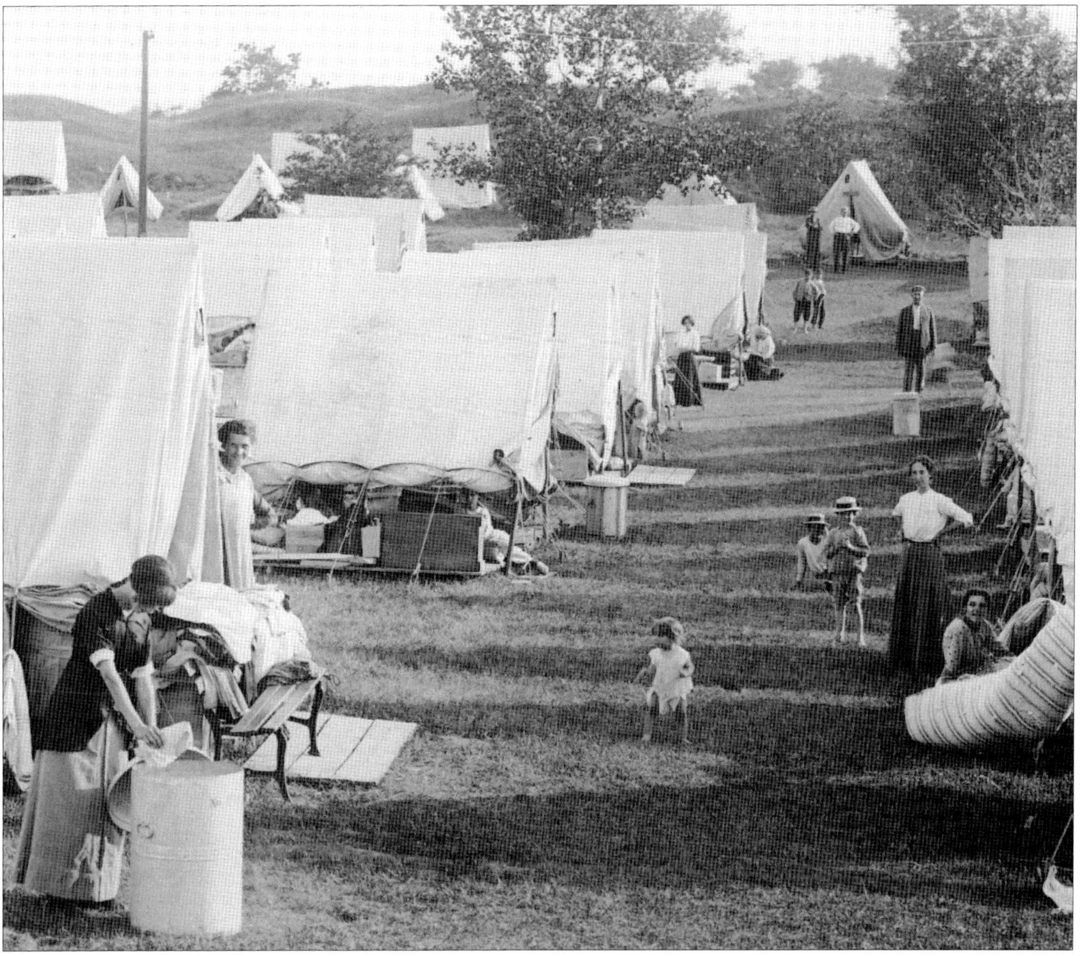

Forest River Camp was the principal camp for refugees. Colonel Frank A. Graves of the Eighth Regiment was in charge of the camp, assisted by the companies of the Ninth Regiment. Members of the cadet artillery were in charge of the camp commissary. Tents at Forest River Camp included wooden floors to protect inhabitants from the rain. Some families built moats around the tents for further protection. Each tent was labeled with the name of the family occupying the tent, identified as a "rude substitute for a doorplate" in a July 30 article in *The Salem Evening News*. Tents housed homeless families and stored the variety of possessions saved as refugees fled burning homes—family portraits, empty bird cages, sewing machines, hat boxes, wash basins, kitchen utensils and rolls of embroidery. Father Binette, acting pastor of St. Joseph's Church (Reverend George A. Rainville, pastor of the church, was in Europe at the time of the fire), performed outdoor prayer and benediction at the end of each day, including a sermon to guide camp residents in their physical and spiritual well-being. When the camp first opened, residents were called to Mass at the sound of a military bugle; eventually Father Binette was given a big bell to call his parishioners to the evening benediction. (*Neg #4606, top previous page; Neg #4606DetailA, bottom previous page; Neg #4606DetailB, this page*)

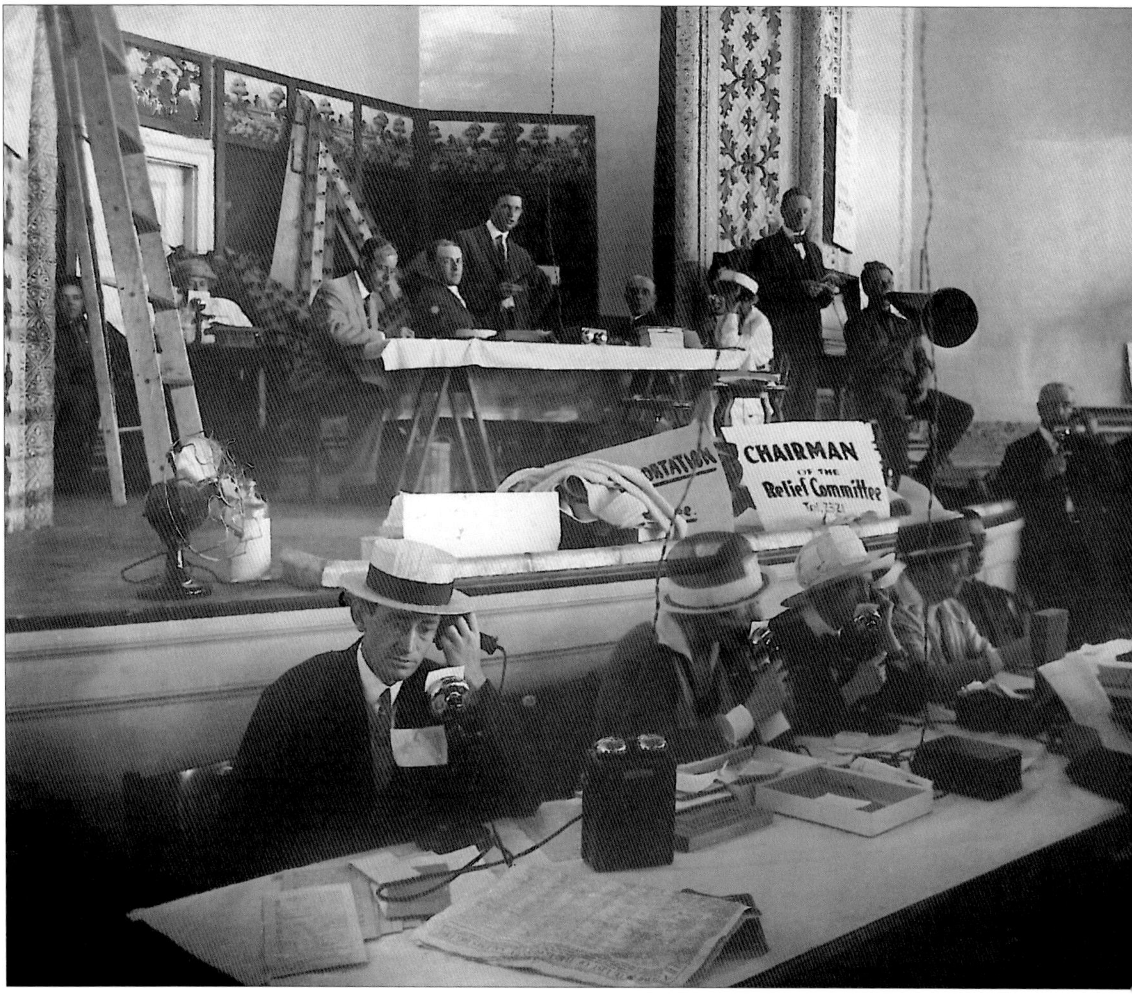

Organized relief efforts began at dawn on June 26. A Committee of One Hundred was chosen by popular vote at a meeting held at city hall later that day. Too large to be effective, the Committee of One Hundred created an executive committee from which seven people were chosen to work with a group of seven others appointed by Governor Walsh. This committee became known as the Committee of Fourteen, and its members included John F. Moors, James J. Phelan, Reverend Father Michael J. Scanlon, A.C. Ratchesky, Bishop William Lawrence, F.L. Higginson Jr. and John F. Fitzgerald, appointed by Governor Walsh; John F. Cabeen, John B. Tivnan, Daniel A. Donahue, Charles H. Danforth, Ethel B. Osborne, Christian Lantz and Theodore B. Bacon, appointed from the Salem executive committee. According to Montanye Perry, author of *The Salem Fire Relief*, the Committee of Fourteen established a committee list that included departments on application, complaint, transportation, insurance, housing, registration, investigation, certification, food, furniture and clothing. Generally known as the Rehabilitation Committee, these committees originally started their relief efforts in the armory but soon moved to the Now and Then Hall on Essex Street. The *Saturday Evening Observer*, dated June 27, reported that volunteers for relief work came from all parts of Massachusetts, as did people from throughout the country who had experience working with the aftermath of conflagrations and disasters. (*Neg #32872, photographed by the New England Telephone and Telegraph Company*)

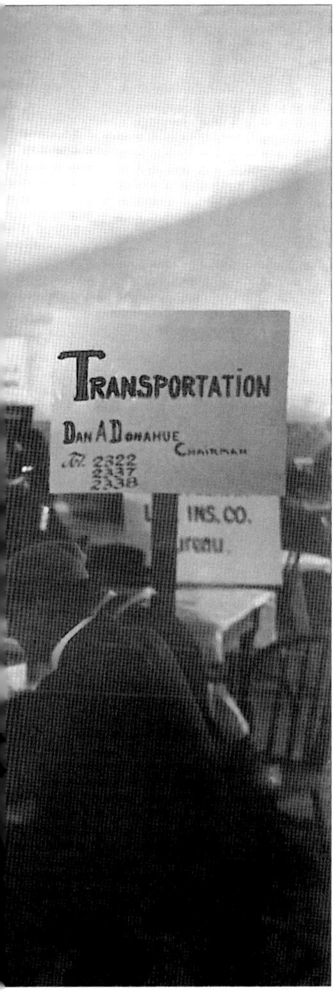

Miss Janet M. Curran, toll chief operator, and Miss Blanche E. Ross, local chief operator, were in charge of the switchboard operators during the fire. Sixteen women were working when the fire started; additional operators returned to work and calls for extra operators went out to neighboring towns as the fire progressed. The fire came within five hundred feet of the telephone building; awnings were removed from the office windows, telephone men were stationed at every window with pails of water and the roof was continually drenched with water. Miss Curran saw her home in flames from one of the windows in the building; she is reported to have said, "There goes everything I own." Undeterred, she continued to work. Over one thousand phones burned, one third of the total in use in Salem before the fire. More serious was the burning of one third of the cables connecting Salem and the North Shore to Boston and the metropolitan area. These lines were lost when the railroad bridge on Highland Avenue was destroyed, as the lines were suspended in a conduit under the bridge. Repairs to the telephone system started at 2:30 a.m. on June 26 under the direction of Division Cable Foreman E.A. Robinson of Lowell. Seven special lines were placed at the relief committee headquarters and over sixty special lines were put into operation at relief camps, city offices, military stations and relief committee stations; a special branch exchange was set up to handle the extra lines. Additionally, twelve extra lines were set up in the offices of *The Salem Evening News* to accommodate the needs of Boston reporters. Such a response from the telephone workers was not surprising, as the telephone originated in Salem. Alexander Graham Bell lived at 292 Essex Street, where he perfected the telephone; he made the first long-distance call at the Lyceum on 43 Church Street. (*Neg #32917, photographed by the New England Telephone and Telegraph Company*)

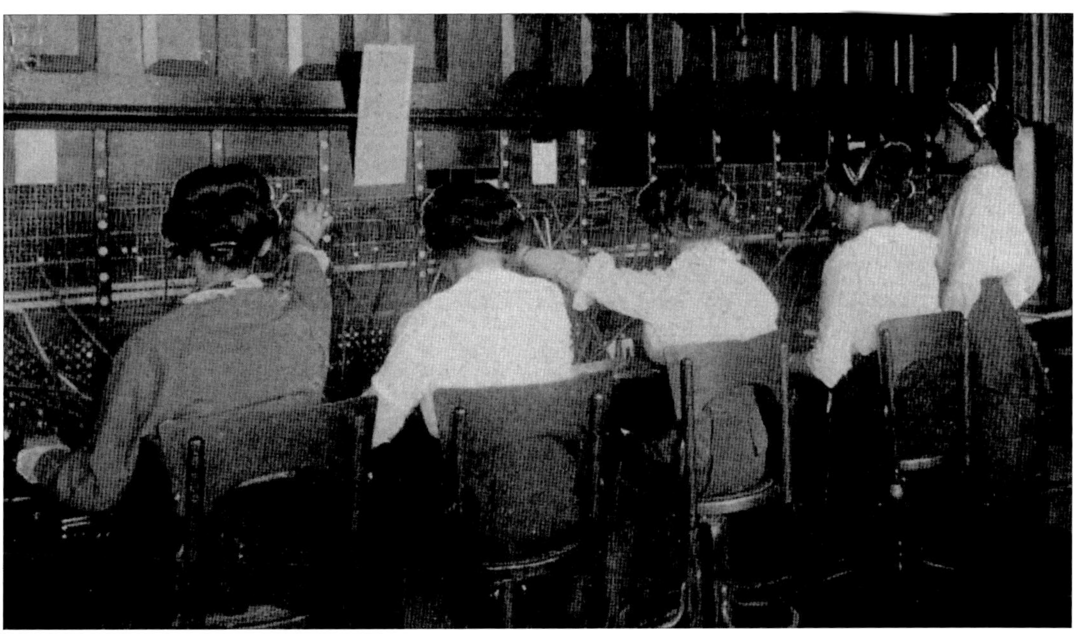

GOODS

TO BE SUPPLIED BY THE
REHABILITATION COMMITTEE.
SALEM FIRE RELIEF COMMITTEE

Name ....

Address ....

No. in family ....

Stove ....

Bed, Double ....

Bed, Single ....

Spring ....

Mattress ....

Pillows ....

Pillowcases ....

Blankets ....

Sheets ....

Table ....

Chairs ....

Bureau ....

Knives, forks & spoons ....

Dishes ....

Tin ware ....

Wash Tubs ....

Wash Boards ....

Ice Box ....

I, ...

of .... Street
Salem, Mass. hereby certify that I have no real estate, or personal property, money in savings bank or tangible assets of any kind, that I have no claim for insurance and that I have not previously received aid except ....
from the Salem Fire Relief Committee.

....

Name ....

Previous Address ....

Present Address ....

No. in family ....

Occupation ....

Real Estate ....

Insurance ....

Money in Bank ....

Other Assets ....

Previous Aid ....

Signed,

....

Mrs. Bik has had from Salem Fire Relief Com

#8 Stove and funnel
double bed and bedding
4 pillows
8 pillow cases
6 prs. blankets
10 sheets
dining table
5 dining chairs
4 bed room chairs
$50.00 cash made payable to Mrs. Hunt
$62.00 food
chiffonier
bureau
Rocking chair
2 bed room tables
Complete bedding for crib
Washing and ironing outfit
Ice box
sewing machine
$12.00 for rent
Clothing for woman and 3 children

The above relief is on the records of the Salem Fire Relief Committee and it is quite possible Mrs. Bik has had additional relief from the Lynn office which we cannot get a record of this morning as Miss Barnett is not at her office.

The law provides that Mothers' Aid shall be adequate!

In *The Salem Fire Relief*, written by Montanye Perry, a question is posed: "To whom is relief given? The answer is so simple as to admit to no misunderstanding: Relief was given to those who had to have it to start again in life." To start the process, refugees had to certify that they were homeless and needed help after the fire. They were required to complete forms listing their previous address and present address, as well as provide information about their current financial status and employment history. (*Neg #32854*) Forms were then filled out to identify the items provided by the relief committee (*Neg #32855*) and records were kept to keep track of those who had already received relief. (*Neg #32834*) Those refugees at the relief camps were also required to have identification cards to move freely onto the campgrounds. As much as possible, male refugees helped to clean up the debris from the fire; these men received additional cards to identify their employment status.

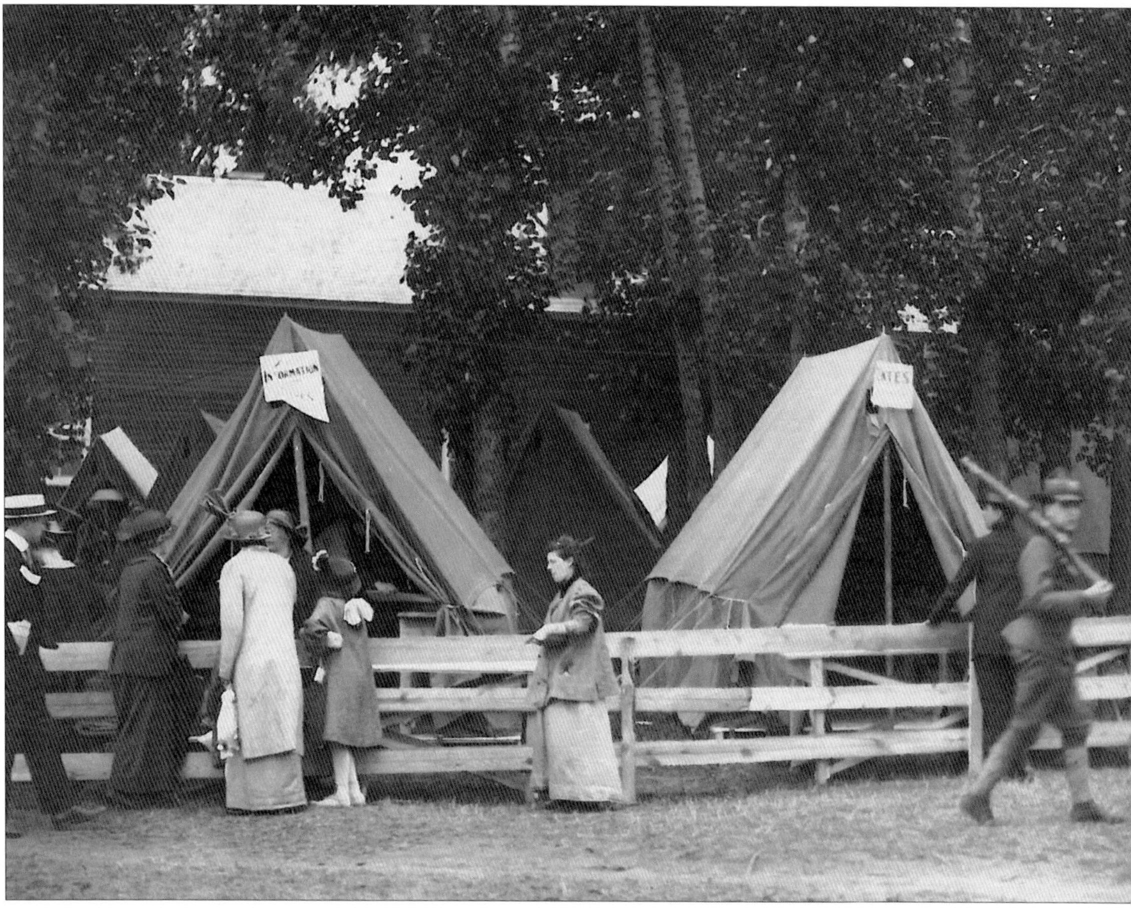

Tents at relief camps were provided for housing as well as for the administrative needs of the camp and its residents. The top image shows tents providing information to the refugees as well as a tent that was the camp post office. (*Neg #4008*) The bottom image shows the tents that were supply stations at the Forest River Tent Camp. The militiaman in the picture is using one of the special phone lines specifically set up to service the relief camps. (*Neg #32836*)

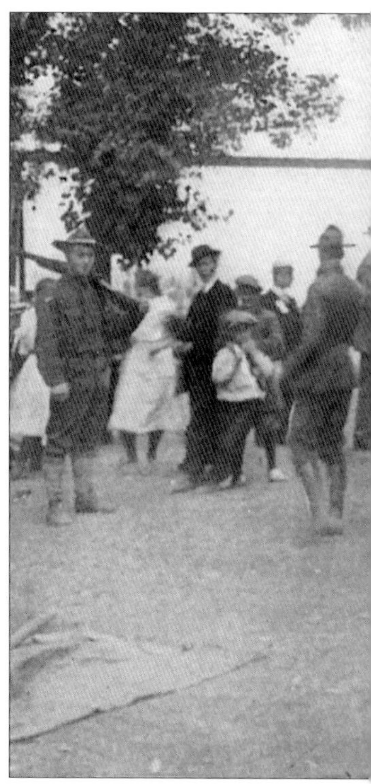

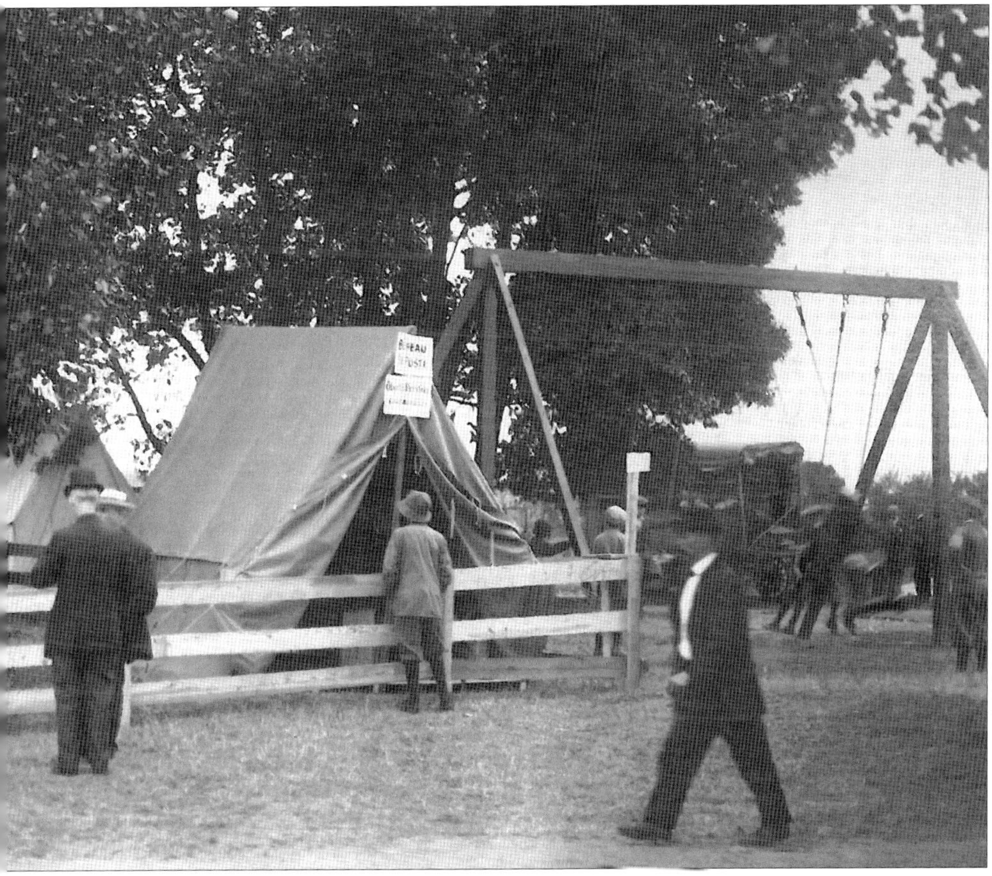

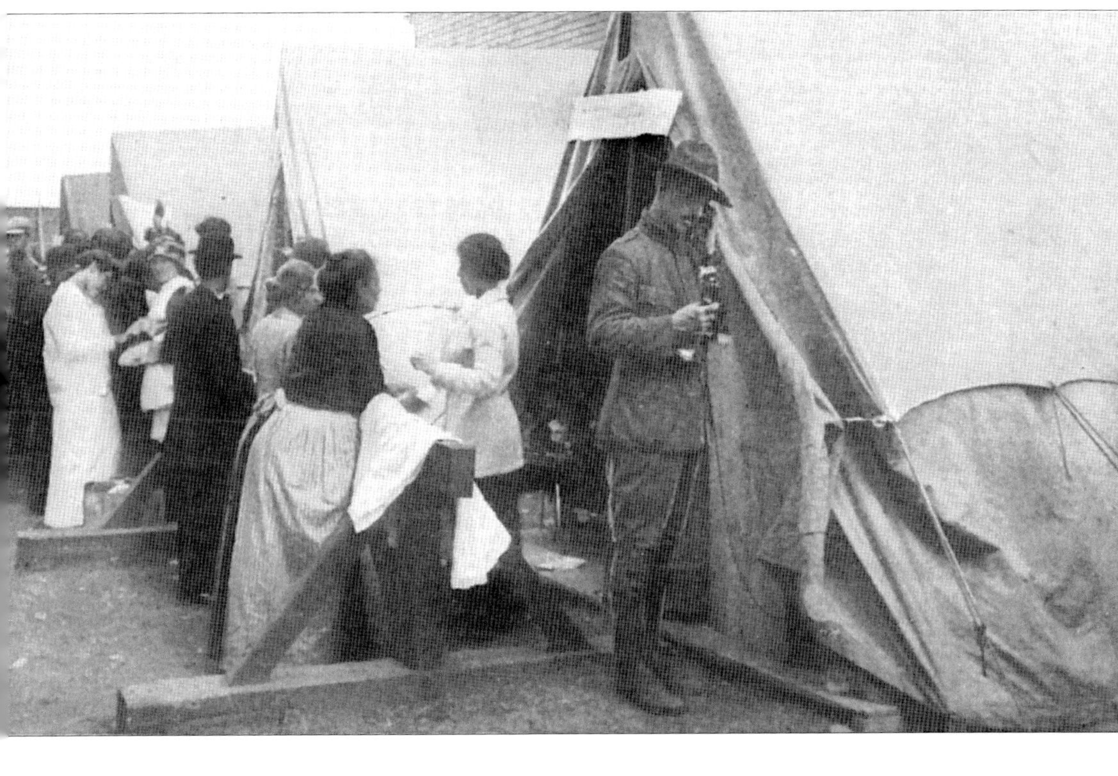

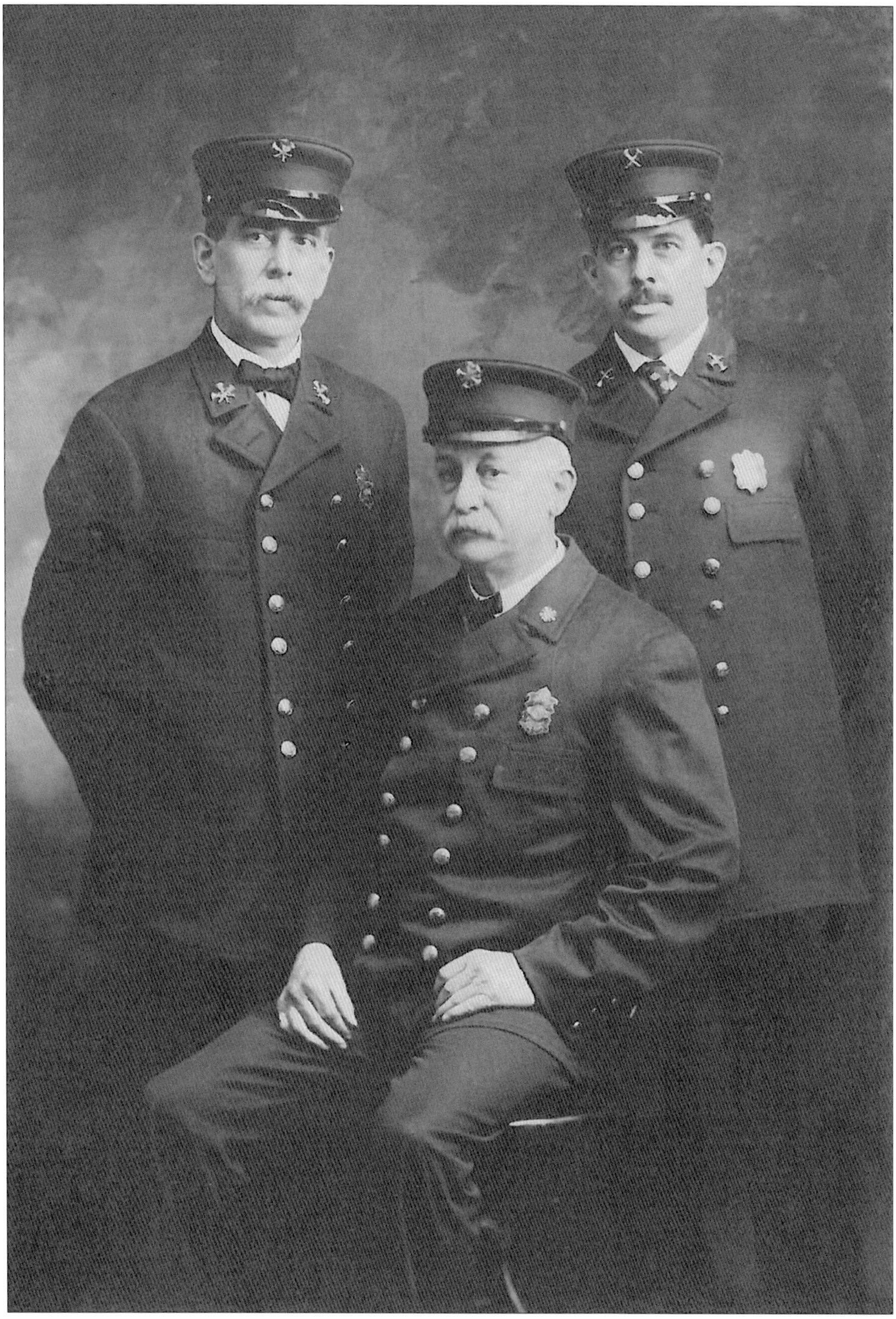

The Board of Engineers in charge of the Salem Fire Department on June 25 is shown in this photograph. Seated is Chief Engineer William O. Arnold. To the left of Arnold is Assistant Chief Howard C. Kimball; to Arnold's right is Assistant Chief Edward H. Hay. The Salem Fire Department consisted of permanent firemen as well as call firemen. Chief Arnold was a member of the permanent staff; both assistants were call firemen. Chief Arnold received a yearly salary of $1,500; assistant chiefs were paid $400 per year. (*Neg #30472*)

Chapter 5

Fighting the Fire

At the time of the fire, the Salem Fire Department consisted of twenty-three permanent men and eight call men, divided into eight companies. Fire equipment included four second-class steamers; one auto combination with two seventy-five-gallon chemical tanks and one thousand feet of hose; one horse-drawn combination with a twenty-five-gallon chemical tank and one thousand feet of hose; five hose wagons with one thousand feet of hose; and one aerial ladder truck and one ordinary ladder truck. The fire alarm system consisted of ninety-six boxes. The equipment was largely new and in good condition.

The fire started in Blubber Hollow. Box 48 sounded at 1:37 p.m. When the alarm sounded, only nine men were on duty to respond to the fire; the remaining permanent men were at lunch. The first alarm called the motor combination, three steamers, three hose wagons and a ladder truck. The second and general alarms following soon after brought out all the remaining apparatus of the city fire department, leaving no protection for the rest of the city.

Engine 4 was the first to arrive at the fire; it connected to a hydrant on the corner of Boston and Robinson Streets, near the origin of the fire. Before the steamer could work effectively, the intense heat caused the company to retreat and haul away the engine by hand. Engine 3 arrived shortly afterward and connected to the hydrant on the corner of Boston and Federal Streets. This hydrant was defective, forcing the engine to retreat as well. Defective hydrants, intense heat and lack of men to fight the fire caused delay in responding to the fire; little water was squirted onto the fire, which quickly spread to the Brennan Leather Company across Proctor Street, the Charles M. Keefe leather factory on Boston Street and the Carr Leather Company on Popes Court, 250 feet away from the start of the fire.

Out-of-town companies were called in as soon as the general alarm was called; they began arriving around 2:00 p.m. and continued to do so until 6:00 p.m. Towns sending fire apparatus included Beverly, Boston, Chelsea, Everett, Gloucester, Lawrence, Lynn, Marblehead, Newburyport, Peabody, Reading, Somerville, Swampscott and Wakefield.

Fire traveled eastward along both sides of Boston Street, crossed Essex Street and continued destroying buildings as it moved toward the Boston and Maine Railroad. The firemen left their first positions and resumed their fight along the northeast side of the fire and stopped its spread within the area bound by Federal, Pine, Warren, Mt. Vernon and Broad Streets. On the south side, the fire burned unhindered as far as the open fields. On the north side, the fire reached the railroad and crossed it, igniting the P.A. Field Shoe Company, which in turn set fire to nearby factories and continued its spread to the north section of the Edwin S. Woodbury Factory on Canal Street. Embers fell on hundreds of shingle roofs beyond the railroad in South Salem and started small fires, which created a larger fire spreading across Lafayette Street to the harbor extending southward beyond Holly and Leach Streets.

At about 6:00 p.m. the wind shifted southward, driving the heat and conflagration to the Naumkeag Steam Cotton Company. The Naumkeag private fire brigade attempted to fight the fire, but the large mills were ignited from the neighboring tenement buildings. At 9:30 p.m., the Naumkeag fire brigade retreated to escape being hemmed in by the fire. A few men remained and continued the fight by running the pumps. At 10:15 p.m., the heat became so intense that these men were forced to flee by a boat that had been brought to the wharf. The remaining sections of the mills were destroyed around 1:00 a.m., except for the fireproof goods storehouse and the round, brick cotton storehouse, which had an asbestos roof. The Salem Fire Department took its last stand along Front and Derby Streets at Derby Wharf, finally stopping the fire about 3:00 a.m. Friday within one block of the mercantile district.

The water supply during the fire was significantly inadequate. Boston Street, where the fire started, typically had water pressure of fifty pounds and a street main that was twenty inches in diameter. However, in one of the factories nearby, the gauges registered only twenty pounds of pressure about fifteen minutes after the fire started, causing the hose streams to be incapable of reaching any higher than the second story of the buildings on fire. Gauges showed pressure dropping to fifteen pounds at 3:30 p.m., at about the time the mill service and sprinkler supply pipes were broken in the Charles H. Keefe and the Carr Leather Factories. The loss of water from the broken pipes caused by the collapse of the buildings produced a severe drain on the system.

Additional water sources were required. At 3:30 p.m., the twelve-inch connection from the water supply in Peabody was let in slowly and the water pressure gradually increased; at 5:00 p.m., the water supply was cut from the Carr Factory, which built up the pressure to twenty-three pounds by 6:00 p.m.; this held for thirty minutes until it dropped again. At 8:00 p.m., the twelve-inch supply from Danvers was let into the system and at 9:00 p.m. the two eight-inch connections from Beverly were also let into the Salem system. These extra supplies had little effect on the pressure as it continued to drop until 10:30 p.m. and reached ten pounds at Church Street. At 11:30 p.m., pressure increased as water streams from fire apparatus decreased and rose until 3:00 a.m. Friday, when it reached fourteen pounds, the fire then being under control. One of the Beverly connections was shut off at this point; the Danvers connection was shut at 11:00 a.m. and the second Beverly connection shut at 1:00 p.m., with the Peabody connection shut at 3:00 p.m. on Friday.

After the fire, the National Board of Fire Underwriters made recommendations for the location of engine houses and recommended the types of fire apparatus that should be stored in each. Additional permanent firefighters were eventually added to the fire department, each being required to pass tests of skill before joining the force. The city also planned to build a high-pressure water system; a new reservoir would be built at Folly Hill in Danvers, 210 feet above sea level, capable of holding ten million gallons of water; and a pumping station would be erected at Wenham Lake. The new pumps would send the water from the lake into the reservoir. Plans were also made to equip the entire city with iron pipes of larger dimension than the cement-lined pipes in place during the fire. Plans included taking in an auxiliary supply of water from the Ipswich River. The high-pressure system was to be completed by January 1916.

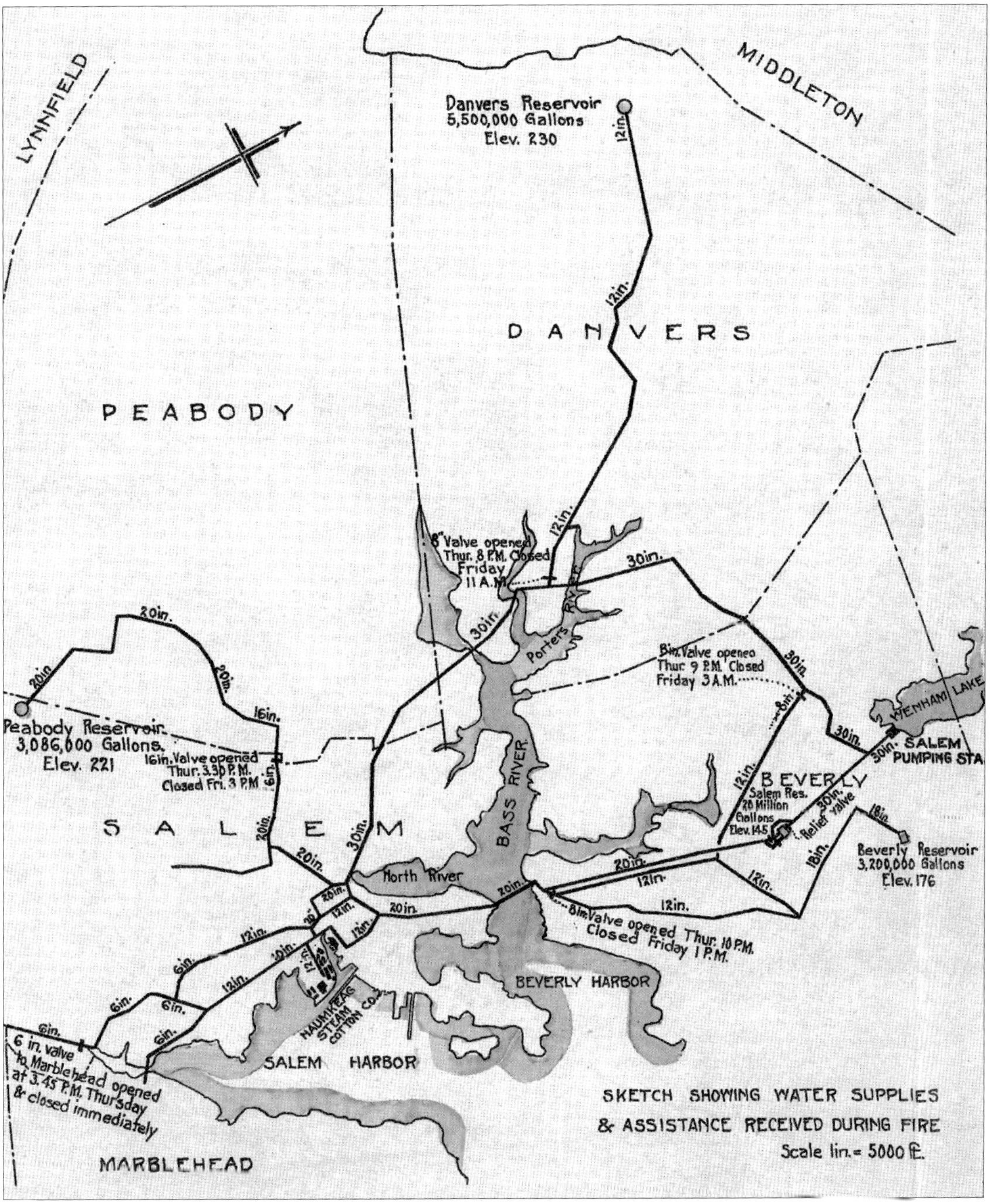

Water pressure and the amount of water available to fight the fire were major factors in its spread. This map, printed in the Fire Report, Index #12375, shows the sources of water supply to the city and identifies the time at which water assistance was received from the neighboring towns of Peabody, Danvers, Beverly and Marblehead. (*Neg #32883*) The fire report was submitted by Fire Inspector Charles H. Smith of the Association of Factory Mutual Insurance Companies of Boston, after he inspected the Naumkeag Steam Cotton Company Ruins on June 26.

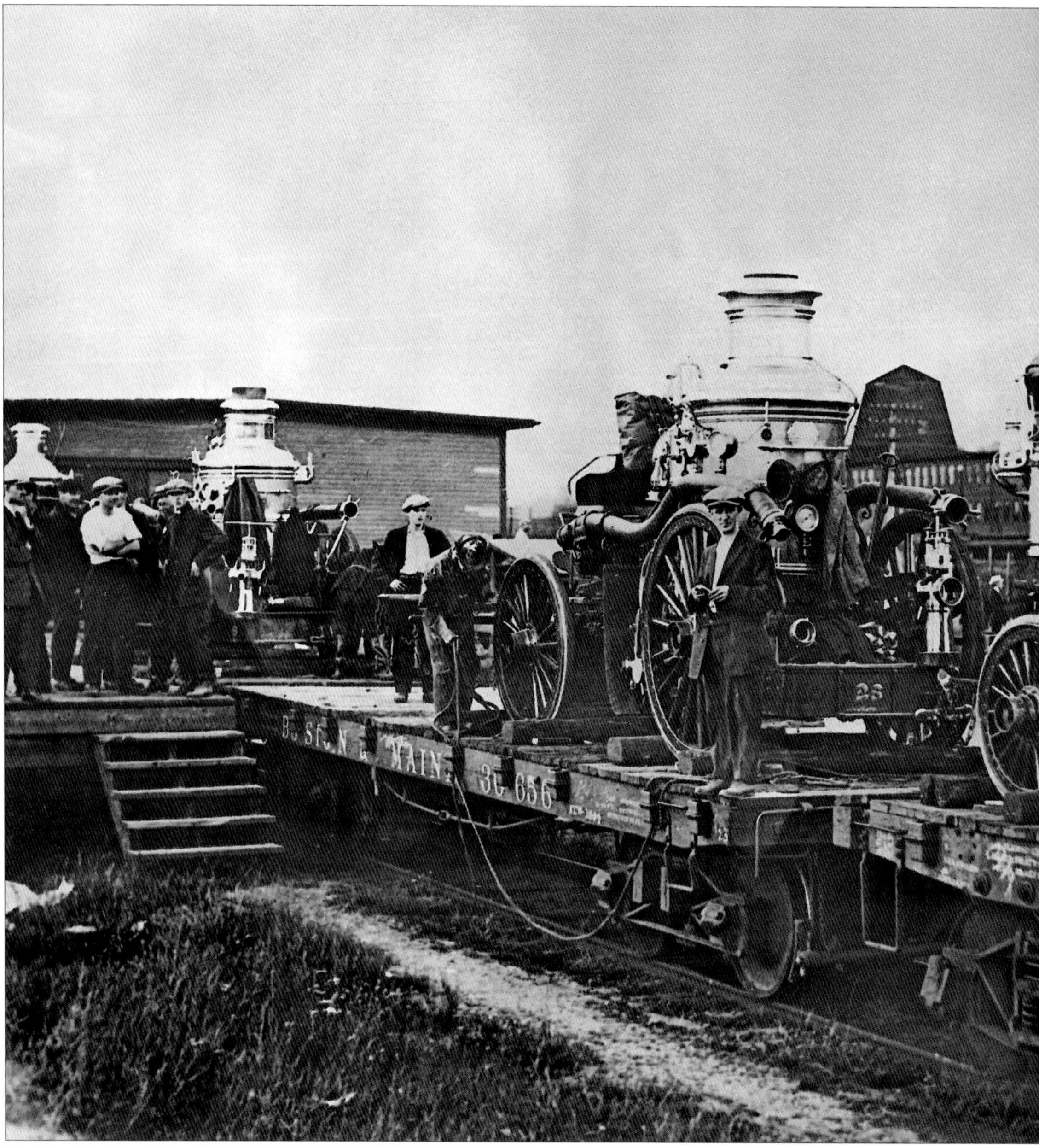

In response to two calls for help, the Boston Fire Department sent five engines and one ladder company to assist Salem. Engines 8, 26 and 39 responded to the second call at 4:40 p.m., arriving by 6:00 p.m. via flatbed rail cars. Engine 8 was commanded by Lieutenant Thomas W. Roose; Engine 26 by Lieutenant Daniel J. Hurley; and Engine 39 by Lieutenant J. O'Brien. Sixty-seven firemen accompanied the engines to Salem. Engines 8 and 26 fought the fire at Summit and Walnut Avenues. Cannel coal ran out within an hour of their arrival, so firemen tore down fences and unhinged blinds to feed the boilers on the engines. When water ran out they assisted the bucket brigade on Misservy Street carrying salt water from the harbor to protect the street from houses burning on Eden Street. Engine 39 was used on Lafayette Street near the Salem Laundry where the final stand against the fire was made. The Salem Laundry survived the fire. (*Neg #15794*)

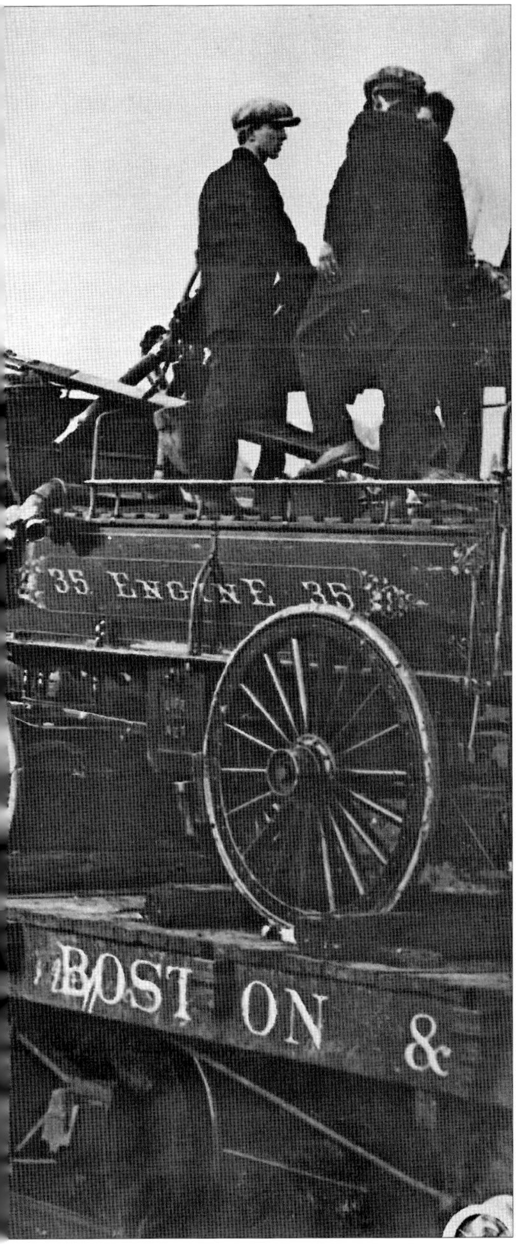

The Lowry flush-type hydrant and hydrant chuck were used in Salem as of 1868. An article in the June 1964 issue of *Firemen* stated that 534 Lowry flush hydrants were in service as of January 1, 1914, 17 of which were in Salem. Unlike post hydrants, these were attached directly to water mains. The chuck was carried on the hose wagon, included one to four independently gated connections for attachment of the fire hose(s) and was set over or screwed onto a large water main after the manhole cover was removed. The multiple connections allowed several steam engines to pump from the same hydrant. This type of hydrant provided a safety advantage in that it removed engines from the danger of falling walls and out of the way of ladder trucks. The hydrant outlet boxes had to be covered with rock salt in the winter to protect against freezing. The Lowry hydrant was a source of controversy in the aftermath of the fire. Some of the assisting fire companies did not have the correct chucks to connect to the flush hydrant; delays of 30 to 120 minutes ensued to find the correct chucks and many buildings burned without any water being applied to them. (*Neg#32202*)

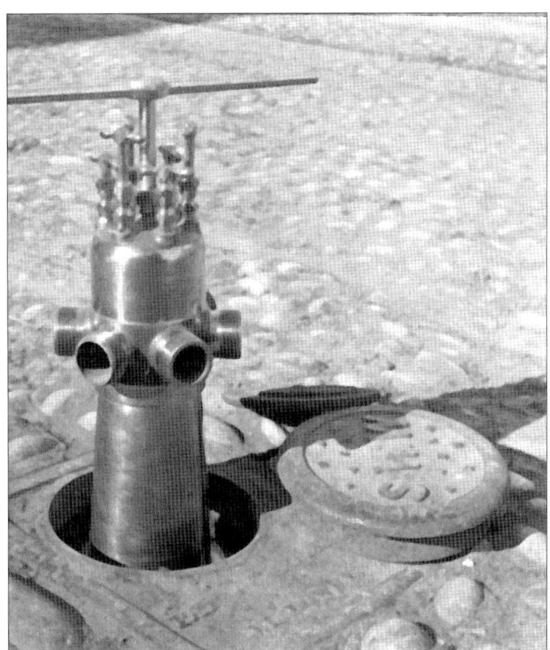

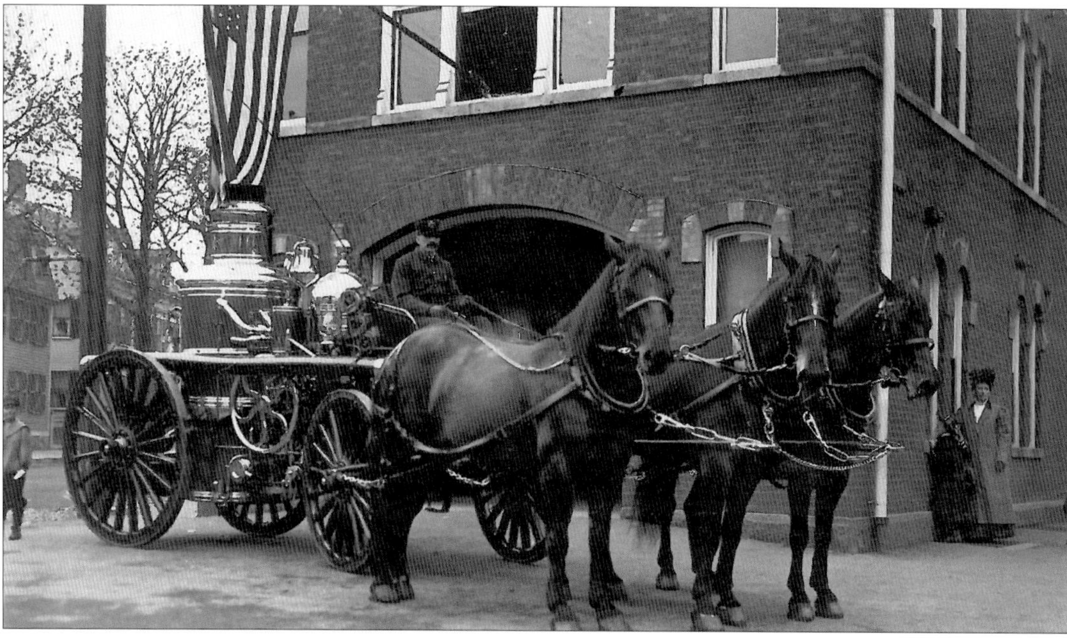

The fire apparatus above is Engine 1, photographed by H.S. Walker. In *A History of the Salem Fire Department*, Edward R. Tufts identifies this engine as an Amoskeag Steamer #812 with a 4 5/8 Base Crane Neck Frame. The engine was unique in that it had a fly wheel on one side only. Depicted in the image is one of the drivers of the engine, John Carter. The steamer engine in the image below fought valiantly to stop the fire at Bridge Street, near the corner of Boston and Goodhue Streets. (*Neg #32824, above; Neg #15155Detail, below*)

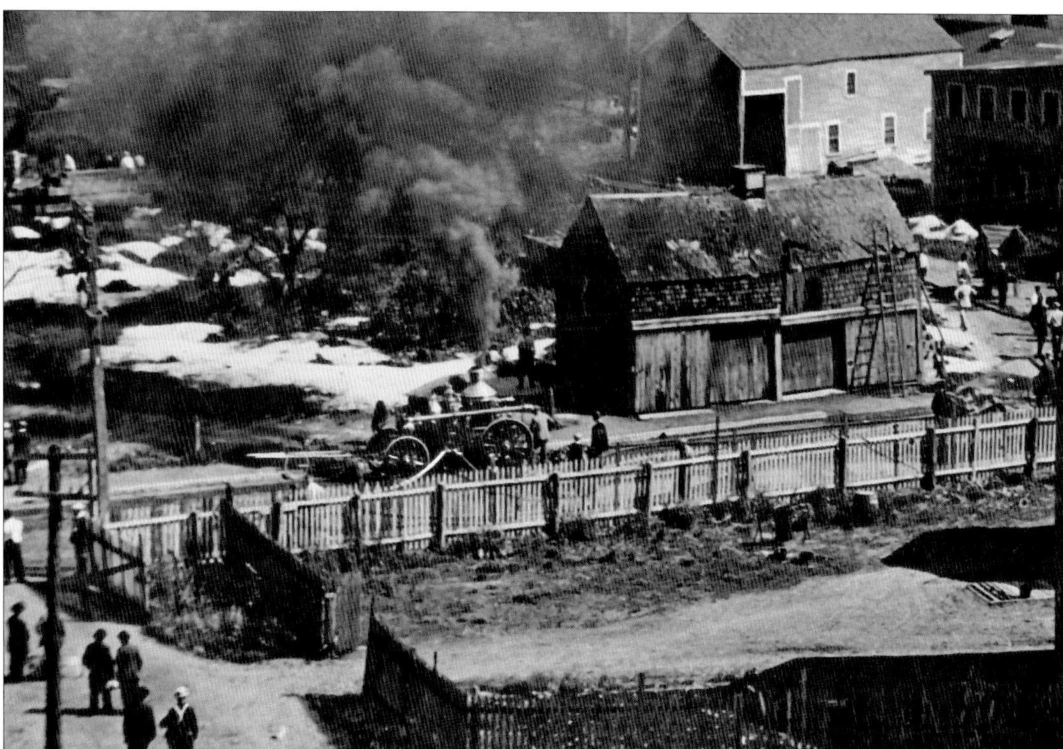

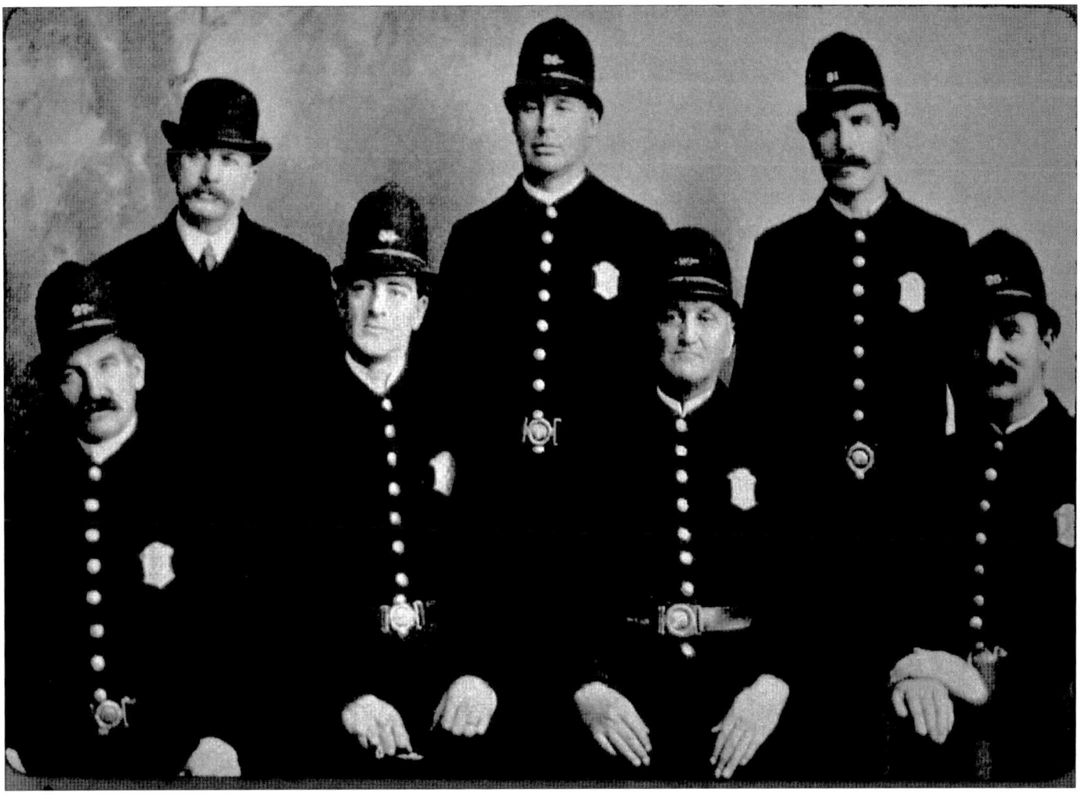

The Phillips Library holds the papers of Eleanor Broadhead, a local resident who lectured about the Salem fire. This image of Salem police officers was one of the slides used in her presentations. The names of the people in the photo are not given. According to the 1914 *Naumkeag Directory of Salem*, the police station was located at 17 Central Street. The city marshal at the time was Patrick J. Lehan. The captain of the department was John J. Carr; his acting lieutenant was Edwin W. Dennis. The police department had 37 patrolmen and 9 reserve police officers along with 138 constables. The city marshal received a salary of $1,700; the captain's salary was $1,250; and the acting lieutenant received $1,100. Patrolmen earned $3 per day, and a reserve policeman earned $2.75 a day when on duty; no salary was supplied for the constables on the force. All off-duty police were required to report to duty when general fire alarms were called. (*Neg #32947*)

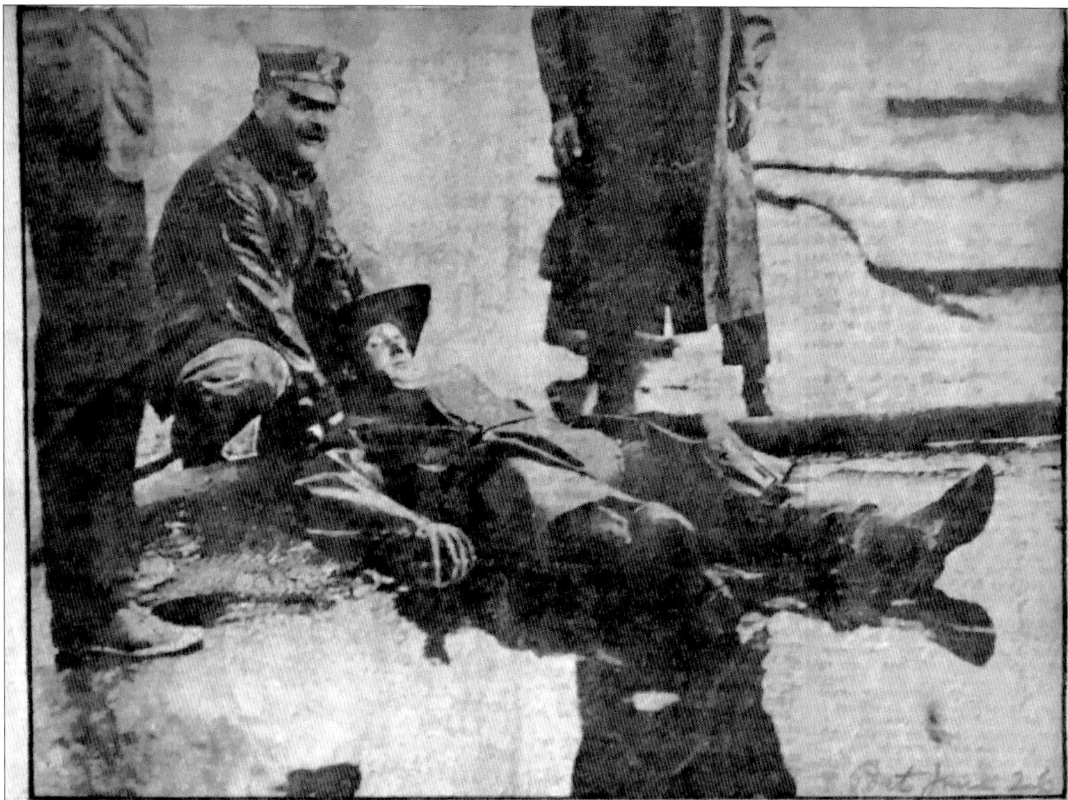

FIREMEN SEEKING RELIEF IN PUDDLES OF WATER. MANY FIREMEN WERE OVERCOME BY THE INTENSE HEAT. THEY LAID DOWN IN PUDDLES OF WATER UNTIL REVIVED, WHEN THEY WENT BACK TO THEIR WORK.

This image, labeled "Post June 26," is from a scrapbook in the library collection. It aptly depicts the exhausting affects of the intense heat on the firemen. The caption under the image reads: "Firemen seeking relief in puddles of water. Many firemen were overcome by the intense heat. They laid down in puddles of water until revived, when they went back to their work." Puddles were also used as a last resort to pump water onto a burning building when other water sources had failed. (*Neg #32913*)

Opposite page: Chief William Arnold is depicted in the image on the bottom of the facing page, along with his driver, James Keon. Seated in the chief's buggy, Keon and Arnold were photographed in front of the Fire Department Headquarters on the corner of Broad and Phelps Streets. (*Neg #30471*) The Engine House on Essex Street, shown at the top of the facing page, was destroyed by the fire; it was originally built in 1898 and was equipped with Steamer No. 4, a hose carriage and ladder truck. A two-story, red brick building was built on the former site of the firehouse; the building is fireproof and was built to accommodate five pieces of fire apparatus. (*Neg #32783*)

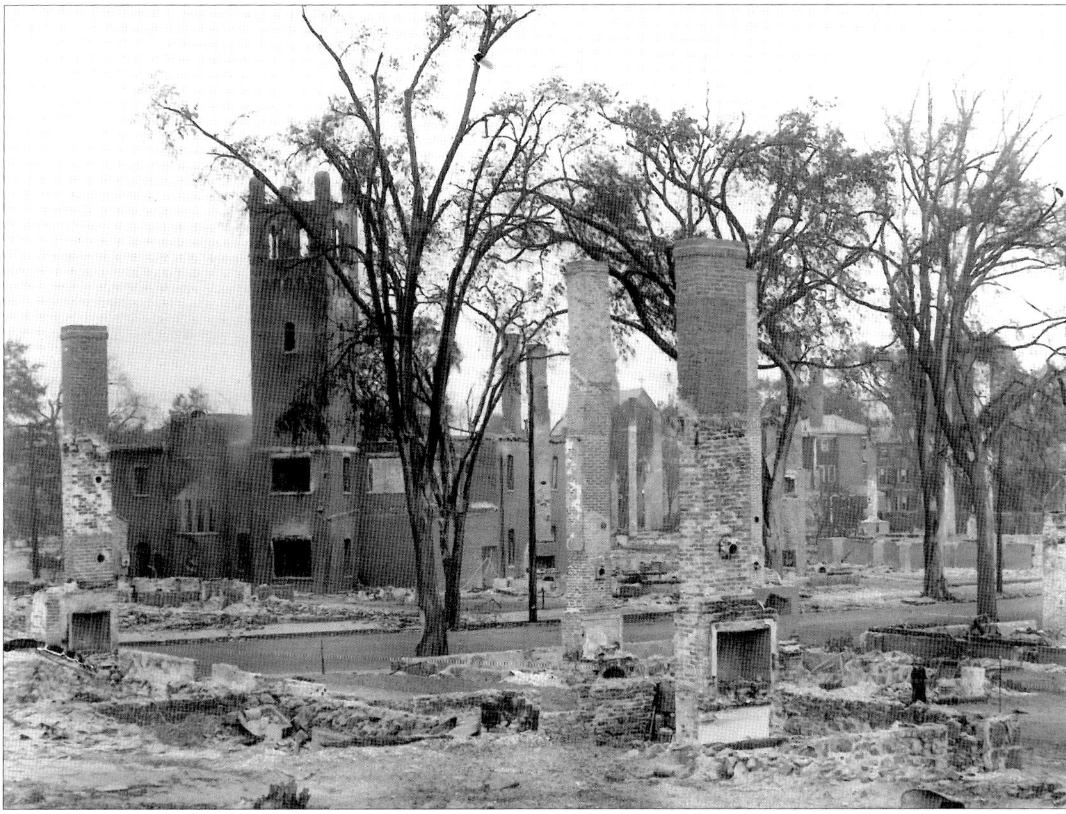

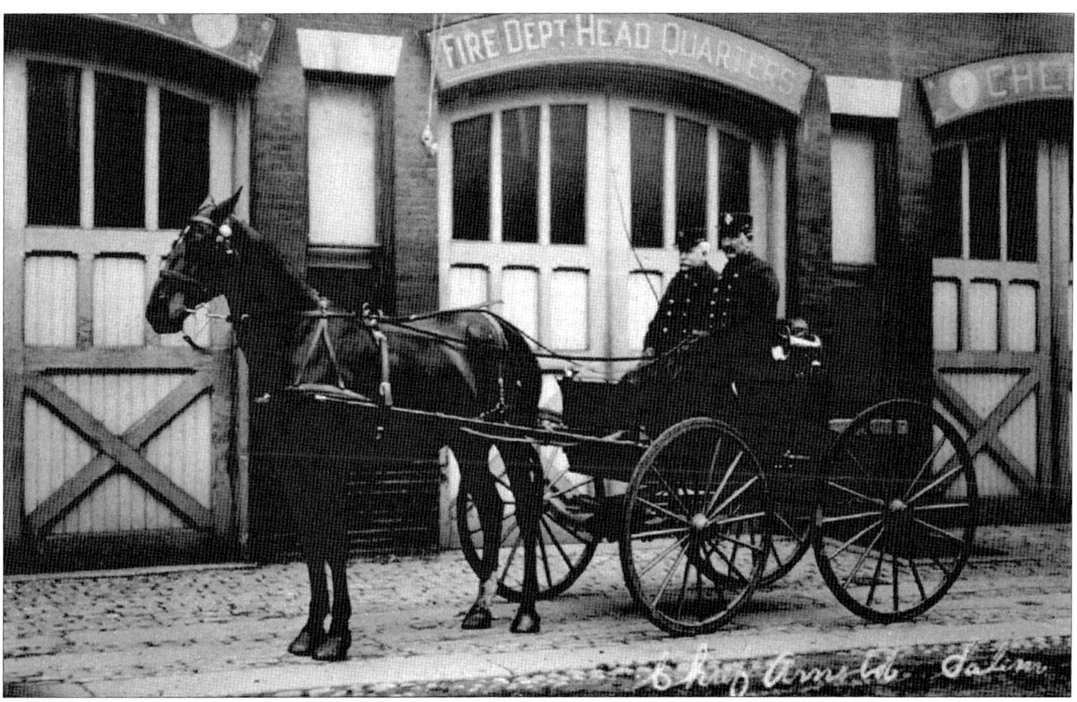
FIRE DEPT. HEAD QUARTERS
Chief Arnold Salem

How Conflagrations Start

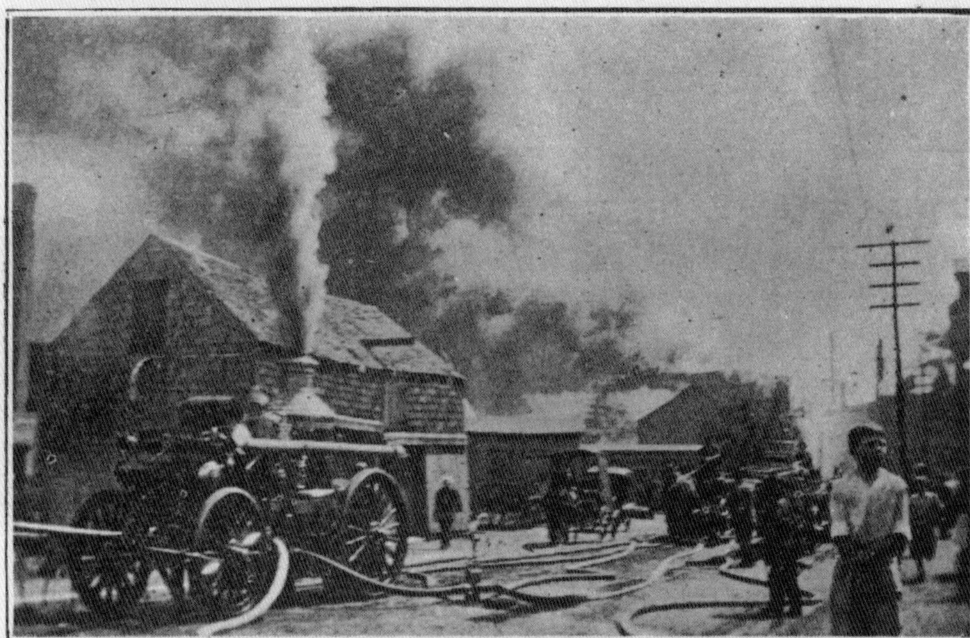

Beginning of the Salem, Mass., conflagration of June 25 and 26, 1914, at the corner of Boston and Proctor streets. The fire engine is drawing from a "flush" hydrant through a portable "chuck".

(Courtesy of Underwriters' Bureau of New England.)

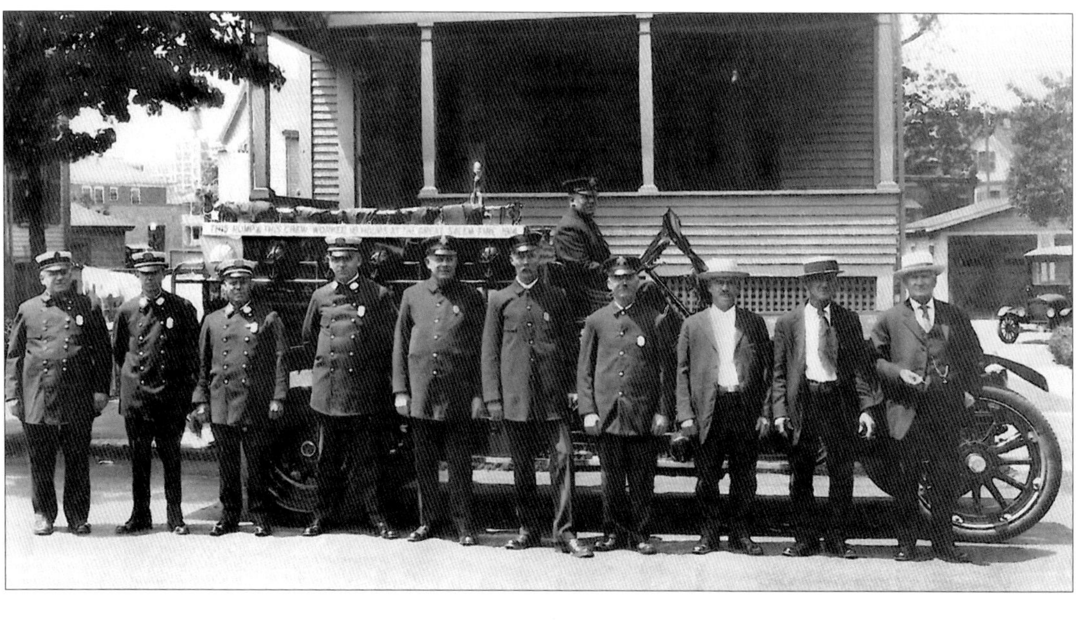

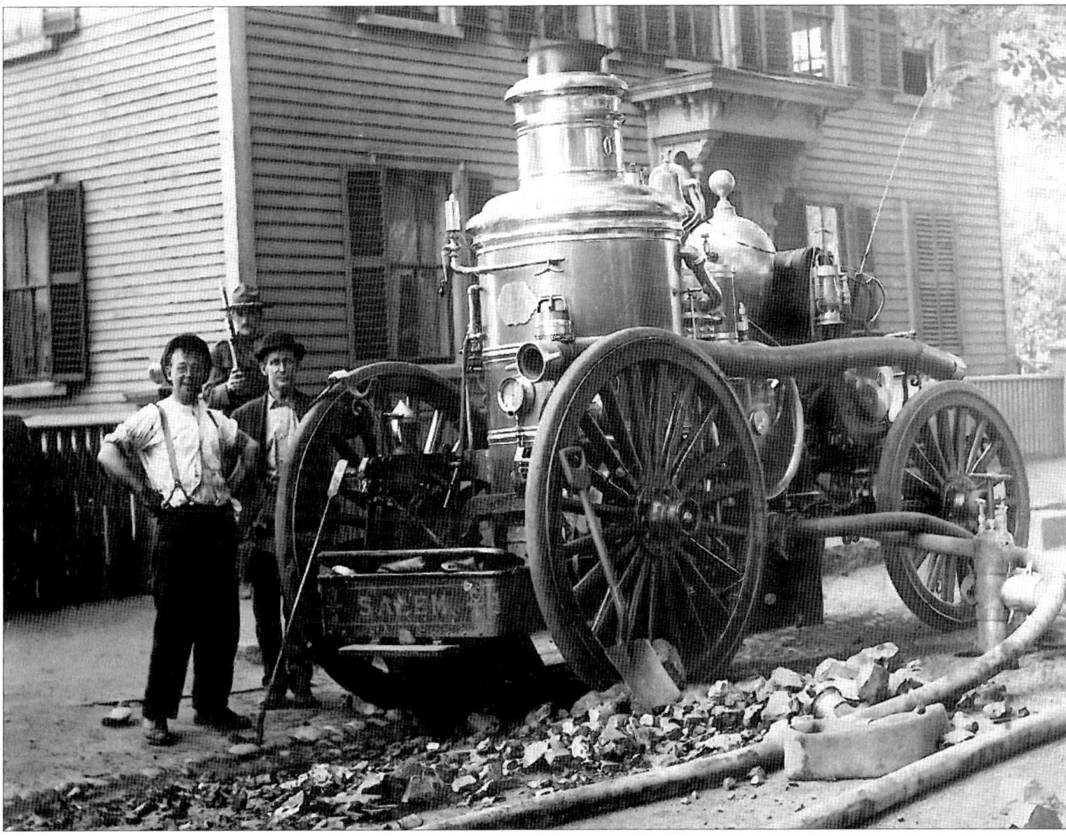

As previously stated, the Lowry Hydrant was a source of controversy in the aftermath of the fire. The images at the top of each page show engines from the Salem Fire Department hooked up to the flush-type hydrant. The caption to the top image on page 112, which was published in *Protection Engineering*, July 1914, states that the "fire engine is drawing from a 'flush' hydrant through a portable 'chuck.'" Some assisting fire companies did not have the appropriate chucks to hook up to the hydrant, which delayed attempts to effectively fight the fire. The top image on the facing page shows firefighters at the corner of Boston and Proctor Streets (*Neg #34070*); the location of Salem Engine One in the image above is not identified. (*Neg #32816*) The bottom image on page 112 depicts the American LaFrance motor pumping engine from Swampscott. The photograph of the Combination "B" engine and its crew was taken in front of the Swampscott Central Fire Station in 1924, on the tenth anniversary of the fire. The banner on the engine reads: "This pump and this crew worked 18 hours at the Great Salem Fire, 1914." According to Arthur Jones, the company held its longest stand at the corner of Broad and Winthrop Streets, after which it fought the fire on both Union and Derby Streets. (*Neg #32825*)

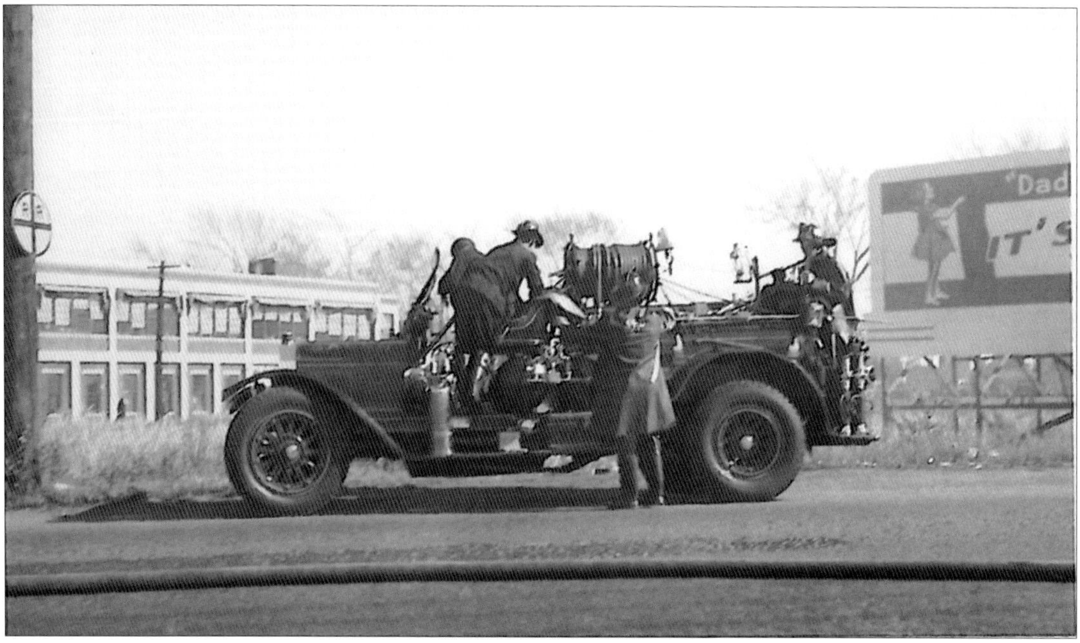

The image at the top of this page shows Engine Company 1, housed at the Church Street Fire Station. It was a 1912 American LaFrance Combination Hose-Chemical engine that was scrapped in 1941. (*Neg #32867*) H.S. Walker took this photograph as well as the photograph on the top of the facing page, which shows Hook and Ladder 2. Hook and Ladder 1 moved to Essex Street but the sign on the building was never changed. Ladder 2 was located at Bridge Street and was a sixty-five-foot Babcock Aerial purchased in 1893. Edward Mason was the driver of the engine. (*Neg #32918*) The Robinson Pumping Engine, similar to the engine from Chelsea that helped fight the fire, was the first gasoline engine purchased by the Salem Fire Department. It was purchased from the Robinson Fire Apparatus Manufacturing Company of St. Louis in July 1914 at a cost of more than $8,000. The engine had the capacity to throw four streams of water, a pumping capacity of nine hundred gallons per minute and was designed to do the work of a chemical engine, horse wagon and pump. It carried one thousand feet of two-and-a-half-inch hose, was equipped with a thirty-five-gallon tank and had a twenty-foot trussed extension ladder along with a ten-foot roof ladder. It was housed at the South Salem fire station upon its arrival in September. (*Neg #32848*)

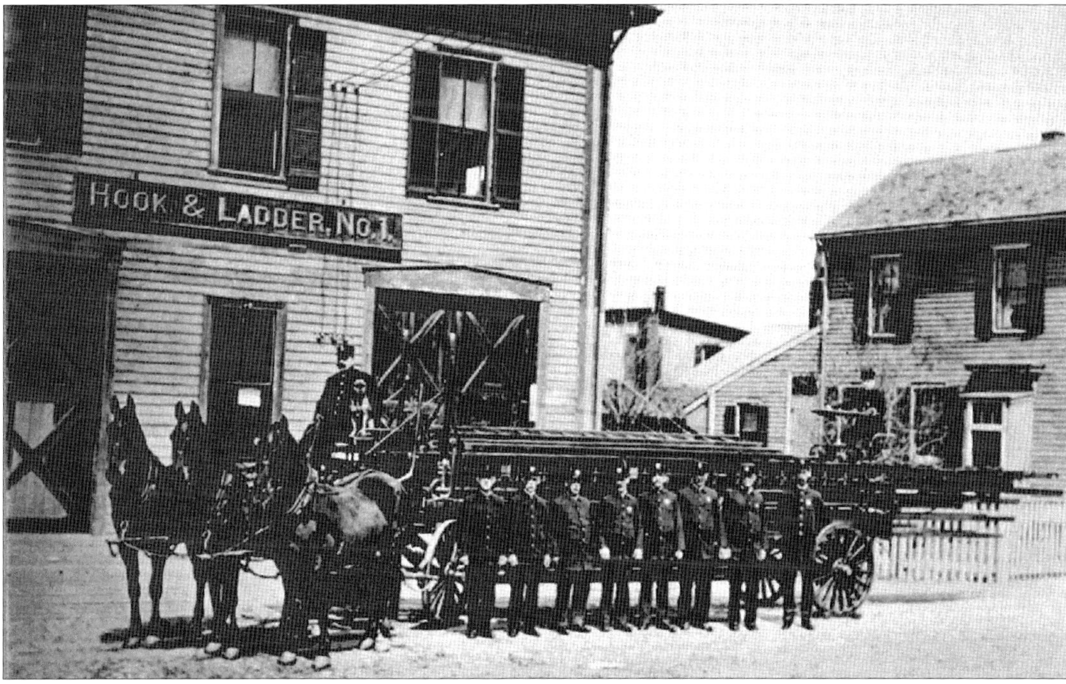
HOOK & LADDER. No. 1.

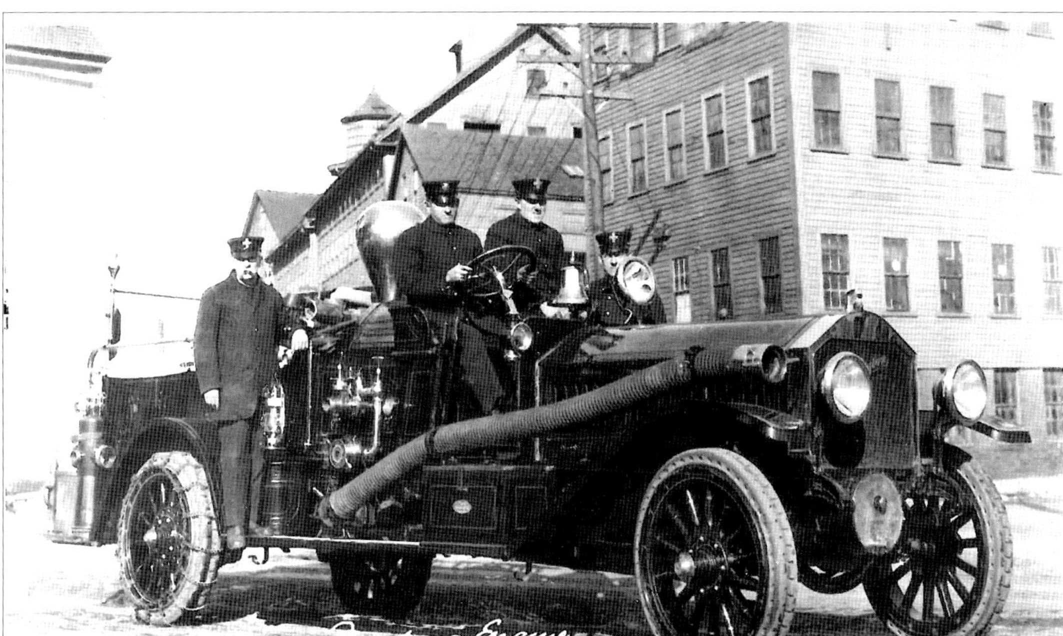
Engine

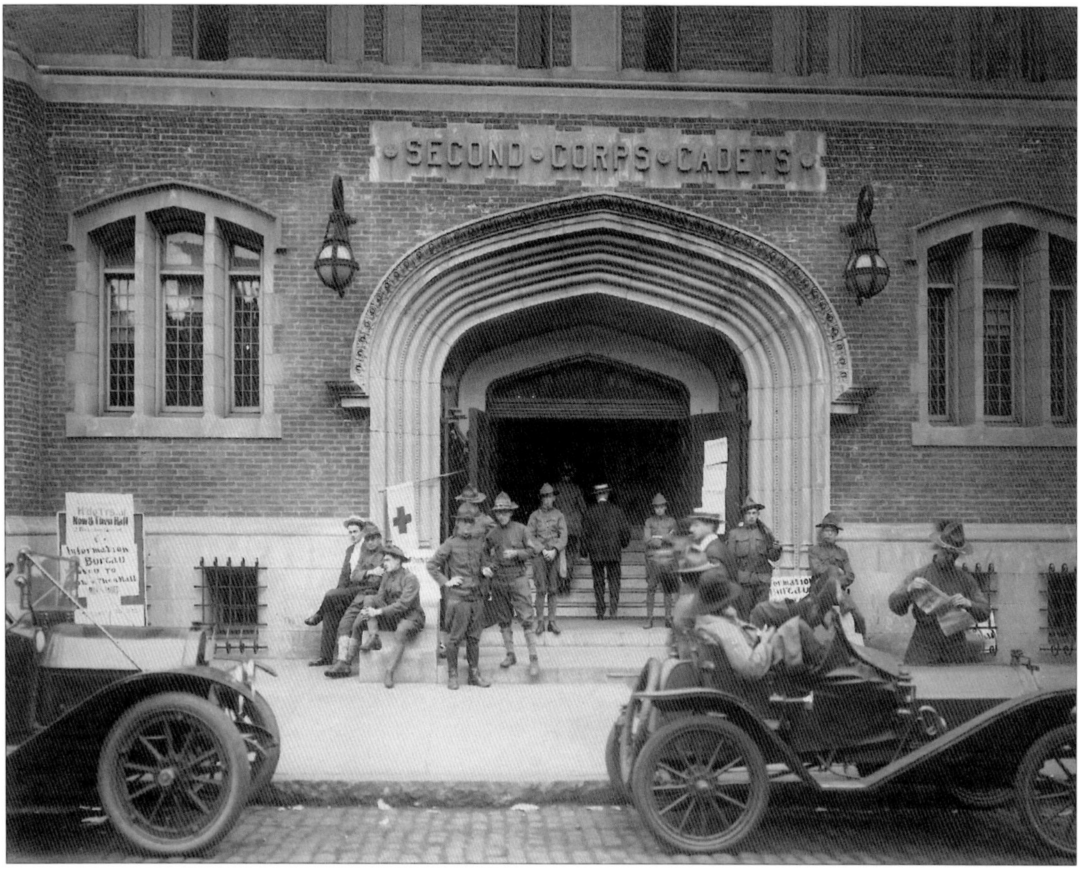

At 2:00 p.m. on June 25, the mayor of Salem established martial law when he called out four companies of the Second Corps Cadets and Company H of the Eighth Regiment for patrol duty. Later in the evening, the entire Eighth Regiment and the Naval Brigade from Lynn were called to duty. Eventually, officers from every militia in Massachusetts responded to the call. An unidentified newspaper clipping dated July 7, 1914, stated, "Without the militia Salem would have been in a chaotic condition. Straightening out matters was a task that none other than the military could have accomplished in such a short time, as the sight of guns and bayonets in the hands of the experienced soldiery had its quieting effect without a doubt." Colonel Frank A. Graves and Adjutant Julian I. Chamberlain were in charge of the Eighth Regiment; Colonel Charles R. Ropes was the commander of the Second Corps Cadets. The militia was on duty until July 7, 1914. (*Neg #4294*)

When the sounding of the military call was made, most of the commissioned officers of the local militia were out of town to attend the Service School session in West Newbury. Privates and noncommissioned officers still within the city limits answered the call. Sergeant Clay, the armorer, and his assistant were at the State Armory; they helped to equip the men as they arrived. The Hospital Corps was part of the military call. They were among the first to arrive; the surgeon from the corps was also at the Service School, but he arrived as the afternoon progressed. The Hospital Corps helped firemen with blistered faces and smoke-filled lungs and also helped the police, other militia and the general populace during the fire. (*Neg #32919, a Robb photo*)

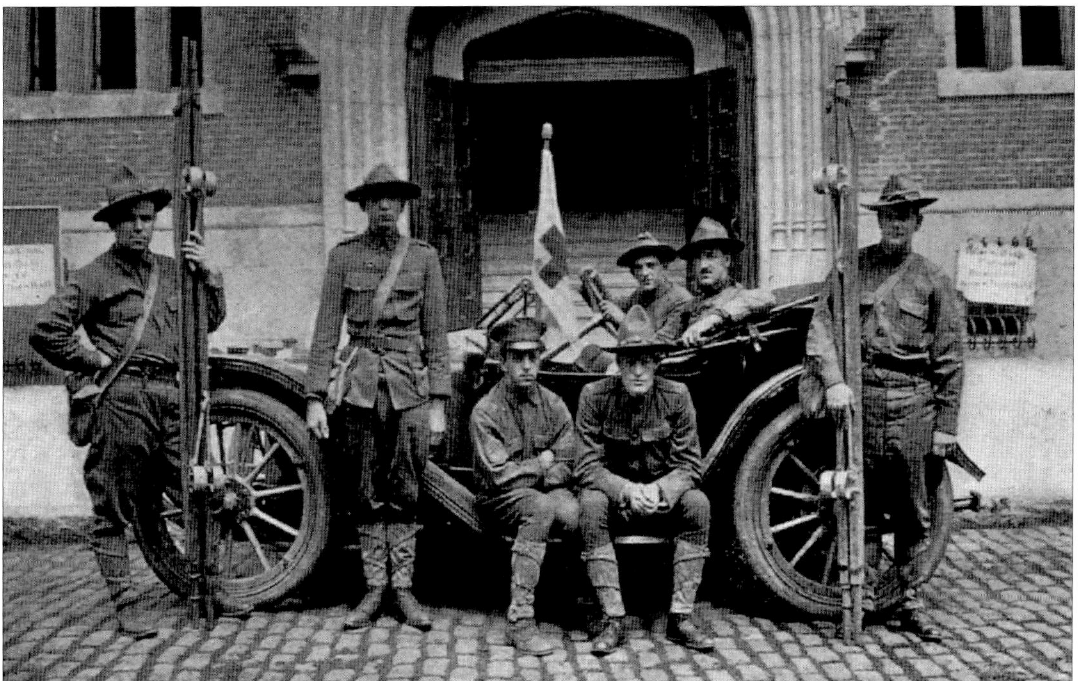

HOSPITAL CORPS, DETACHMENT OF THE SECOND CORPS OF CADETS

Photograph by M. E. Robb, Salem

Priv. Carl P. Brewer
Lieut. E. A. Rushford
Serg. C. D. Brown   Priv. Steven Richards
Priv. Percy Ray   Priv. Armie Michaud   Priv. Myron Hutchins

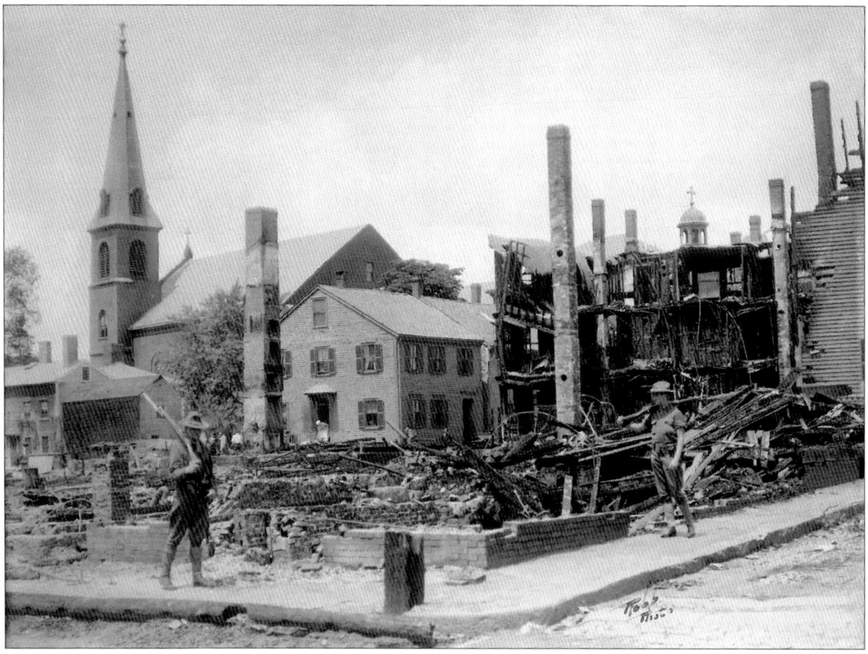

One of the major duties of the militia was to patrol the ruins to protect damaged property and prevent looting. The Robb photo above was taken at the corner of Herbert and Derby Streets. (*Neg #4003*) The lone militiaman in the photo below patrols an unknown location. (*Neg #32584*) Immediately after martial law was declared, the militia received orders to shoot looters. Quoted in *The Salem Evening News* the following day, Adjutant General Cole said, "Our men were given strict instructions to shoot on sight any looters or burglars. And the orders carried shooting to kill. We don't propose to stand for any lawlessness in the burned area while we are in control."

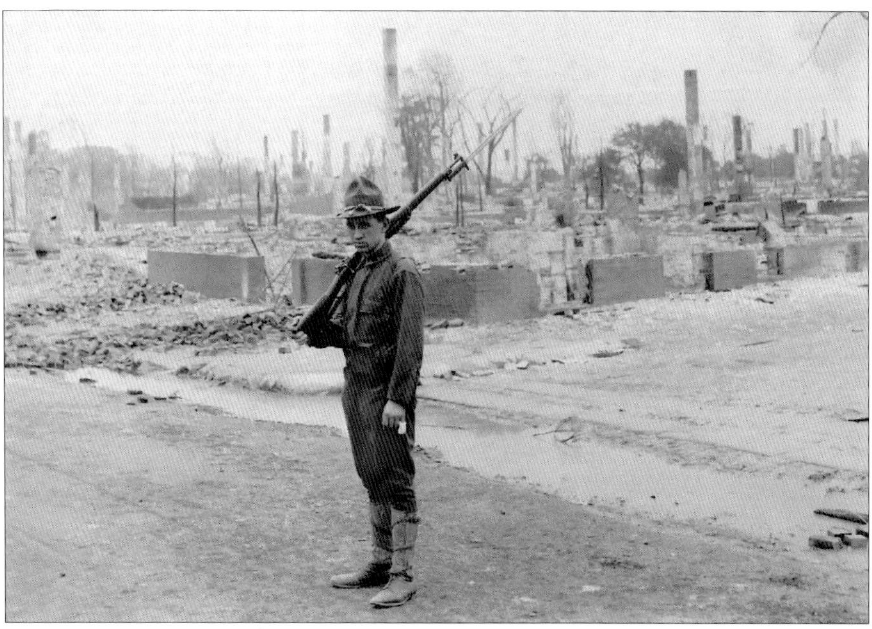

Transcript: Oral History of Salem Fire, 1914—Captain Henry Nichols

Captain Henry Nichols was seventeen years old in 1914 and a member of the private fire brigade of the Naumkeag Steam Cotton Company. He lived at 22 Cross Street. Captain Nichols normally worked on the day shift at the mill, but stayed throughout the evening on that fateful day to fight the fire until he was able to escape the blaze. An oral history of his experience was taken by a student at Salem State College in 1974, a copy of which was recently donated to the Phillips Library by Ben Arlander. Captain Nichols was seventy-seven years old at the time of the interview and was Ben's great-grandfather; he earned the title of captain in the navy. Asterisks separating the two excerpts indicate that the complete transcript is not provided. Blank lines identify inaudible sections of the interview; this includes the intervier's name—she is identified as "Interviewer."

Nichols: *The entire section of South Salem was destroyed, and I was one of seven men trapped in the Naumkeag Mill. Fire was completely surrounding us. Now we had in that mill, we had two huge fire engines. They were steam-driven pumps. They had a normal capacity of 2000 gallons a minute, but when they were forced to the maximum capacity they were pumping nearly 3000 gallons per minute. And this was pumped into a lot of the main system—the mill system and into the sprinkler pipes. But with the fire bearing down on some of the mills, the mills caught on fire on the outside, and the insurance inspectors, who came from Boston, took charge of the mills. And they shut off these valves and shut off certain sections of the mills and let them go. We made the fight with one, two, three mills—we made the fight with the sprinklers on and the pumps operating.*

* * * * *

Nichols: *Now this is quite an experience for a 17-year-old boy. Furthermore, I had charge of 26 boilers and not a one of those boilers was damaged. And the smoke and the sparks were coming through the open windows of this boiler room, so you couldn't see the gauges; you couldn't tell where the water was in the water glass; you had to go by sound.*

Interviewer: *How many men were with you?*

Nichols: *I was the only one with this. The others were outside, see. And once about 10:00* [p.m.], *these insurance underwriters shut off valves going to certain sections of the mill. And when they did, I had no knowledge of it and the pumps were pumping very high, see. They naturally build up high pressure and the release valves on the pumps, the huge things, and they'd discharge all of this water into the room where the pumps were. Fortunately, I heard it and I ran in there and the means of illumination was a small steam engine generator. ___ had gotten up within that side of the generator when I stood the pumps down and stopped the valves from discharging water.*

Interviewer: *Four or five inches?*

Nichols: *Yes. So this kept us in the light. So we stayed for an hour and figured at the last resort we would go out through this mill and drop overboard and swim.*

Interviewer: *Who is we?*

Nichols: *Myself and the other six people. See? Now, the management knew we were there, so they sent a motorboat around on the ocean alongside of the dock, and like a miracle—especially if you're an old sailor like I have* [sic], *you've seen dozens and dozens of miracles, see. For some reason, this wind change and the solid mass of flame that shut off our exit to the water changed and the ____ divided up. So we went down through there and there was the boat waiting for us. And two minutes after we got through that space, the wind changed again and that was one solid mass of fire.*

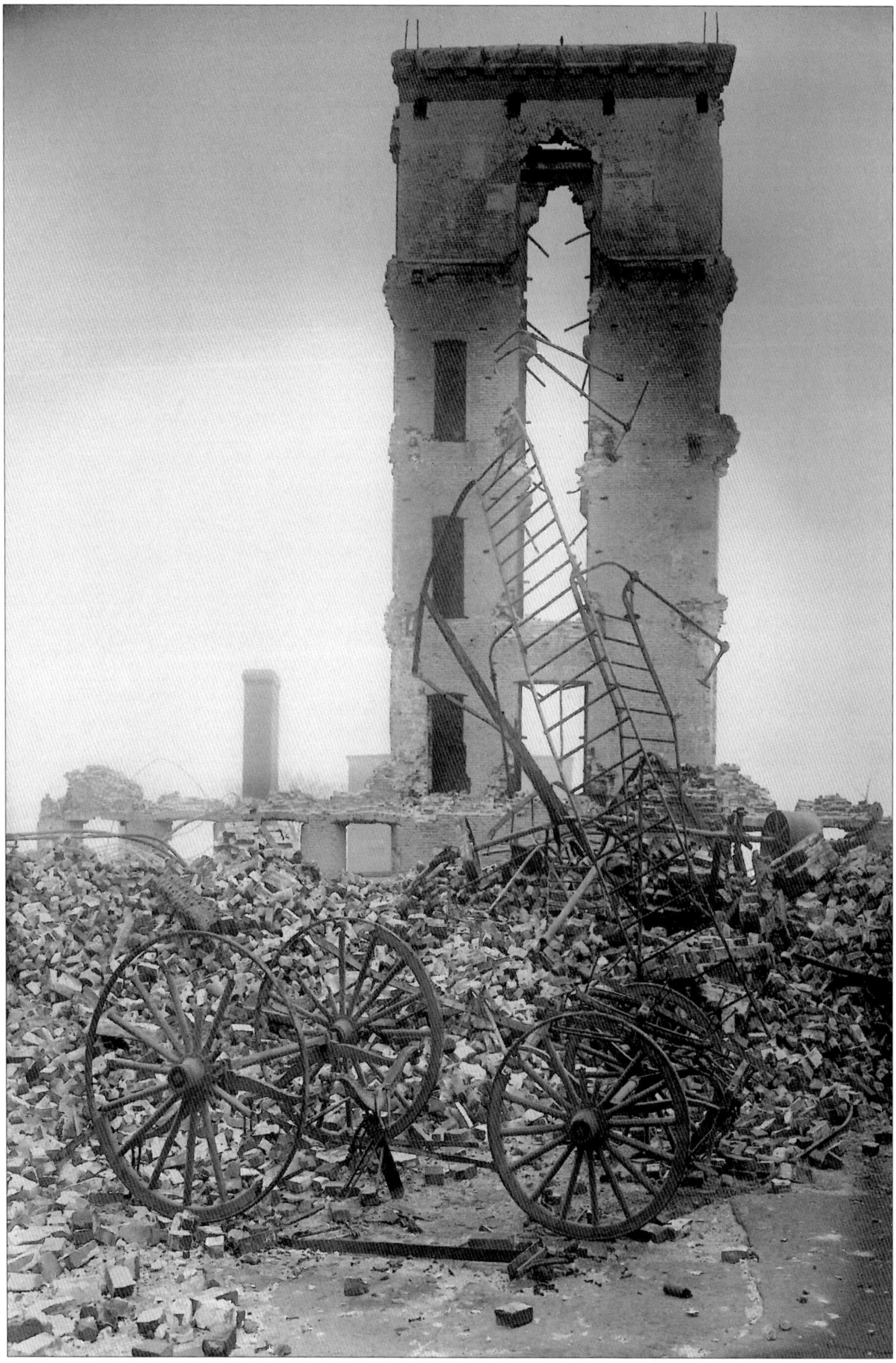

Only the two fireproof buildings in the Naumkeag Steam Cotton Company survived the fire. The tower of the main complex and the towers of the mill buildings did not. The mill complex was constructed of brick and housed machinery made from metal. After the fire, thousands of bricks and tons of metal debris had to be removed before reconstruction could begin. (*Neg #32798*)

Chapter 6

Ruins

One can read articles and reports full of statistics about the many properties that were destroyed by the raging blaze. But the impact of such an event is not truly grasped until one sorts through images that portray the utter destruction and desolation in the aftermath of the fire.

These images intrigued me in their randomness—one shows almost complete destruction of a building that may have been a general store, yet the building next to it is completely unharmed. Two images depict ruins of large brick buildings, but statues of Saint Joseph and an angel survive the fire. The chimneys, the only evidence that a home existed on a particular piece of property, are strong in their ability to survive the fire and haunting because the warmth and comfort they gave a home was stolen by the intense heat of the fire. The amount of debris is phenomenal in the pictures at the textile mill—hundreds of thousands of bricks were left in the wake of the fire, as were tons of metal, which included the building supports and the machinery destroyed by the intense heat.

Many, many more images of the ruins from the fire exist in the collection at the Phillips Library. Photographs chosen for this section show the progression of the fire from its source in Blubber Hollow to its demise at the water's edge along Derby Wharf and the property of the Naumkeag Steam Cotton Company. It is impossible to show the full brunt of the fire in this short chapter; for those who want to explore the fire and its impact more fully, please visit the library and peruse the Photo Archives collection.

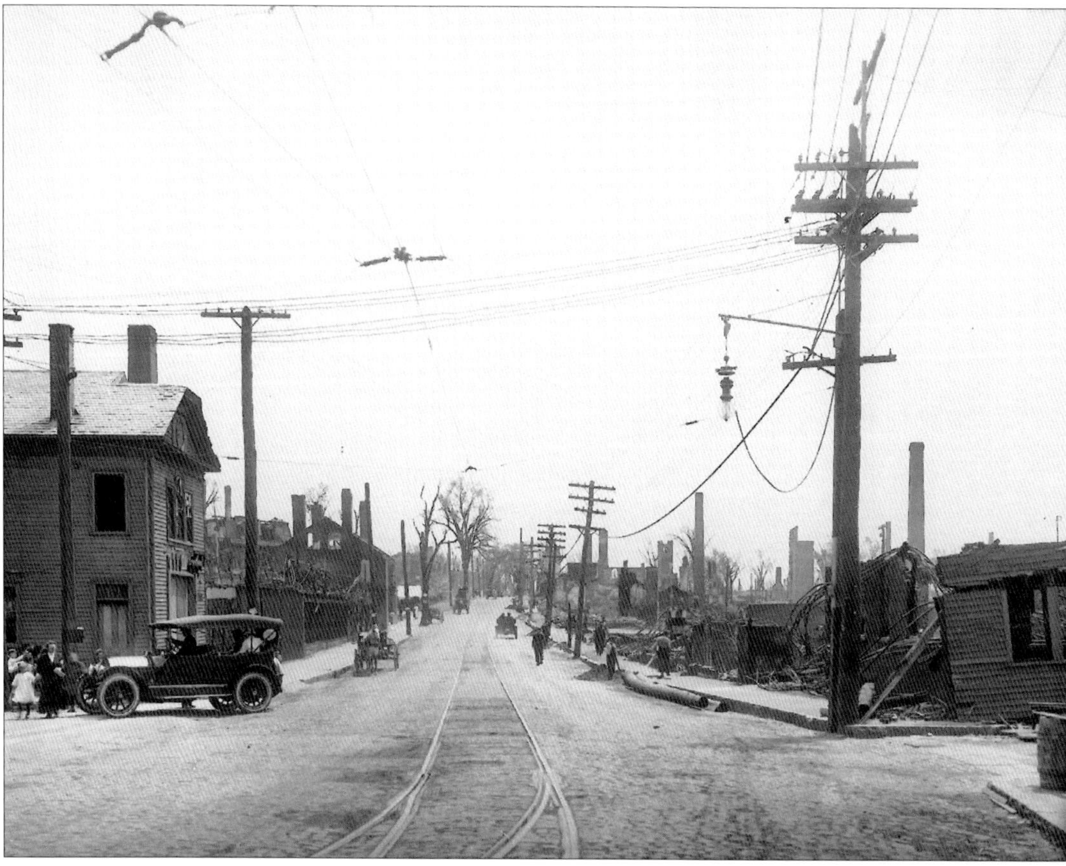

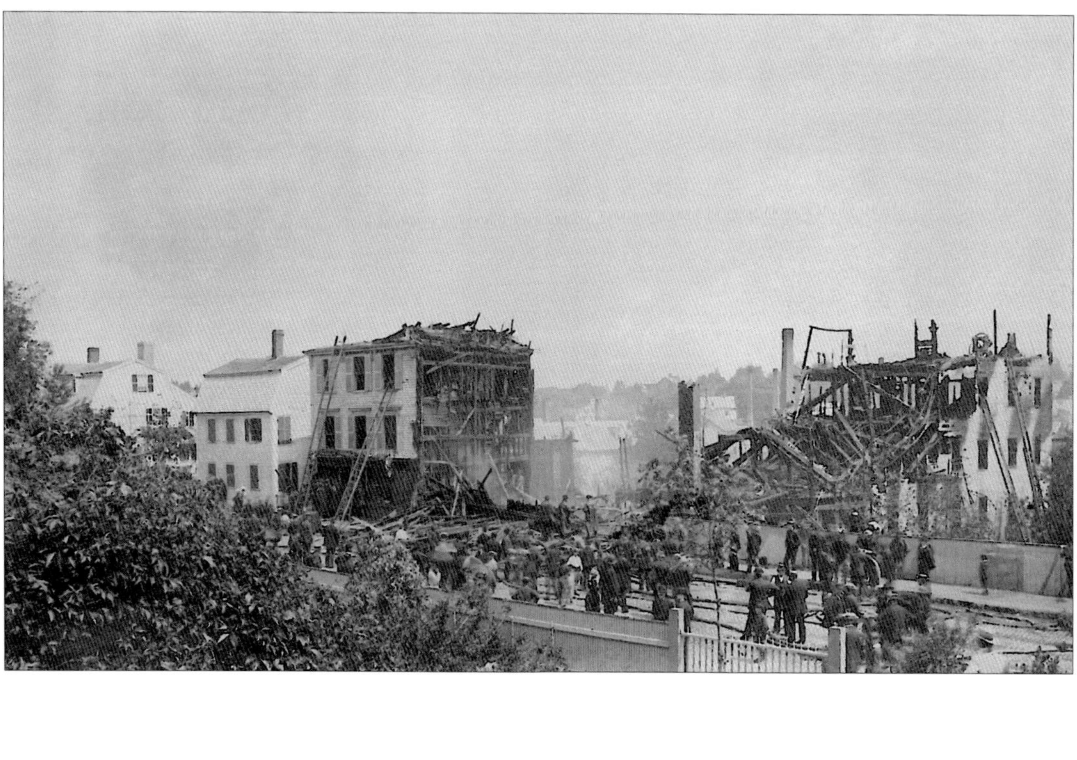

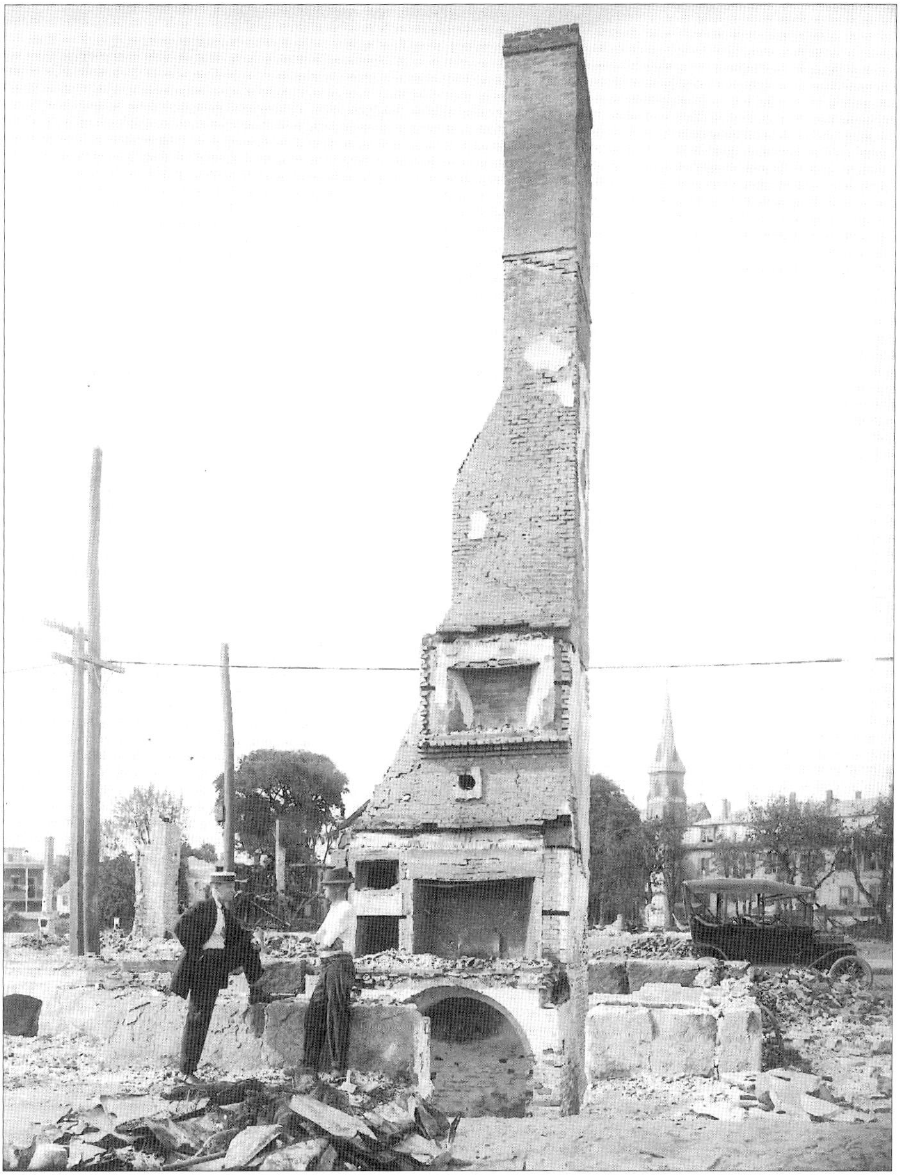

The image on the top of the previous page shows the view up Boston Street from Blubber Hollow toward Essex Street and Highland Avenue. The damage done to the Keefe Leather Factory, across the street from the Korn factory, at 46 Boston Street, is on the right side of the picture. (*Neg #19323*) The Keefe factory was sprinklered, but the intensity of heat from the fire and explosion in the Korn Leather Factory was too strong for the sprinklers to protect the building. The loss of water from broken pipes caused by the collapse of these buildings produced a severe drain on the water supply, allowing the fire to spread more easily, as shown by the factory ruins in the bottom image. (*Neg #32866*) Homes on Boston Street were destroyed along with the factories. This chimney shown stood on Boston Street, near its intersection with Federal Street; the steeple of the Immaculate Conception Church is in the background. The wind direction and extensive efforts by firemen prevented the fire from spreading down Federal Street to the business district. (*Neg #4200*)

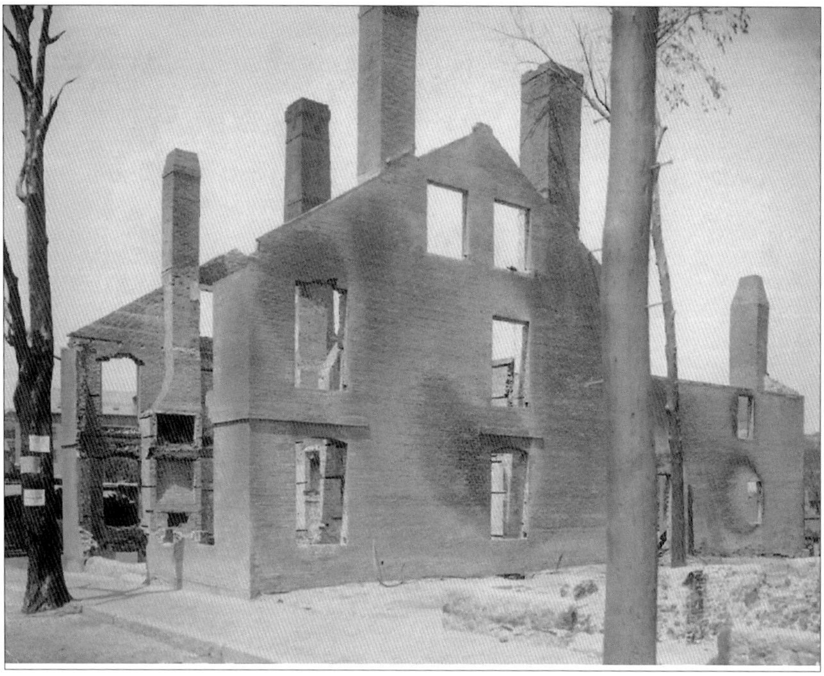

The ruins of this home at 44 Boston Street were photographed by Frank Cousins from several angles. The top image (*Neg #2699*) shows the exterior walls of the house while the bottom image (*Neg #2695*) displays an interior view. The *Lynn Item* of July 7 discusses Cousins's impressions of a Boston Street dwelling that was destroyed by the fire. Cousins indicated that the structure was one of three houses that were "gems interiorily" and that the Boston home, "a frame structure of two and one-half stories, had a beautiful fireplace of the type eagerly sought by modern architects as suggestive of modern dwellings." The ruins on the top of the next page also depict a home on Boston Street; although no address was given for the photograph nor was the photographer identified, it appears to be another view of the same Boston house. (*Neg #4222*)

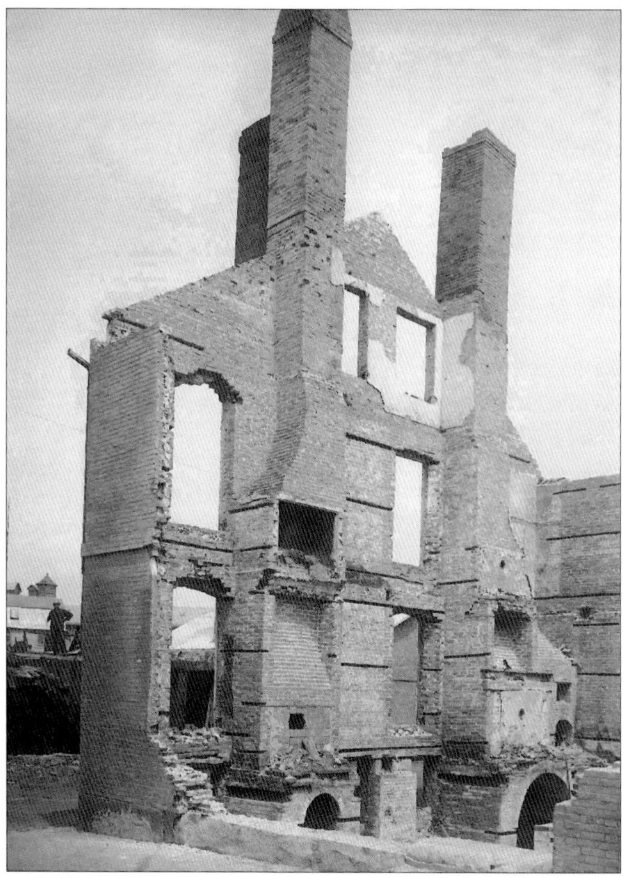

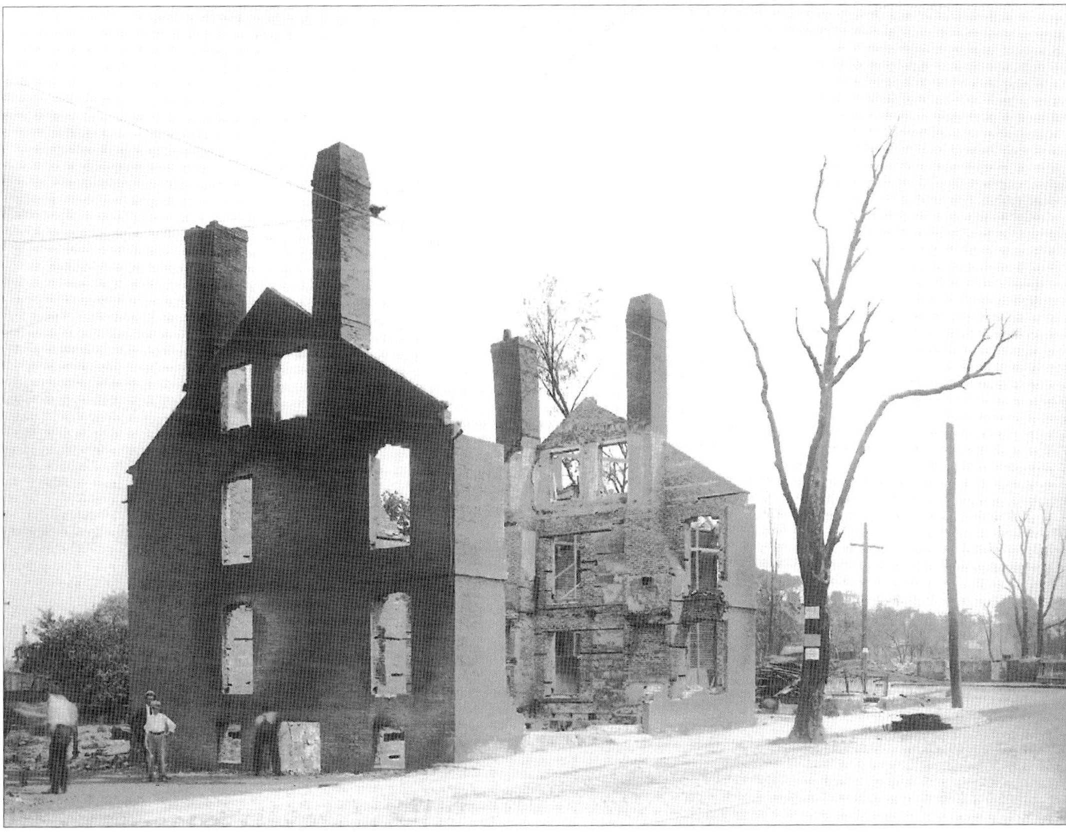

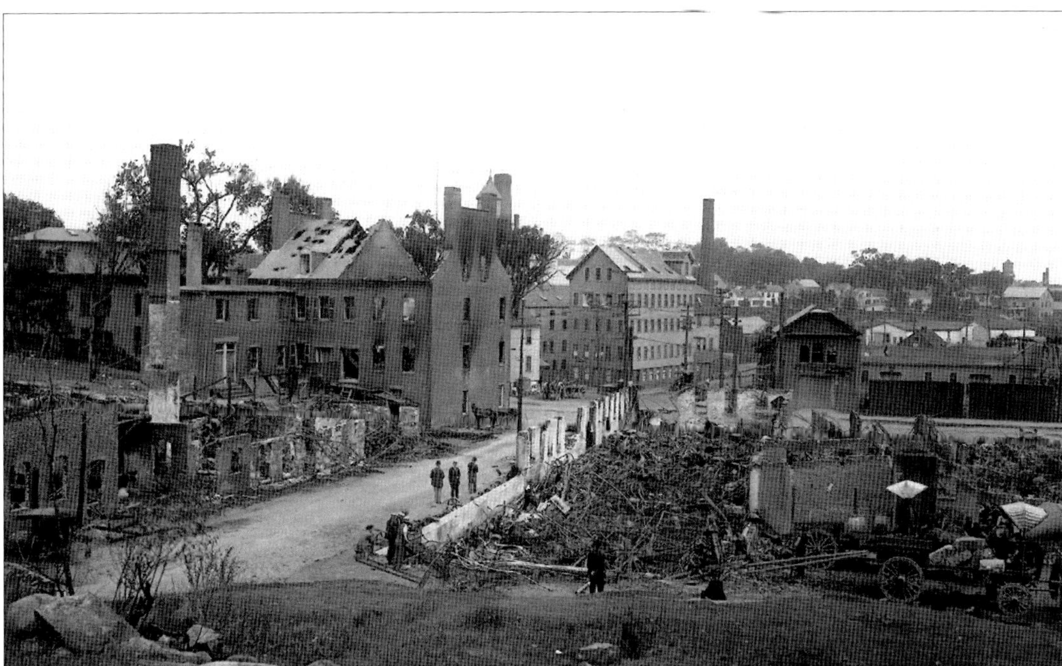

The fire started at the intersection of Boston, Proctor and Goodhue Streets. This image shows unidentified ruins on Proctor Street. (*Neg #4104*)

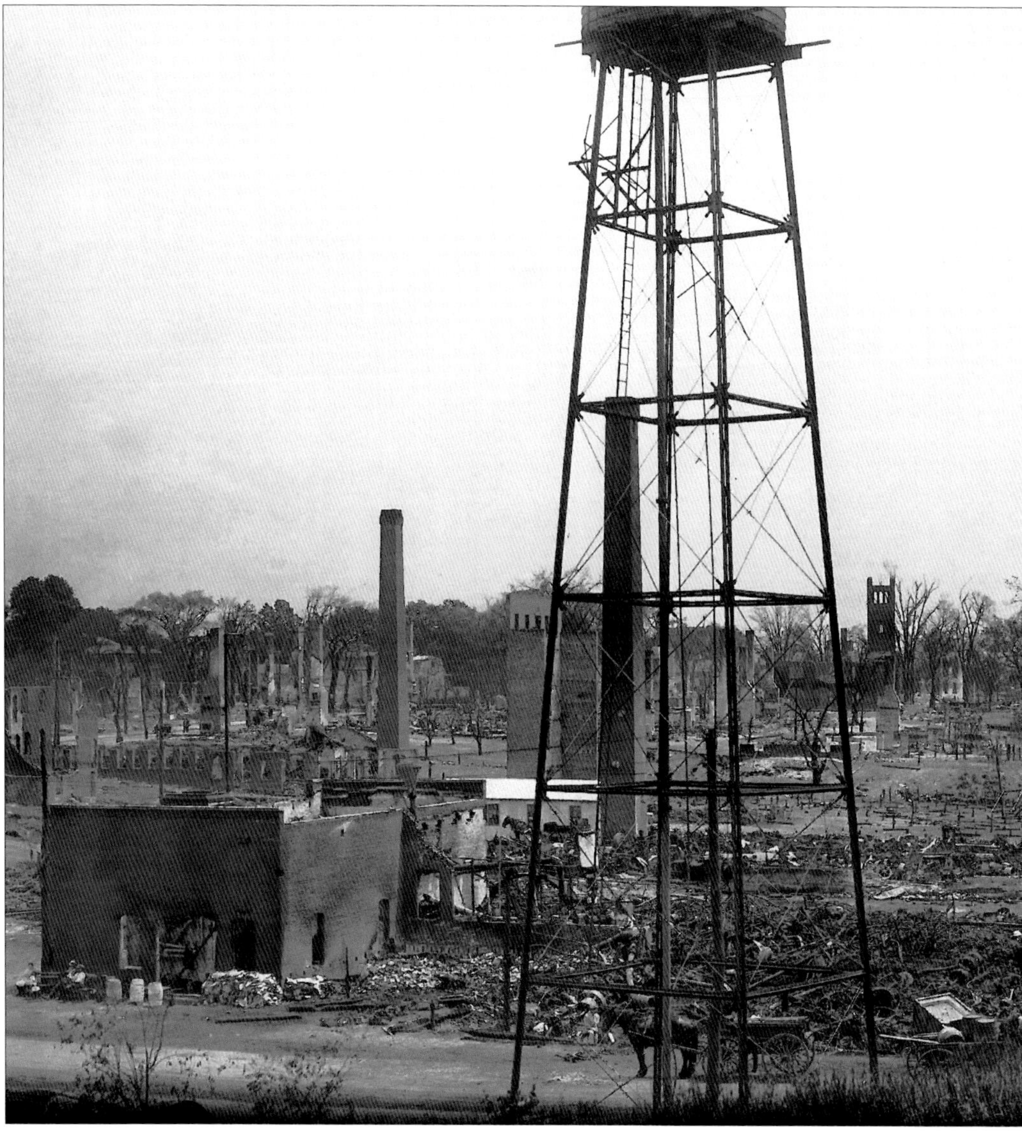

The images of the thriving leather industry at the beginning of the book were taken in the Carr Leather Factory before it was destroyed by the fire. The 1914 *Naumkeag Directory for Salem* locates the Carr factory at 1–8 Pope's Court, two hundred feet away from the start of the fire. The factory had previously been destroyed by another fire. This new factory was a mill-constructed building made of both brick and wood materials with installed sprinklers, which were connected to the city water system and to a large water tank on top of an iron tower near the factory. Originally protected by a water curtain from a single line of sprinklers under the cornice, the factory was destroyed in this fire because the water supply to the sprinklers became unavailable when the bottom of the water tank burned from the heat of the fire. The iron tower supporting the water tank remained unscathed. The Robb photo on this page (*Neg #4021*) shows the damaged water tower; the detail image on the top of the next page (*Neg #4021DetailA*) shows the damaged racks used to dry the leather along with the debris from the fire; the bottom image shows the gutted factory. (*Neg #4021DetailB*)

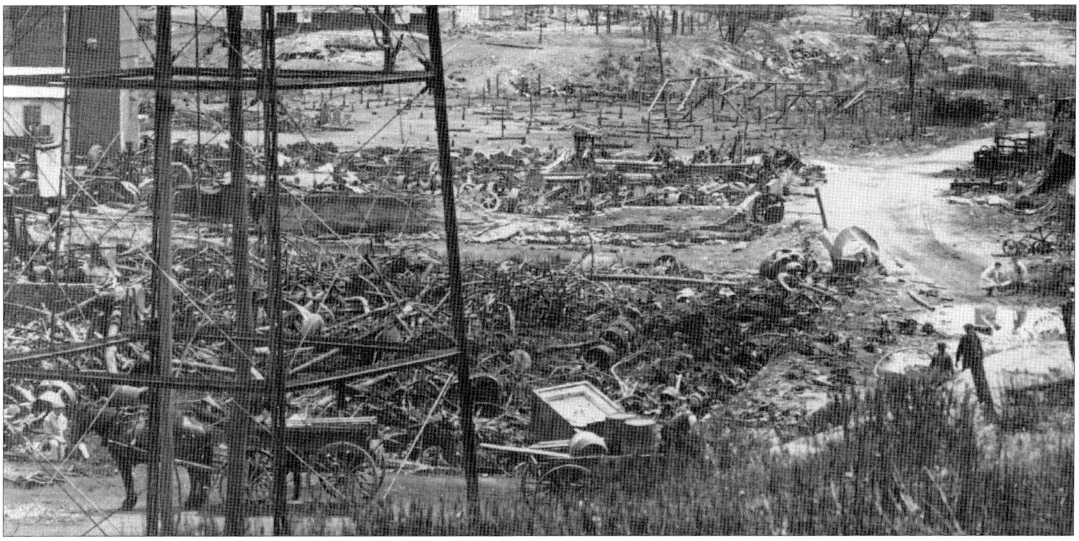

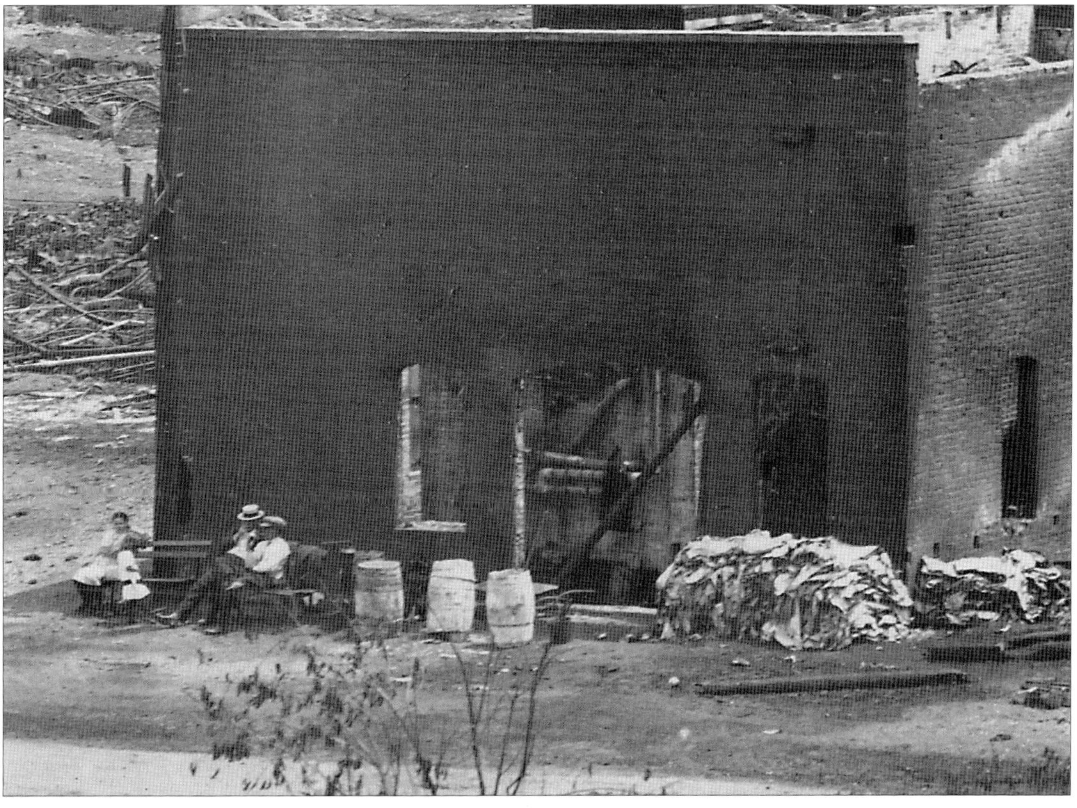

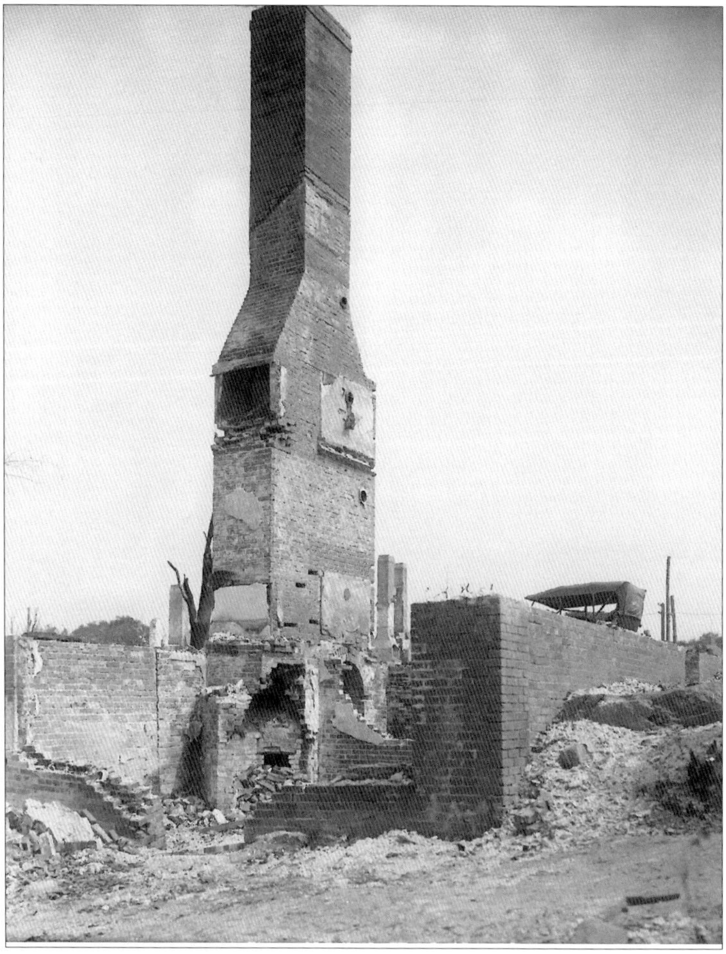

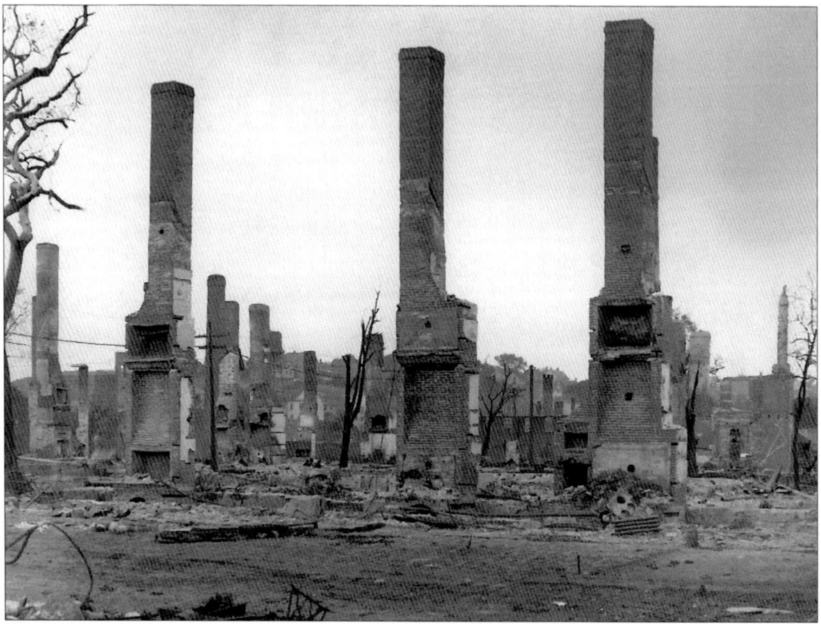

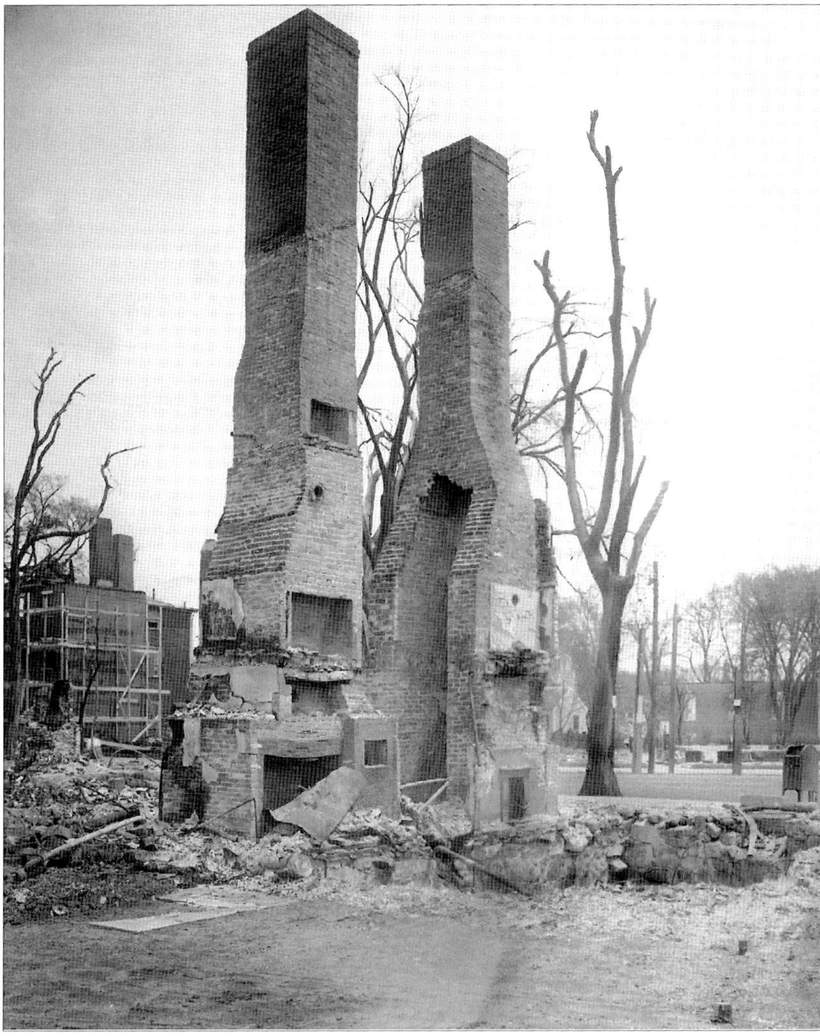

Frank Cousins was a local authority on the architectural treasures of Salem. The *Lynn Item*, dated July 7, quotes Cousins as saying that "10 per cent of Salem's colonial architecture was destroyed in the conflagration of July 25–26." The same article says that the "century and more old houses may be noted through the burned district by the massive brick chimneys which stand like sentinels watching over the remains of the old house which for years they furnished the means of heat, and in the great fireplaces which broke the vertical lines and added to the comfort and beauty of the old rooms." An unidentified news clipping dated June 27 states that one could "believe that the many gaunt fire stained brick chimneys were sentinels watching over the graves of their comrades." The bottom image on the previous page depicts several of these sentinels. (*Neg #32804*) The top image on each page was photographed by Cousins; the photo on the previous page was taken on May Street and the image on this page was shot on Essex Street. (*Neg #2694, previous page; Neg #2691, this page*)

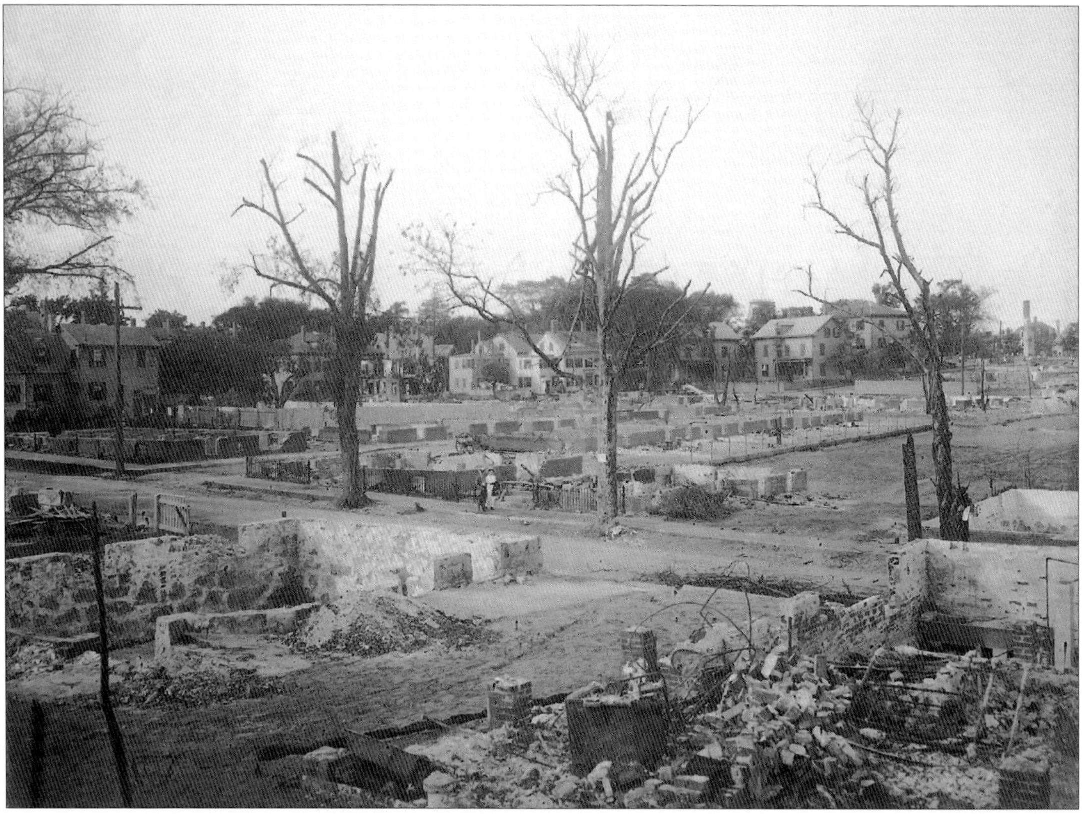

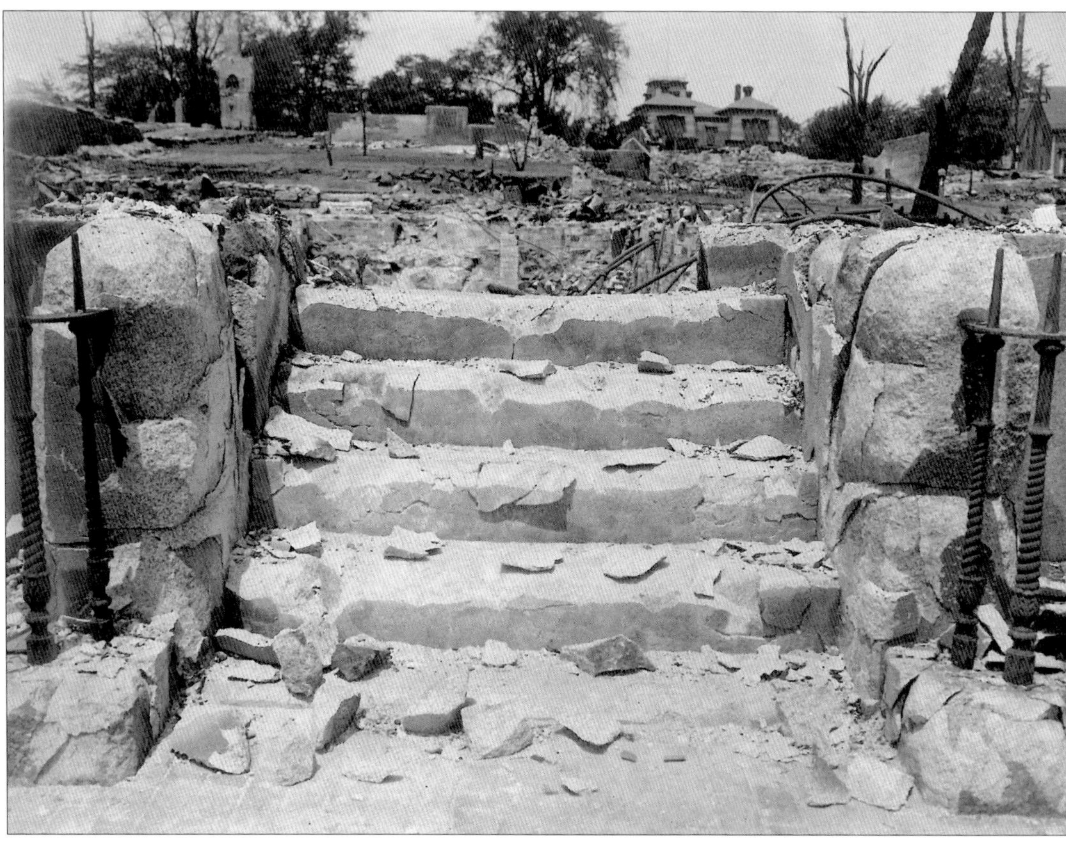

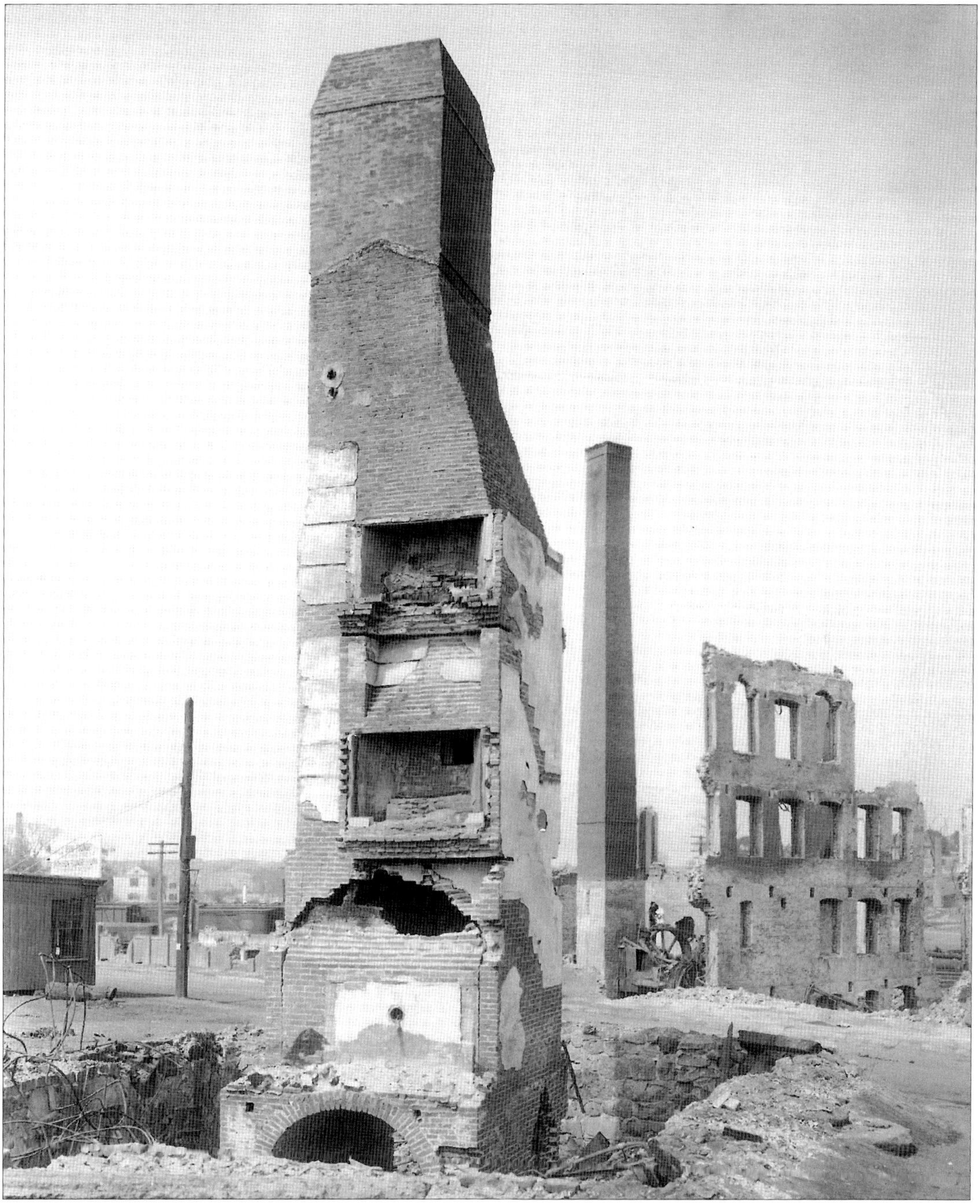

The top image on the previous page shows the destruction in the Orne Square section of Salem bordered by Hathorne, Winthrop and Mt. Vernon Streets. (*Neg #4223*) The heat of the fire was intense and reached a high point that caused the spalling of granite, limestone and brick. The results of such spalling are revealed by the chips of concrete on the stairs on Endicott Street, shown in the image on the bottom of the previous page. (*Neg #32800*) Frank Cousins shot the image of the chimney with fireplaces on multiple levels at the remains of the Kenney House in the Lower Mill Hill section of Salem. The chimney also shows evidence of spalling. (*Neg #2693*)

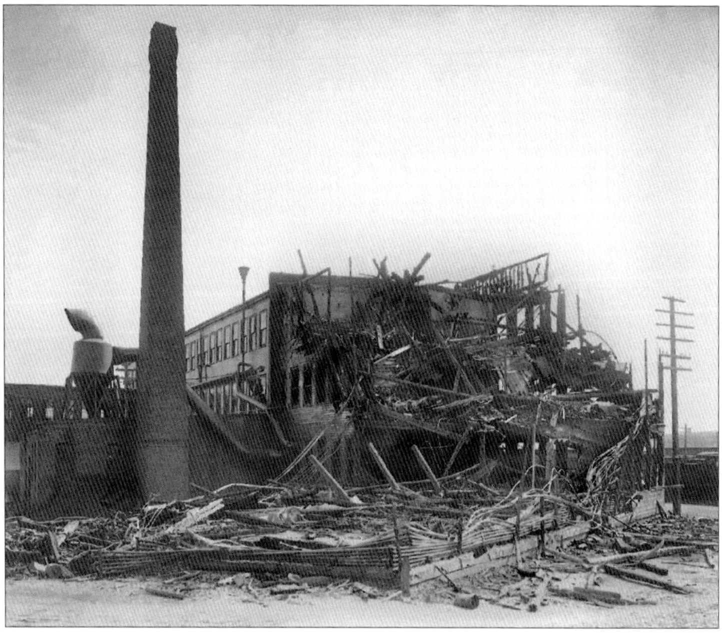

The E.S. Woodbury Shoe Factory stood on the corner of Canal and Ropes Streets. Approximately 60 percent of the building, divided into multiple factories, was saved from complete destruction because of a sixteen-inch brick fire wall with double fire doors at the openings. The Hodgdon Durand Shoe Company was destroyed while the Woodbury Factory survived because of the fire wall. (*Neg #4619, a Robb photo*) The upper image on the next page shows the Washington Street ruins from the vantage point of the train depot. (*Neg #4464*) Frank Cousins's photograph of the ruins of St. Joseph's Church on Lafayette Street also shows the destruction on the northwest side of the Boston and Maine Railroad tracks. (*Neg #32927*)

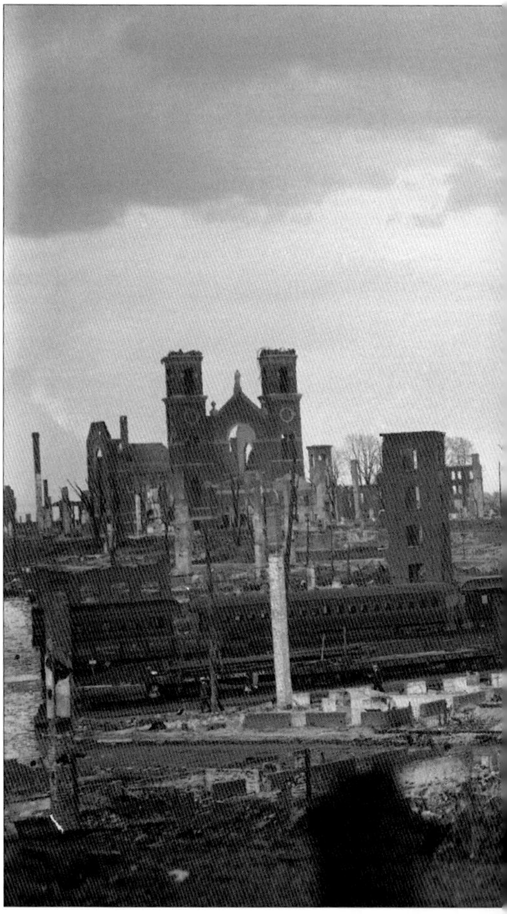

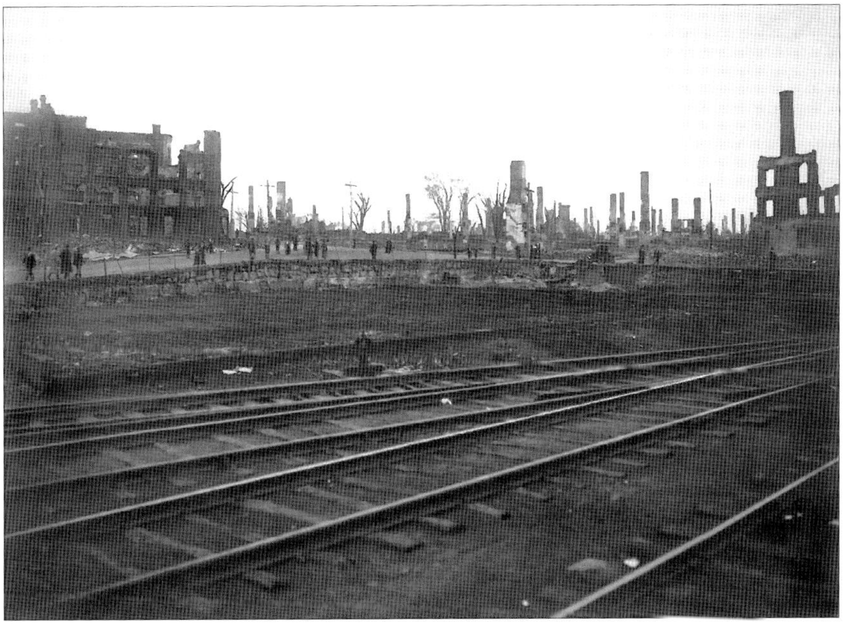

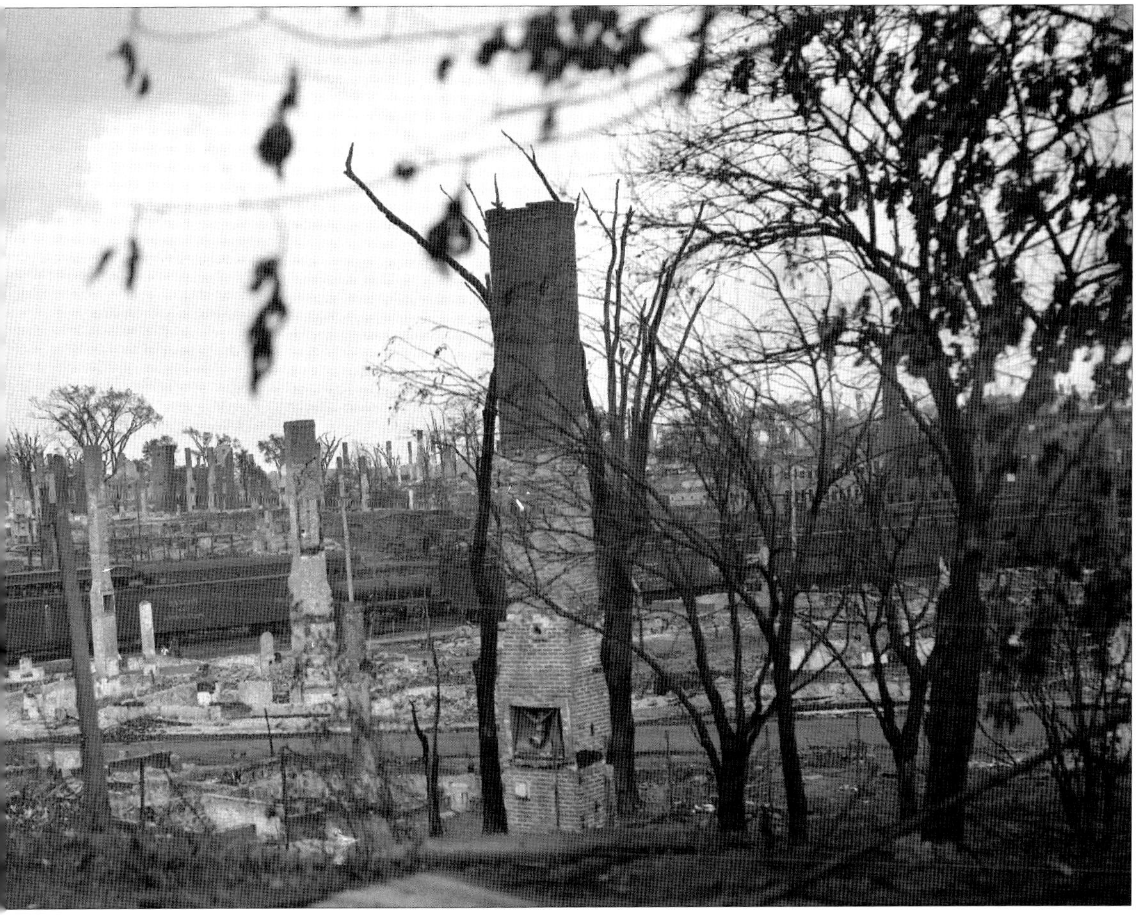

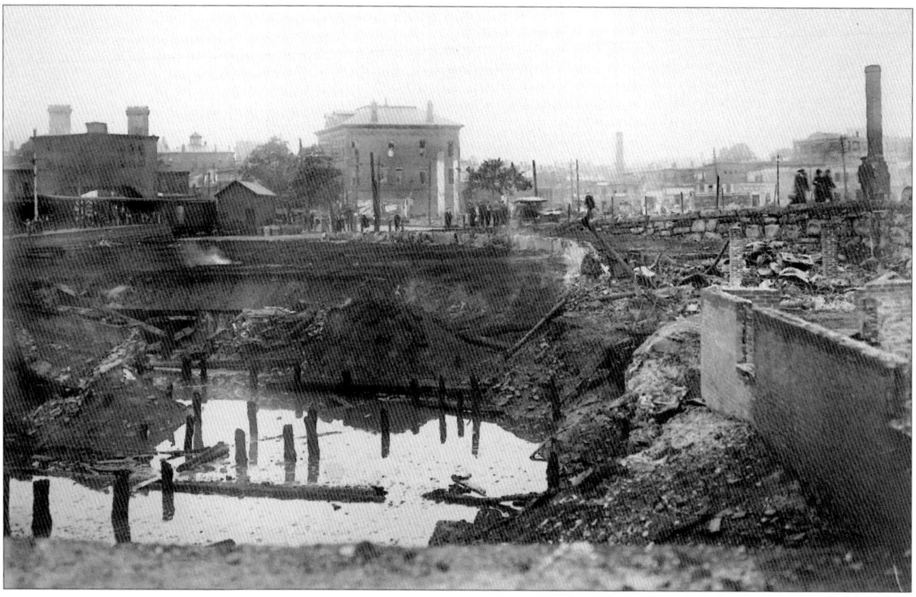

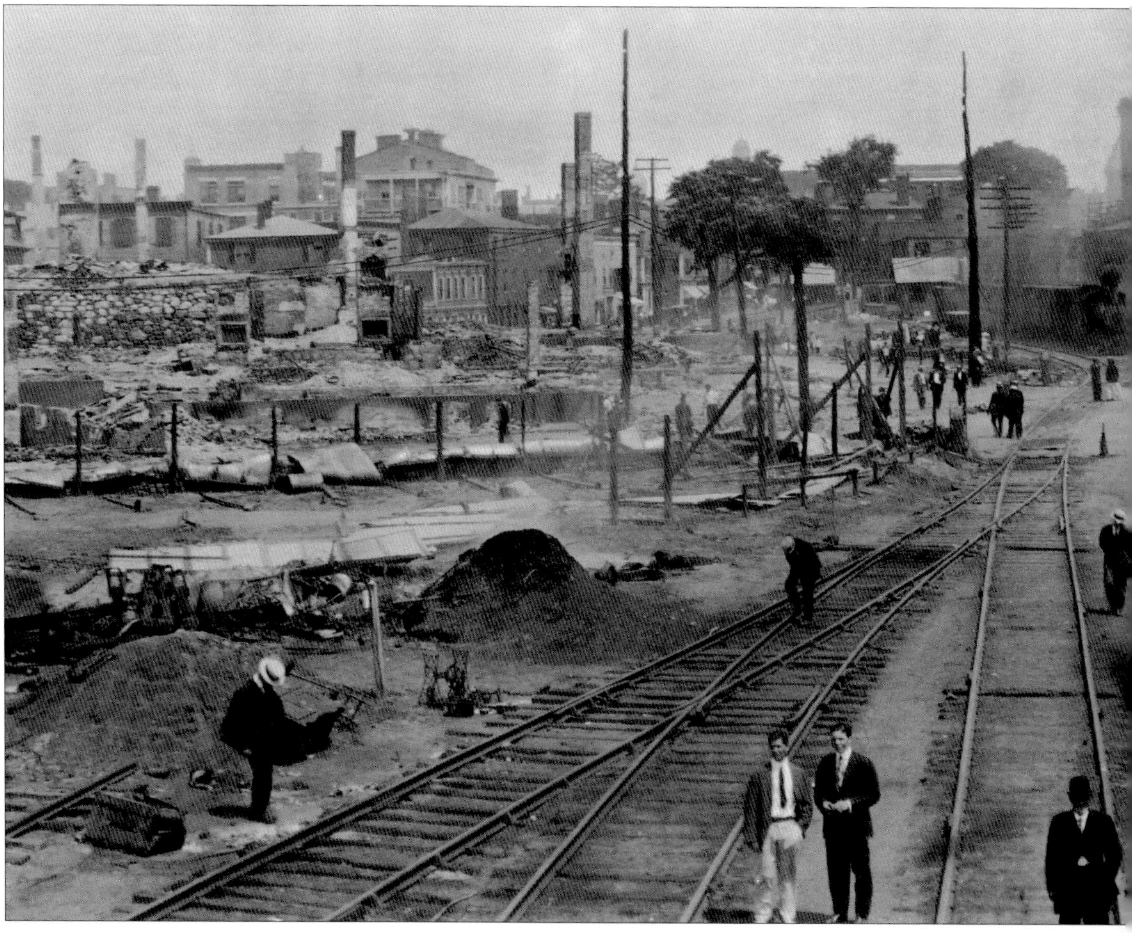

Both of these images depict the ruins near the Salem Depot. The smaller image shows the destruction to the Boston and Maine Roundhouse behind the Salem Depot, the building with the two tower structures in the upper left section of the photograph. The destroyed engine house was torn down a week after the fire. Newspaper articles reported that fire loss to the B&M would not exceed $5,000. Agent Harry A. Whitehead was on duty when the alarm was sounded; John B. Hannon, claim agent, arrived at the station immediately after the alarm was sounded and gave orders for the removal of the freight cars in the yard. Three hundred cars were removed, one of which contained dynamite. Whitehead and Hannon assisted James. T. Tyter, the general superintendent of the B&M, after he arrived from Boston; they removed ticket cases, baggage and furniture left by residents along the tracks. The furniture was moved to Beverly. (*Neg #4001*) The Phillips Library collection includes several panoramic views of the aftermath of the fire. The image of the train station is a section from one of the views photographed by Rankins and Ward of Lynn. It was also published in the *Boston Post* on June 27. The accompanying article mentions that the debris on the tracks was the remains of four Pullman coaches destroyed by flames before a trainman could get a locomotive onto the siding to save them. On the right was one of the first relief trains to arrive in Salem from Boston. The ruins of residences that were dynamited after the fire are to the left of the train; the ruins in the center of the picture are the remains of a shoe factory that was newly built before the fire. Beyond the track was a block of brick buildings completely encircled by fire that escaped destruction. The walls of a factory to the extreme right of the train were dynamited as a precaution against falling debris. (*Neg #32941*)

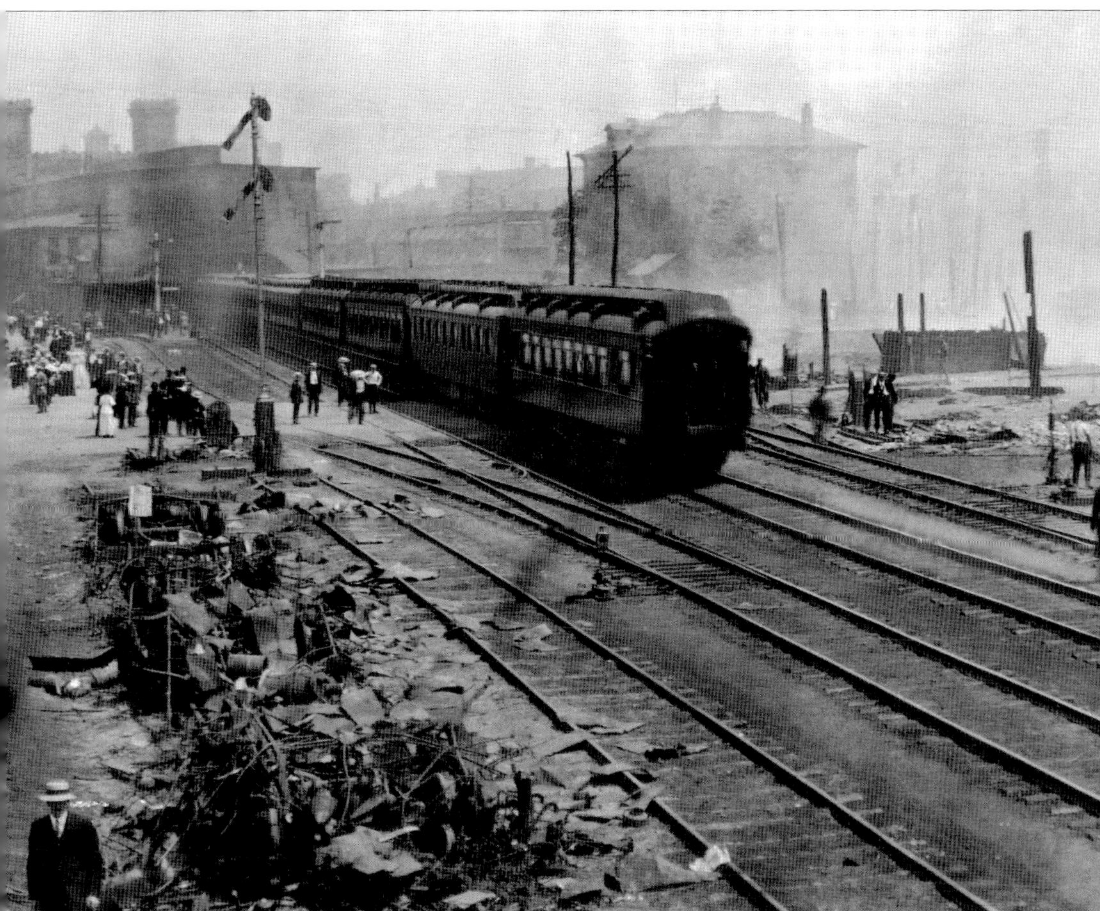

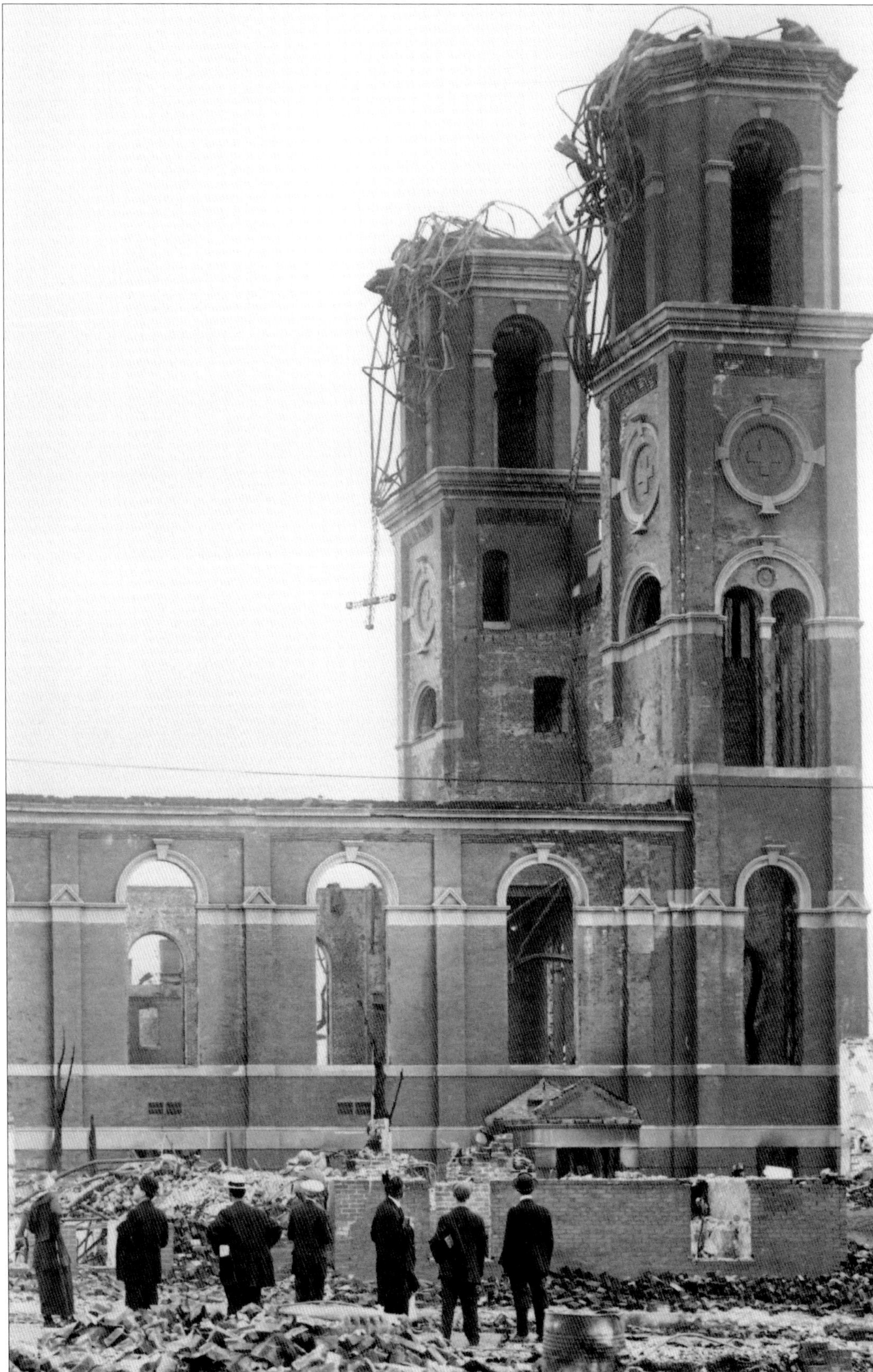

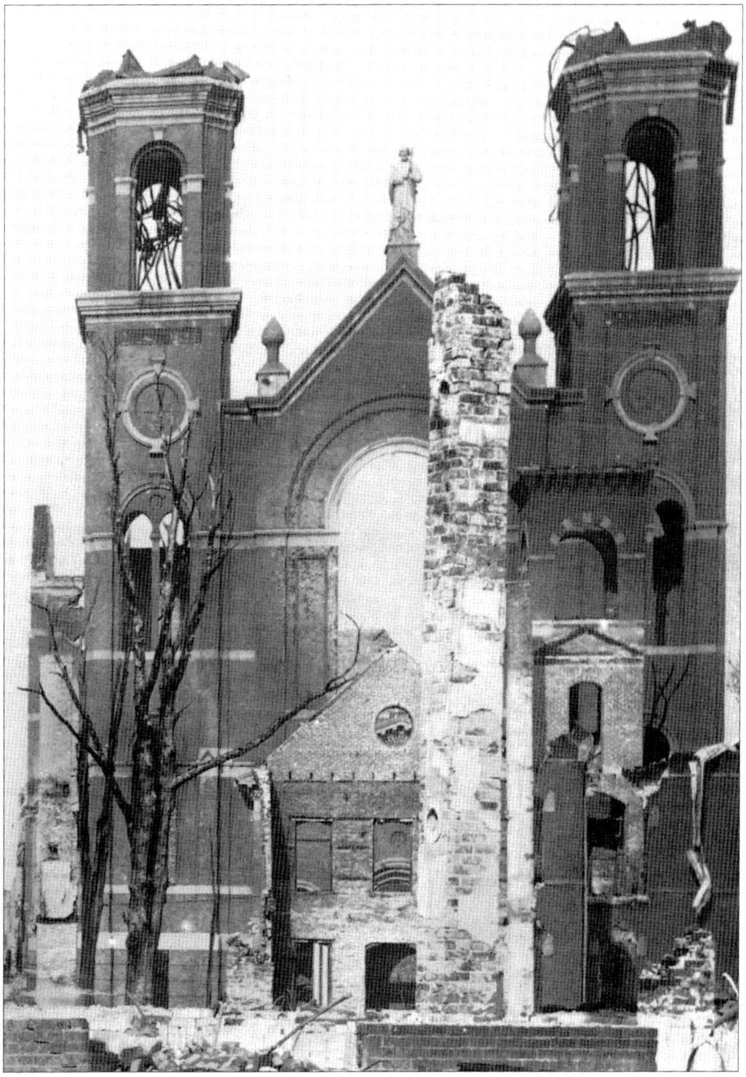

Saint Joseph Parish, Salem Mass 1873–1973, written to celebrate the centennial of the parish, states that St. Joseph's Roman Catholic Church in Salem was founded on May 17, 1873. The Luscomb Estate on Lafayette Street was purchased in 1881 to build the church; a few years later the adjoining Elwell Estate was purchased for the rectory. A church school was built, with separate entrances for boys and girls, in 1892. The original wood church was moved to Salem Street to build a larger church on the Lafayette Street property. Dedicated one year before the fire, the heat of the blaze gutted the inside of the church. The devastating heat bent the metal crosses, creating an interesting juxtaposition to the stone crosses carved on the church towers. (*Neg #5105Detail*). Newspaper reports state that the surviving statue of St. Joseph looked like a "carven figurehead on a tombstone." The basement of the burnt church was remodeled to be used as the church for many years after the fire. Classrooms to replace the school were built on the roof of the basement; eventually a new school was built and the surviving statue was moved to the school grounds. The cornerstone for a new church, designed in cruciform, was laid in 1949. When construction of the new church began, the old statue of St. Joseph was buried in the parking lot of the church to signify the new future for the church. During its reorganization, the Archdiocese of Boston closed the parish in 2004. (*Neg #32928*)

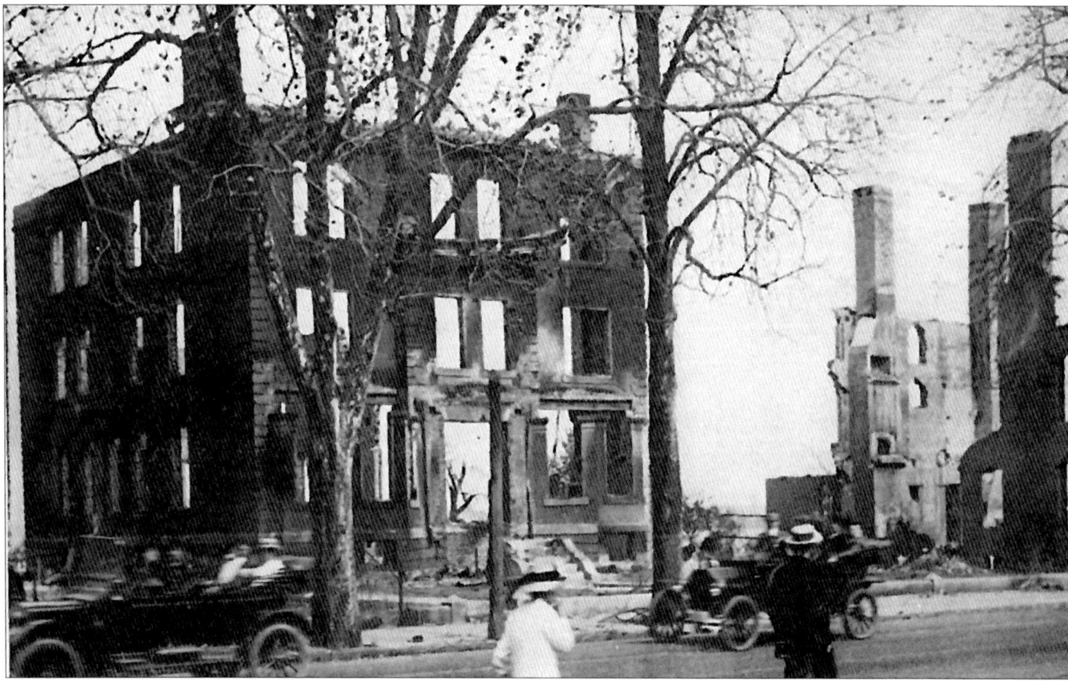

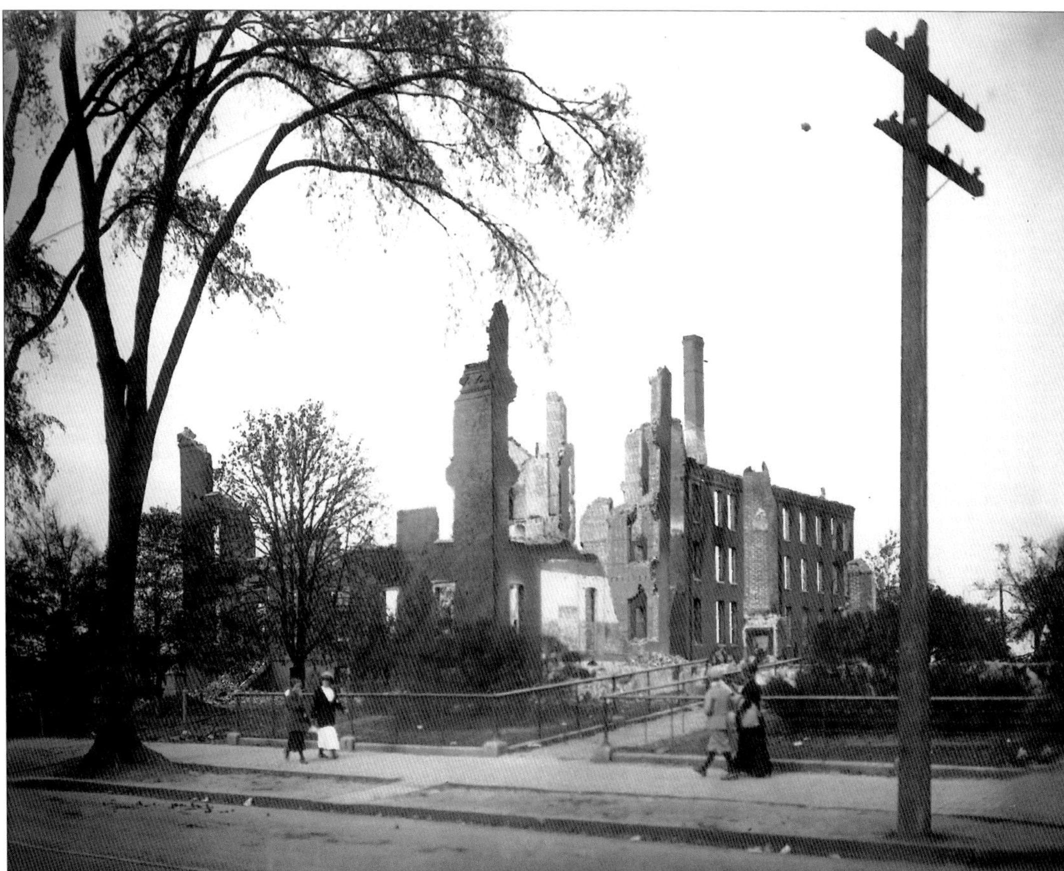

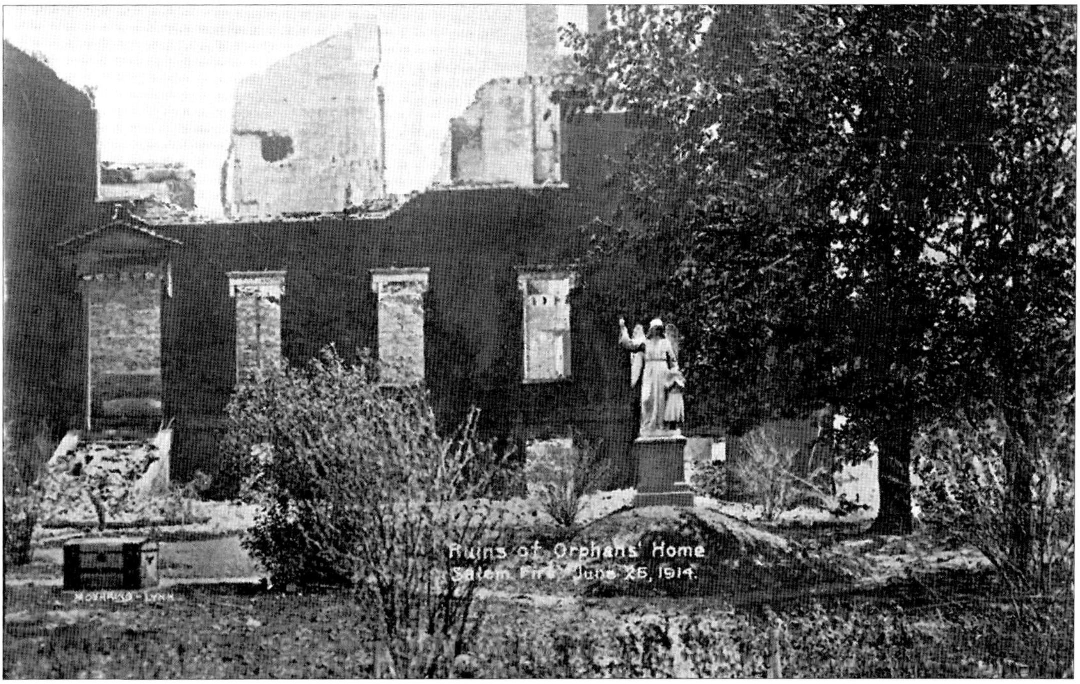

The top image on the facing page shows the destruction to the St. Joseph Parish House. Eventually a new rectory was built on the corner of Harbor and Lafayette Streets. The French Canadian parishioners attended a Mass in the field of Palmer Cove the day after the fire to give thanks for their lives and to pray for the future of their church. (*Neg #32862*) The bottom photograph depicts the ruins of the Orphan Asylum on Lafayette Street, run by the Sisters of Charity. (*Neg #4473, Robb photo*) Reverend Father J.J. Herlihy and Sexton Patrick Quinn from St. Joseph's Church were instrumental in removing the sisters, the orphans and the infirm women from the asylum before it was destroyed by the fire. The occupants of the building were removed with the help of Cahill's Peabody bus and John. T. Farrel's undertaking vehicle. The building was severely damaged in the fire but the statue was unharmed, as shown in the picture on this page. (*Neg #32852*)

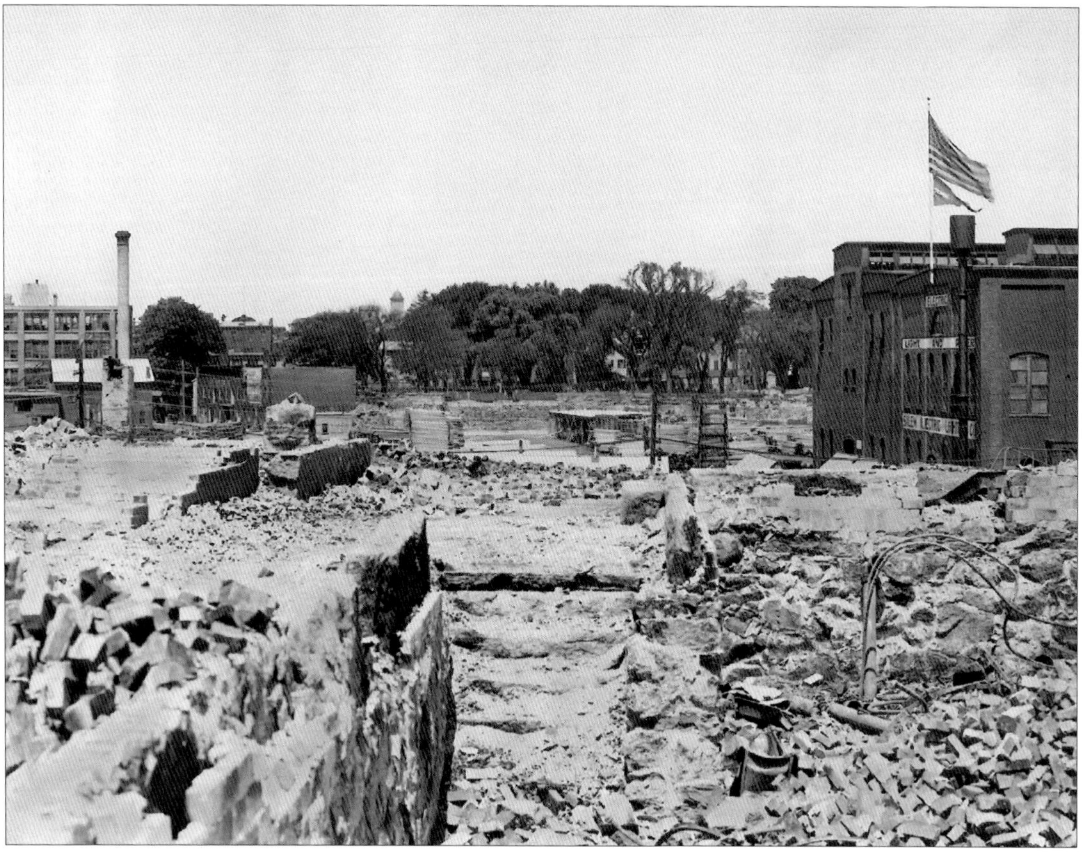
LIGHT AND

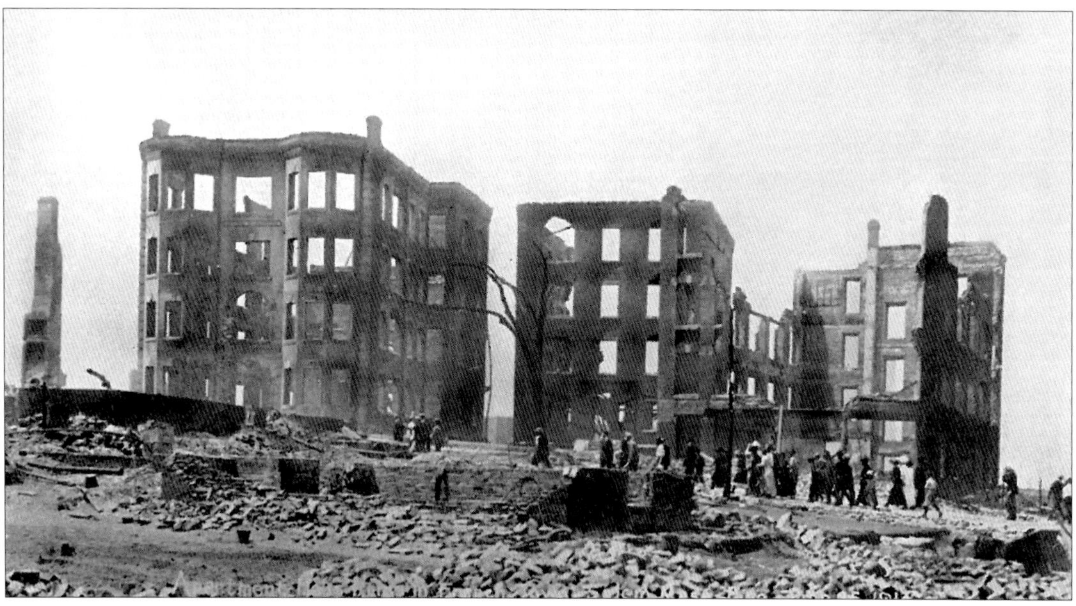

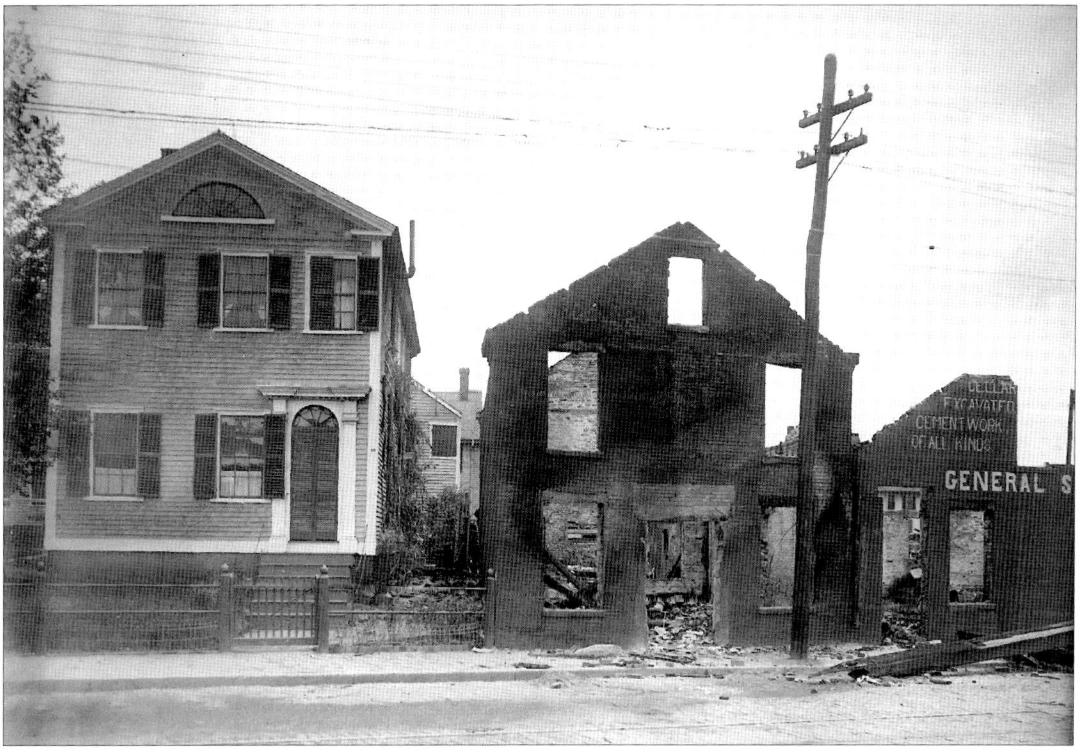

One of the scrapbooks in the library collection contains a news clipping, with no source or date, about the American flag flying in the midst of the ruins near Ward Street. The first sight many saw the day after the fire was this American flag "strung to the pole on the top of the power house of the Salem Electric Company, right in the midst of the burned district, waving unscathed over clouds of smoke, the only bit of color in the entire picture of desolation." (*Neg #32775*) An article in *The Salem Evening News*, dated June 27, with the headline "French Population Destitute," states that in 1914 the French district in Salem housed six thousand residents. The byline of the article was "A News Man." This newsman wrote, "The Point district was the scene of the labors of a lively, aggressive and hardworking people" whose "homes…had been laid in smoking ruins." These French Canadians owned their own homes and businesses, many worked in the Naumkeag mills and many were physicians and lawyers. Another article dated the same day writes that "three brick apartment houses at the top of the hill were wrecked as quickly as wooden buildings." These apartment houses were near the Mill Hill section of the city. (*Neg #32851*) The picture of the Walnut Street ruins above illustrates the role fate played in whether a structure survived the fire. (*Neg #4014*)

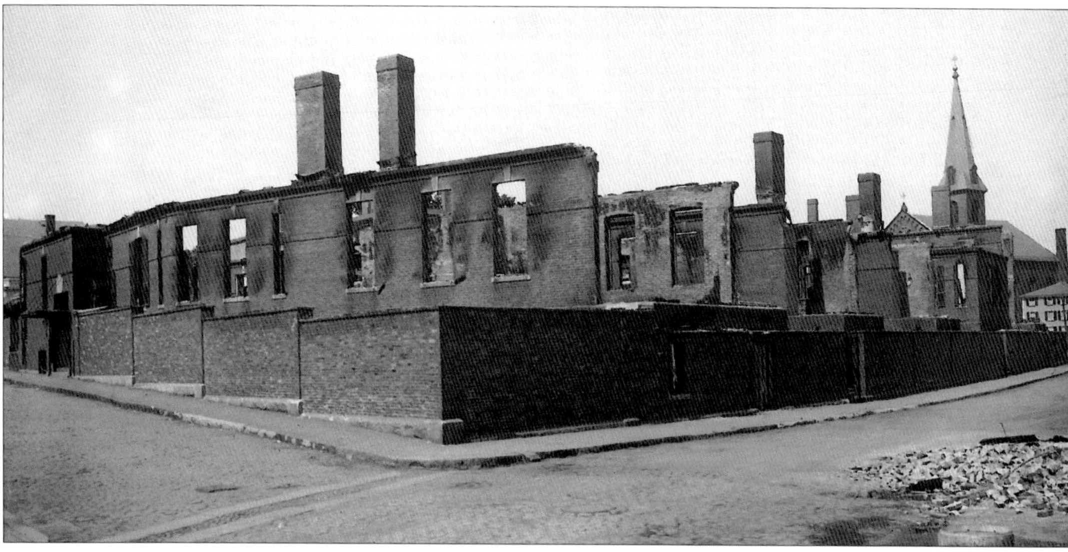

Originally a private residence, Salem Hospital was founded in 1873 with the help of Captain John Bertram. Located at 31 Charter Street, it opened in 1874 with twelve beds and a staff of thirteen doctors led by David Coggin. The hospital expanded to include eight buildings on Charter, Derby and Liberty Streets, which served maternity, surgical, outpatient and pediatric patients. Only the administration building survived the fire. Burned and injured people were taken to the hospital at the start of the fire but were eventually moved when it was realized the hospital lay in the path of the fire. Patients were transferred to Beverly and Peabody hospitals, and to the local almshouse and auxiliary hospital at Salem Neck. A relief hospital was also established at the armory under the supervision of Dr. Cutter, surgeon of the Eighth Regiment. *The Salem Evening News*, dated June 25, wrote: "The transfer was made without mishap and an hour later the rear end of the hospital buildings were in flames, the fire working its way through all the new buildings up to the old structure on Charter Street." (*Neg #16419, above; Neg #4013, below, a Robb photo*)

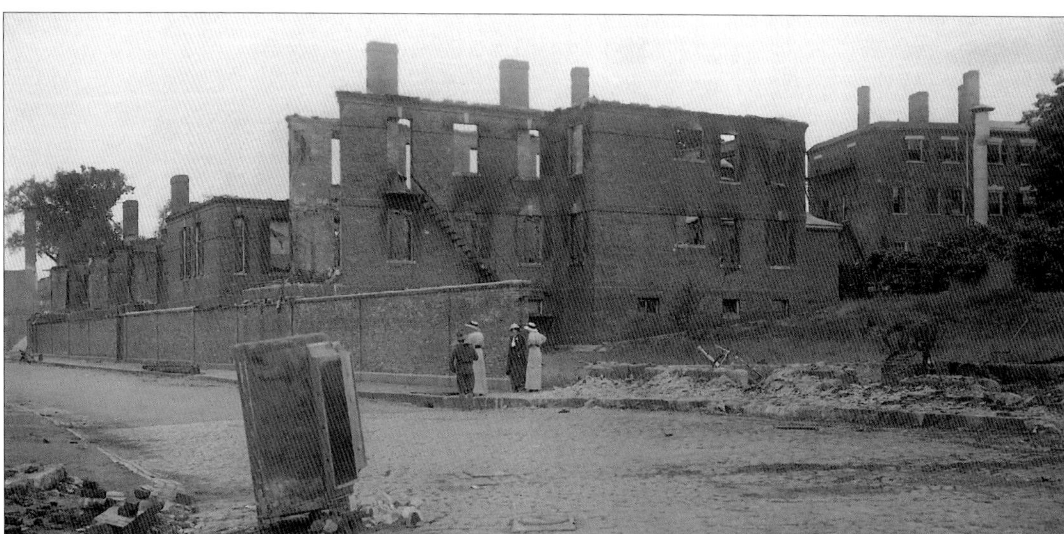

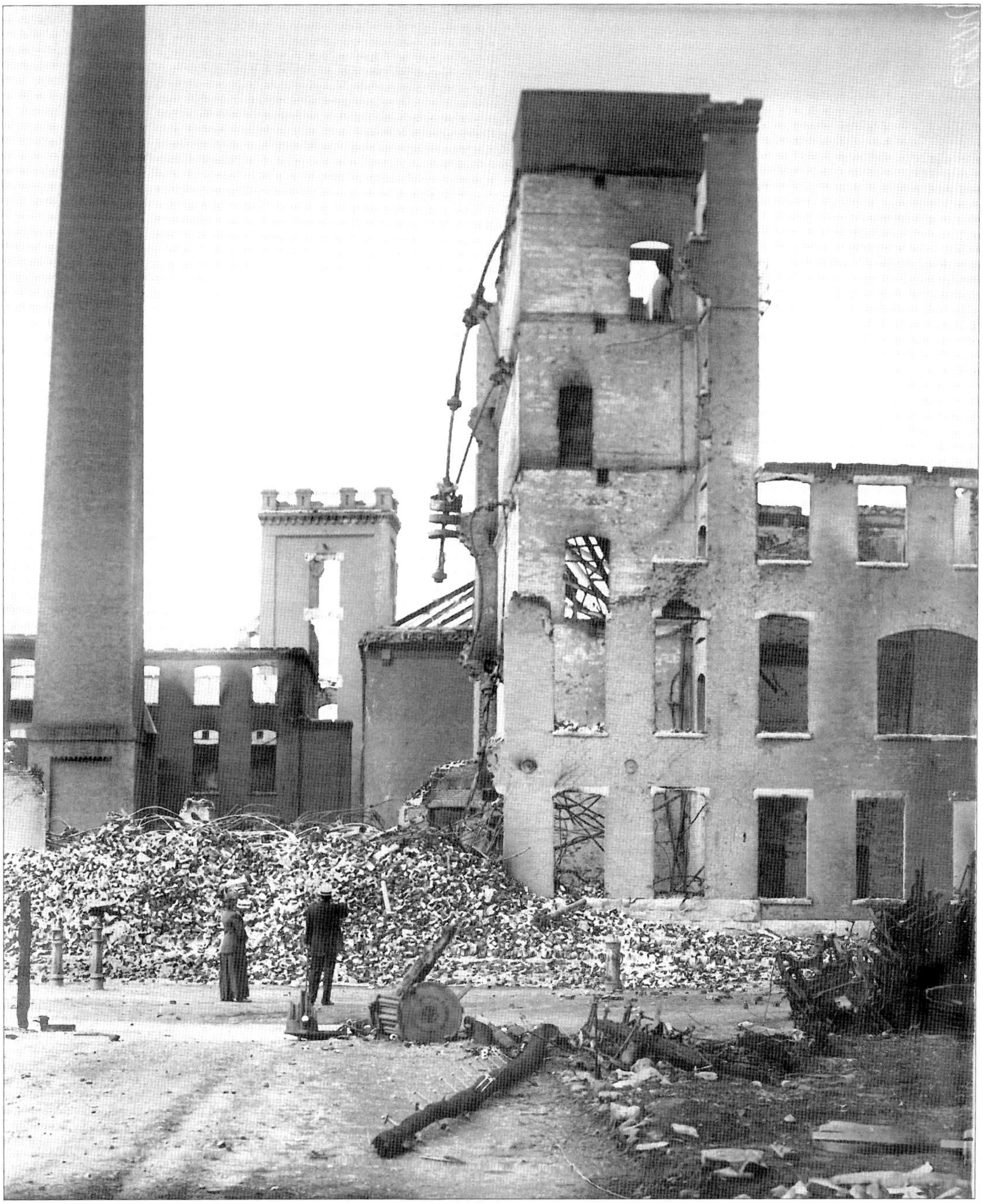

None of the mills of the Naumkeag Steam Cotton Company survived the fire. The ruins of this unidentified mill are indicative of the damage done to the company, which sustained the largest loss in terms of property and payroll. Over $1,000,000 worth of raw cotton was destroyed. Just prior to the fire, the company had purchased over $50,000 worth of new textile machinery from the Draper Works to increase its output. The mill property was across from Derby Wharf and was bordered by Union, Peabody and East Gardner Streets. (*Neg #32799*)

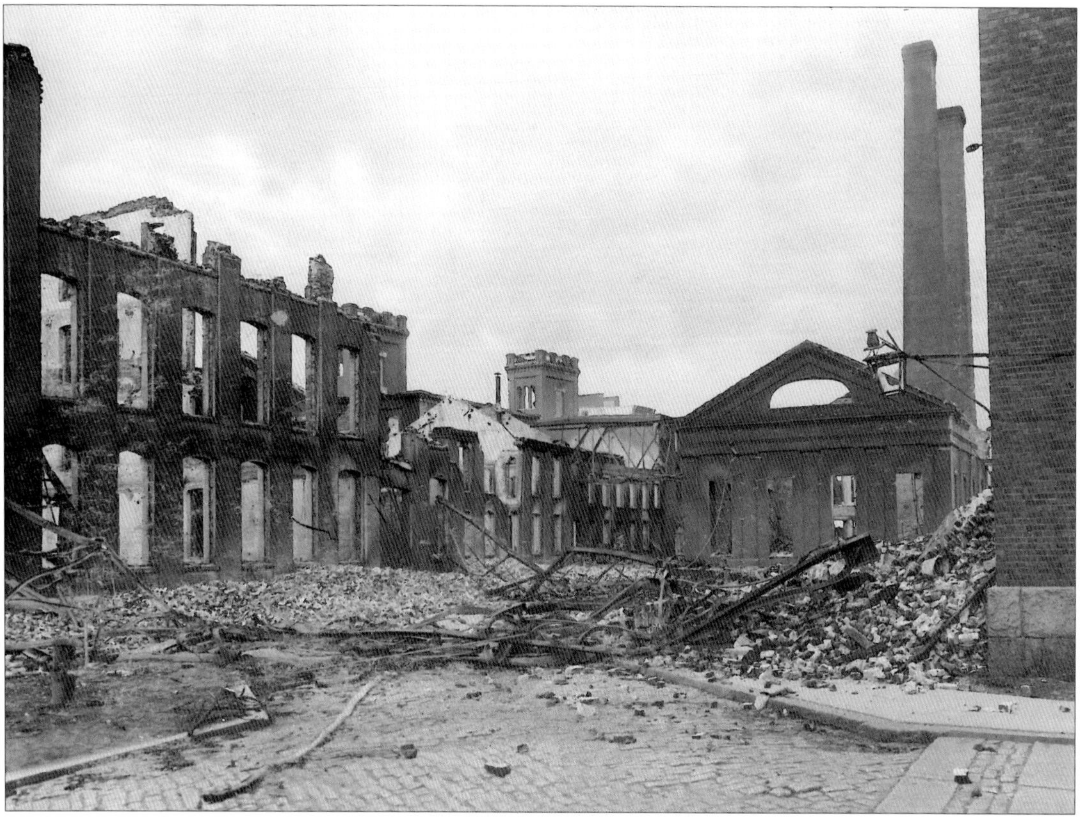

It was originally felt that the fire would move south of the mill district. A southerly shift in the wind around 6:00 p.m. moved the fire just west of the mills. The increased wind and intense heat from the burning multistory tenements of the Point made it impossible for the mill to escape destruction. No public fire apparatus was seen in the area and the mill was left entirely to its own protection. The fire brigade at Naumkeag was under the command of Chief Frank P. Cook. The complex had its own high-pressure water system, a fire alarm system that included twenty-three alarm boxes and steam engines, which burned cannel coal. Water pressure was originally sufficient with two one-thousand-gallon steam fire pumps drawing salt water from the ocean to add to the water supply. The mill complex was protected by external and internal sprinklers. Hoses had been laid to fight the fifteen hundred feet of property and the external sprinklers had been turned on. When the wind shifted, the hoses were turned toward the tenement buildings but the fire progressed, causing several of the large mills to catch fire, which turned on the internal sprinklers, draining the water pressure. The only means of escape was by water because of the advancing fire along the remaining land exit, the bridge on Union Street. By 10:15 p.m., all remaining workers at the plant had escaped via small boats in the harbor. Before they left the boilers were stoked and the pumps were left running in an attempt to save part of the mill; the entire facility, except for two fireproof buildings, was destroyed. The photograph on the top of each page depicts the destruction to the mill complex. (*Neg #32776, top this page; Neg #4010, top next page, a Robb photo*) The corporate boardinghouse in the bottom photo on the next page was also destroyed. (*Neg #32809*)

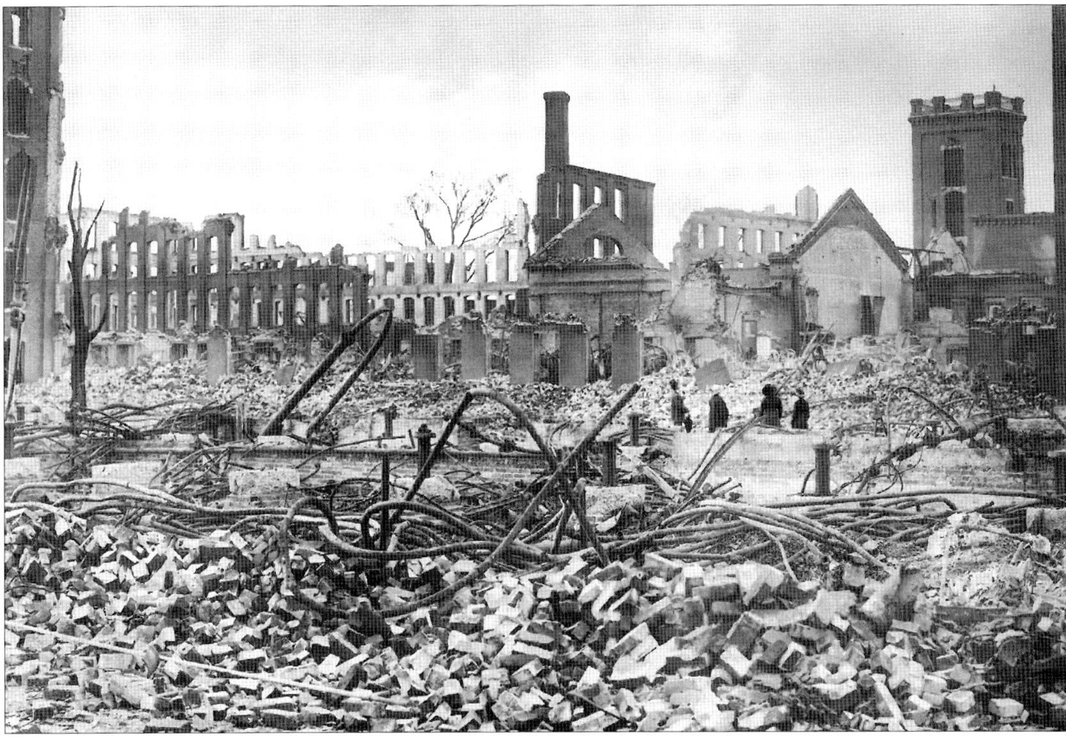

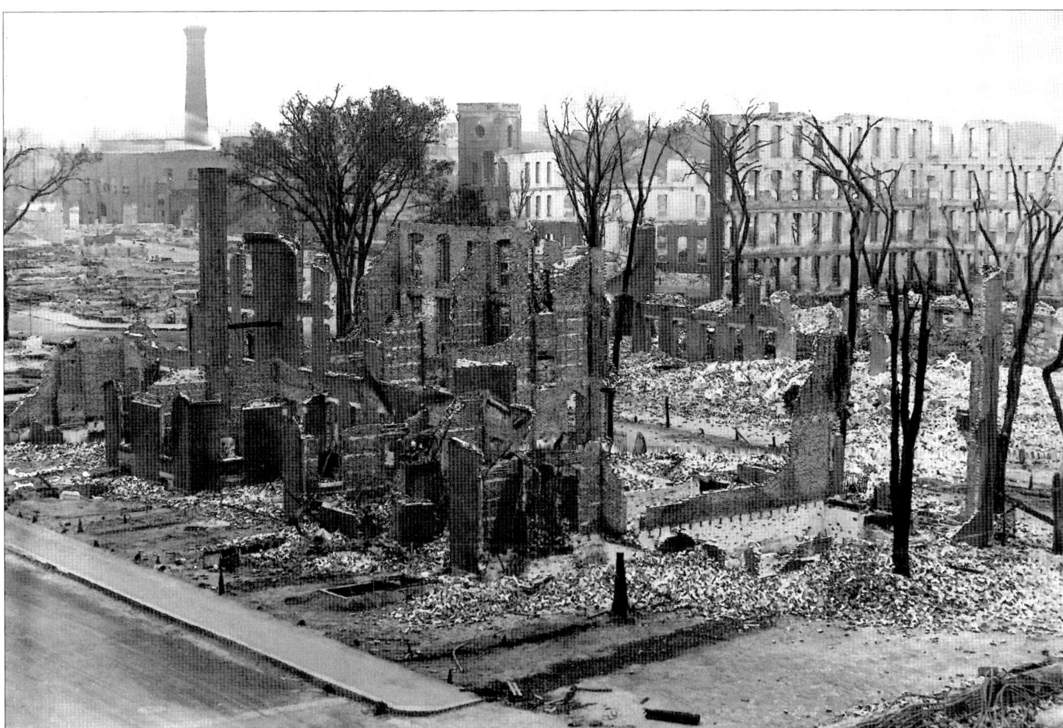

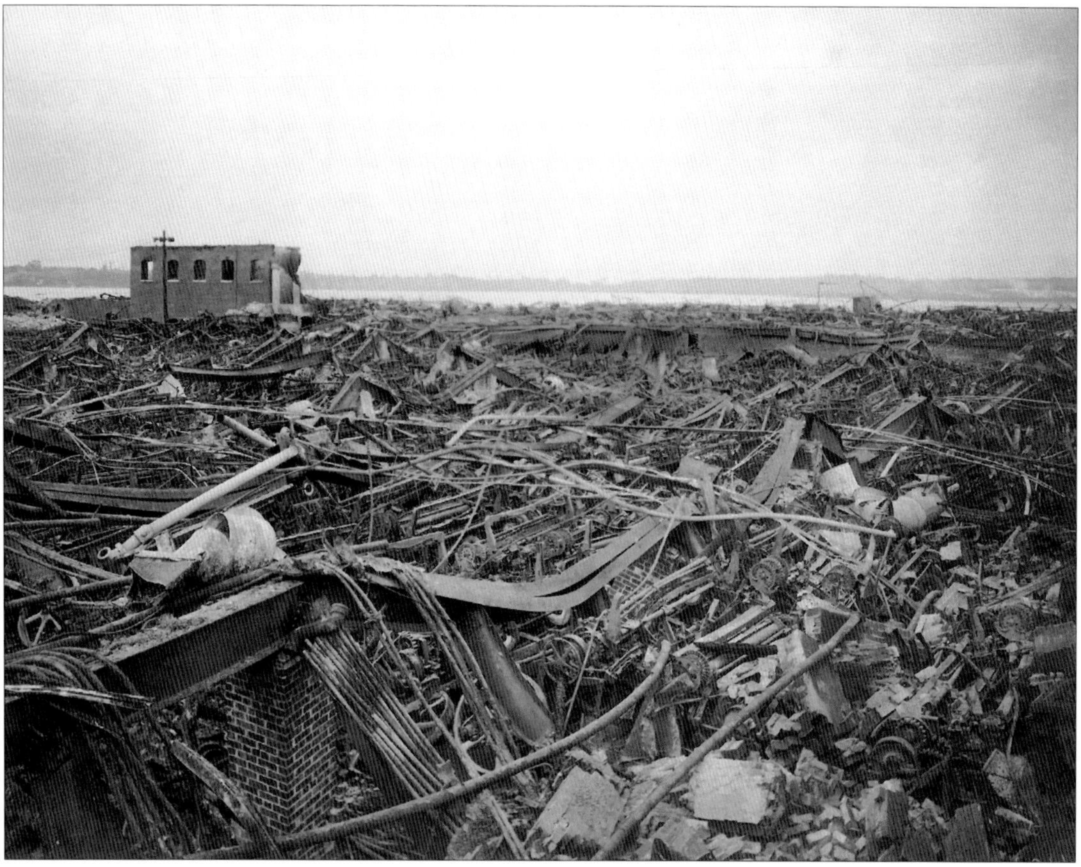

These three images provide additional examples of the total destruction to the Naumkeag mill complex. Debris included materials from the damaged buildings and tons of metal from the destroyed machinery inside the buildings. The photo on this page shows the wreckage of the No. 6 Weave shed, known for its saw tooth roof. Included in the wreckage are floor beams from the building. The weave shed was the first building to be rebuilt when the mill announced plans to reopen. (*Neg #4115, a Robb photo*) The two photos on the next page are unidentified ruins on the property. (*Neg #32781, top; Neg #32790, bottom*) On June 27, *The Salem Evening News* reported that the Naumkeag safe had been found among the ruins. Employees from the Damon Safe Company opened the vaults and everything inside the safe was intact. The article said that all employees would be paid when they reported to a local school the following Thursday. The *Boston Globe*, dated June 26, wrote, "William P. McCullan, agent and manager of the company, is to be given unrestricted authority to put the plant in the condition it was immediately preceding the fire." McCullan hoped to have one of the buildings in commission by September 1. Temporary offices of the company were moved to the Salem YMCA on Essex Street.

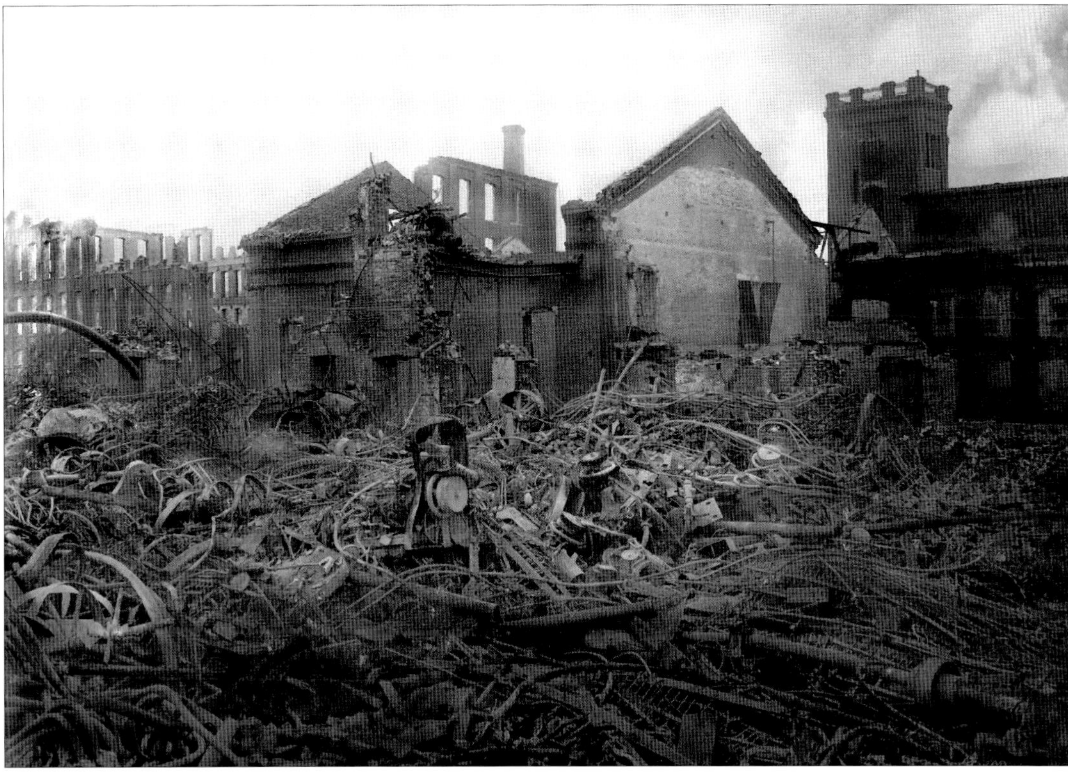

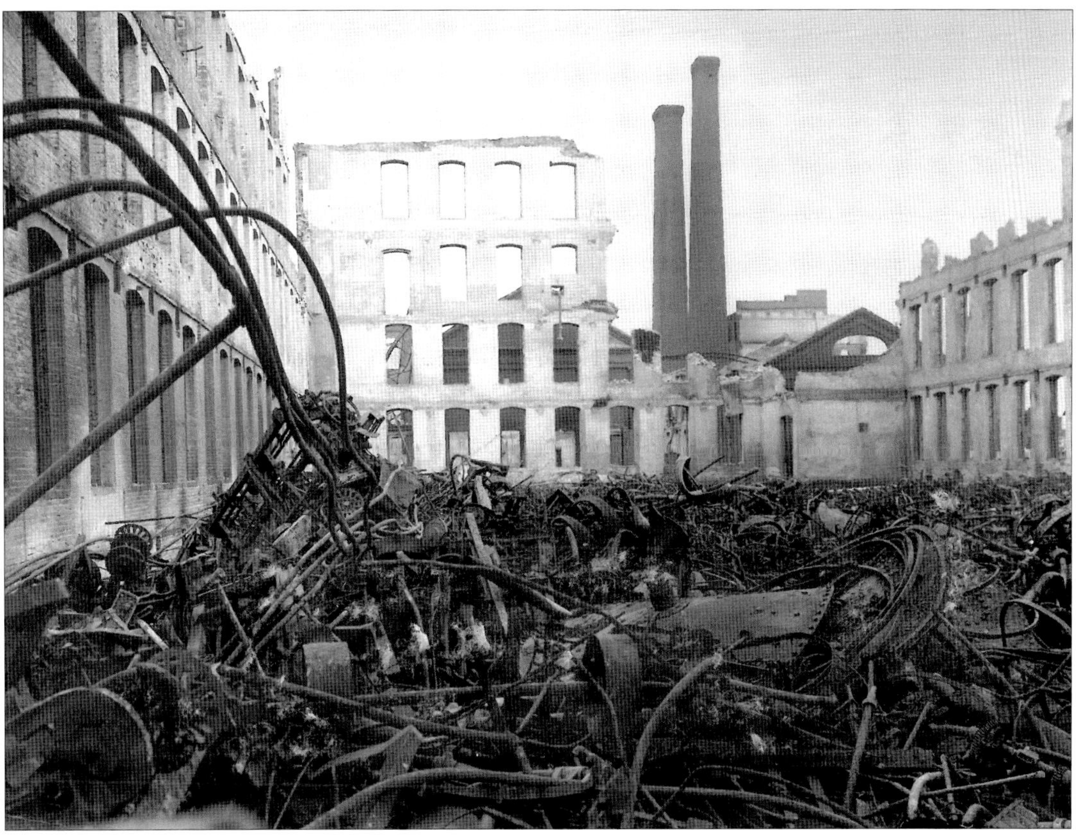

The Naumkeag Steam Cotton Company housed 106,000 spindles and 2,620 looms. The mill complex was powered by steam engines: two Corliss engines and one pair of Cooper-Corliss engines, which produced 3,500 horsepower requiring over 13,000 tons of coal a year to operate. The coal reserves were stored on Union Wharf across the South River; the mill and the wharf were connected by the Union Street Bridge, all of which were destroyed in the fire. The exterior of No. 5 Mill is in the center of the photo. (*Neg #4000*) After the fire, heaps of tons of burning coal smoldered, emitting choking fumes and gases. An unidentified newspaper article dated July 13 reported that the coal piles of the George W. Pickering Coal Company were still burning. "These great heaps with their fringe of gold and blue loom up in a manner to vividly bring back to mind the fire which caused so much destruction here on that fatal June 25." (*Neg #32779, bottom this page; Neg #32784, bottom next page*) Newspaper articles also reported that coal reserves in private homes, which had not been used to stoke the steam engines fighting the fire, were also smoldering for days afterward.

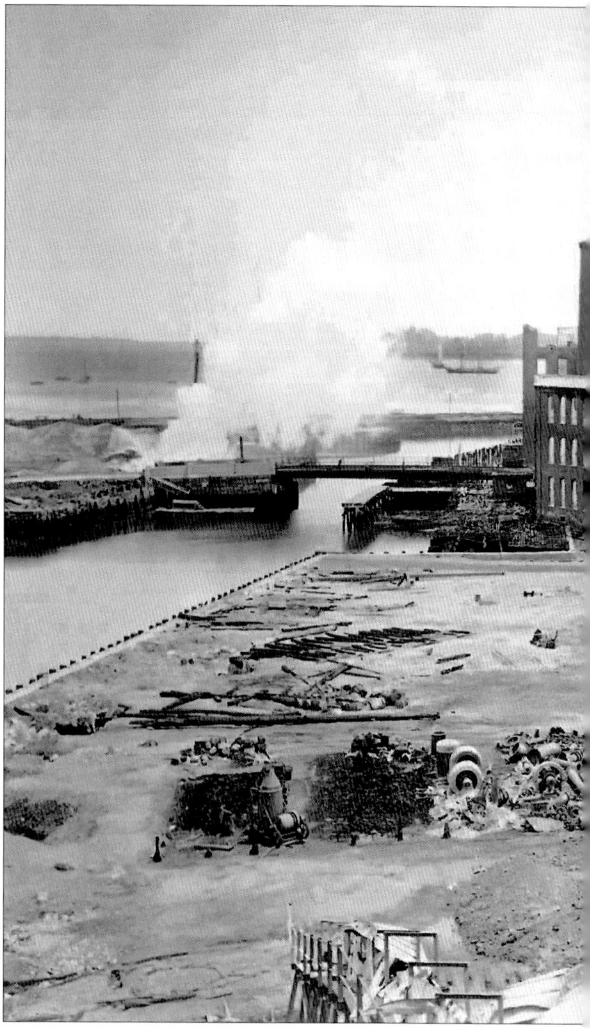

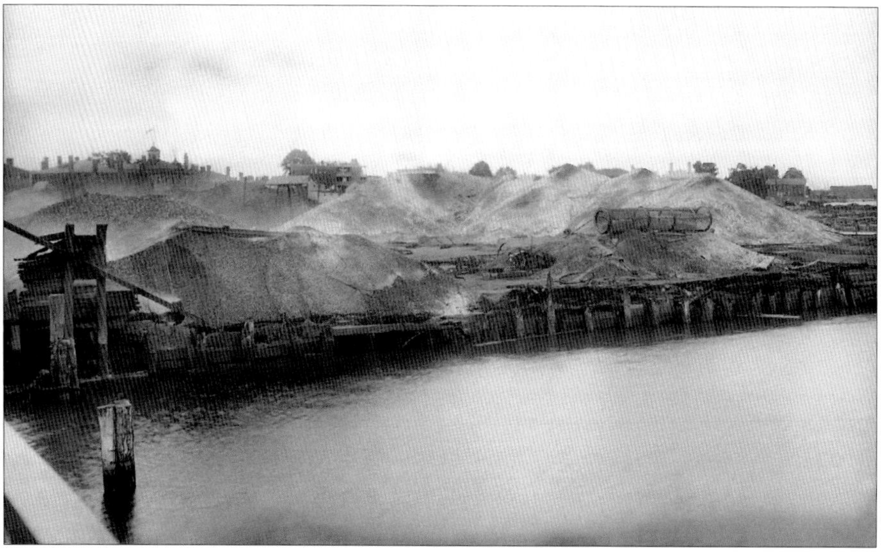

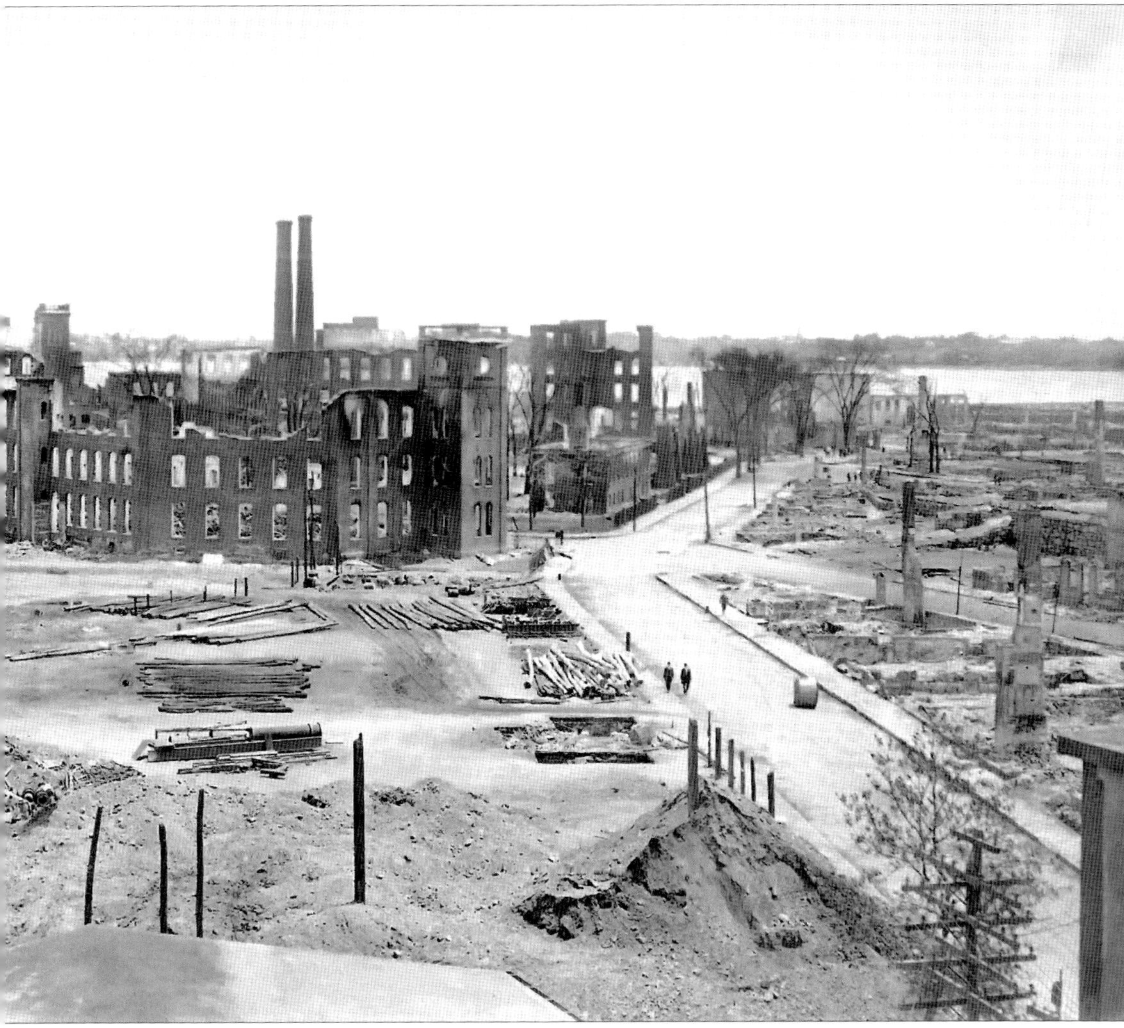

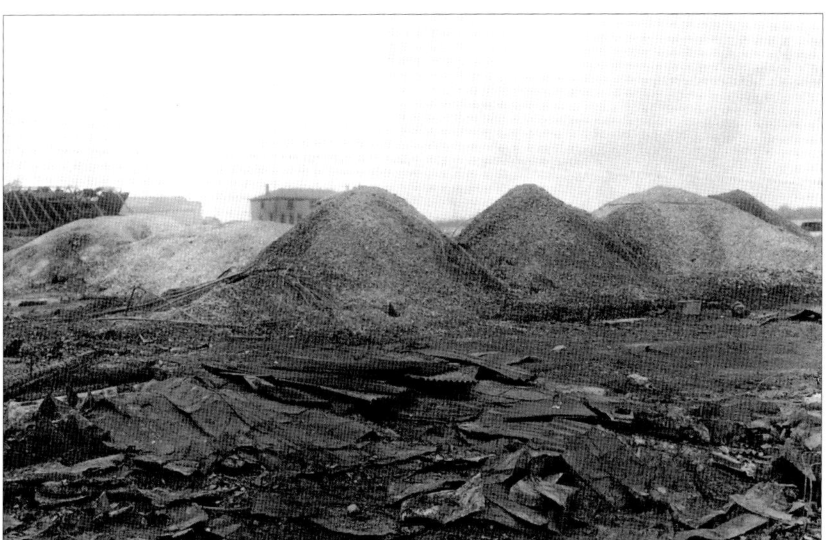

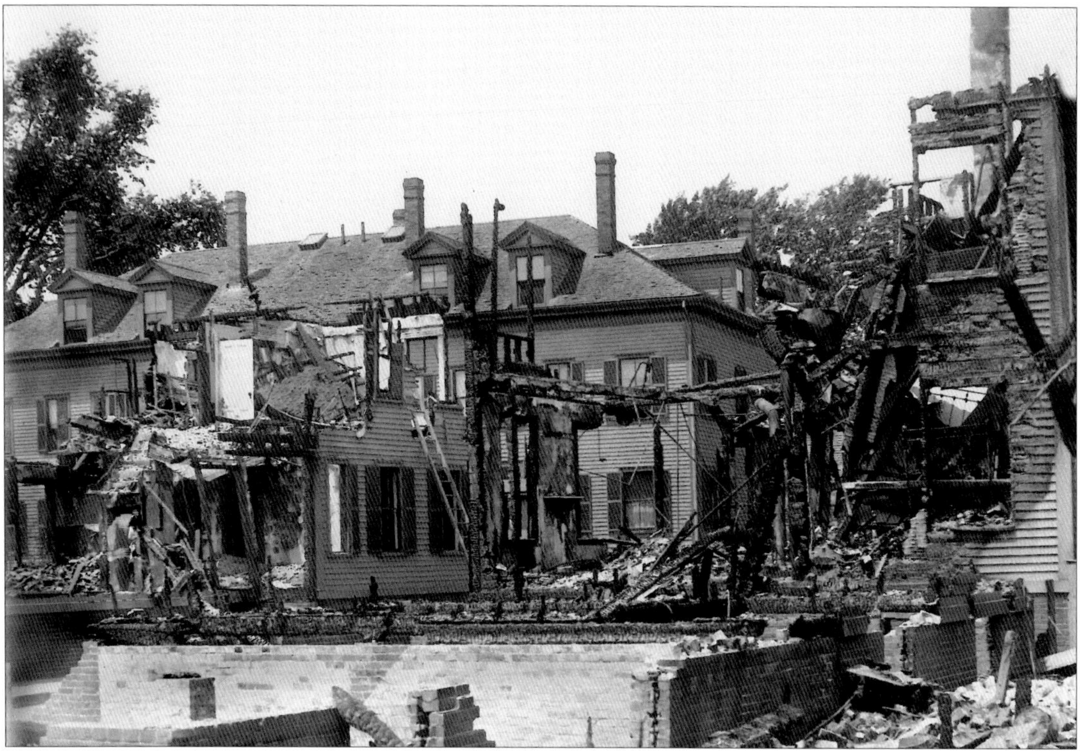

The unidentified ruins above show that although many homes were completely destroyed, others were only partially damaged. (*Neg #17930*) The ruins depicted in the bottom image were on Derby Street. (*Neg #32795*) Both images show the utter randomness of the fire's destructive power.

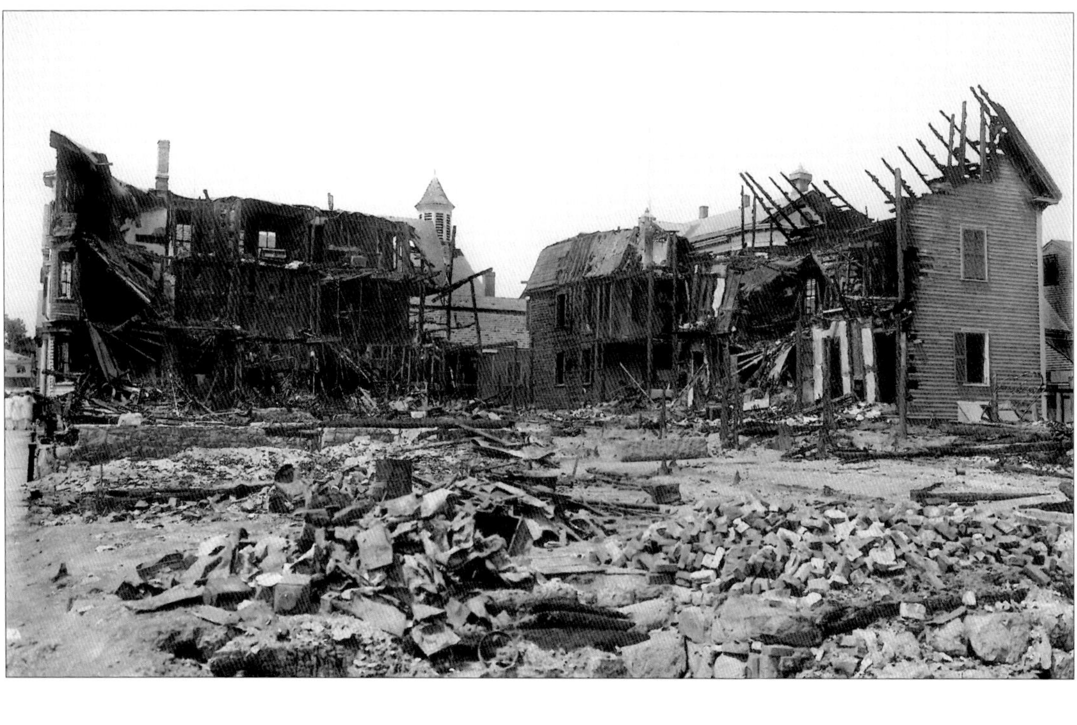

Chapter 7

Reconstruction

The Salem Rebuilding Commission was created by an act of the Massachusetts legislature, known as Chapter 777, Acts of 1914. On July 7, 1914, Governor David I. Walsh appointed five men for a period of three years: Edmund W. Longley, Dan A. Donahue, Eugene J. Fabens, Emile Poirier and Michael L. Sullivan. The commissioners were residents of Salem; they received no compensation for their work and personally paid for any expenses incurred by their duties. The members of the commission remained the same throughout the three-year tenure. The commission believed that Salem was one city, not two, and although its charter was to rebuild the burned area, the commissioners felt it was important to work closely with all the organizations that worked for the city. In 1917, the commission published a report summarizing its efforts.

One other person played a role in the reconstruction of the city. Three days after the commission was created, C.H. Blackwell was hired as an advisory architect and provided significant input into the creation of the new building code, in which the use of wooden shingles was banned. Every new building had to have a fireproof roof and gutters so that sparks and embers that fell on them could not start a fire. Buildings for one or two families had fewer restrictions on them than did buildings built for more than two families because there were fewer people to protect. Those designed for two families or more could not be more than two and a half stories high. None of the multifamily dwellings could occupy more than 75 or 80 percent of the land on which it was built. Chimneys had to be lined, cellar walls had to be made of mortar and heating apparatus in the cellars had to be protected. The walls of wooden buildings had to be stopped off at every floor so that a fire could not travel from floor to floor through the side walls.

Salem was the first city in the Commonwealth to have a City Plans Commission. Appointed in April 1911, it presented a plan to the city for improvements by December 1912. The chief engineer of the city, George Ashton, also worked as the chief engineer for the City Plans Commission, helping to suggest areas of improvement: widening city streets, extensions of certain streets and improvements to specified streets. A map showing the plans created by the commission is shown at the end of this chapter. The plan was put into effect as part of the rebuilding process; according to the report of the Rebuilding Commission, having this plan prior to the fire saved the city considerable money in the rebuilding process because no engineers needed to be hired to determine the best plan of action regarding city streets.

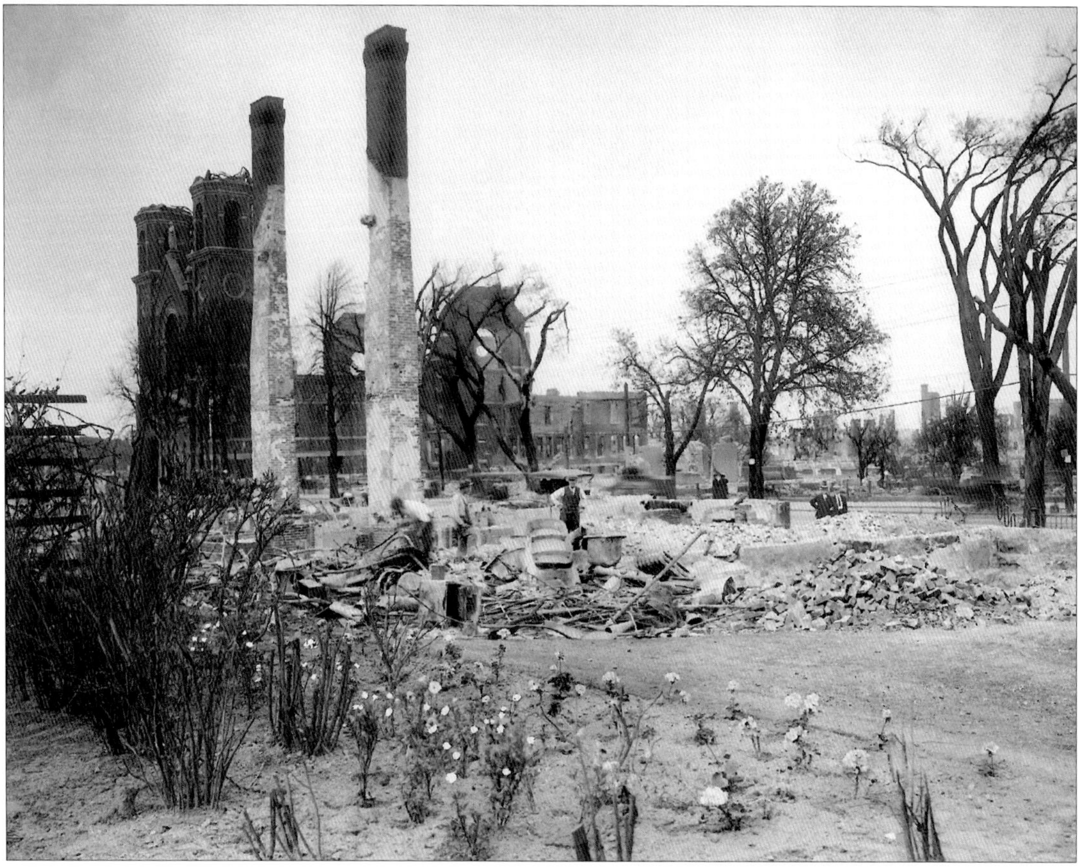

The fire burned for over thirteen hours from its initial start in the leather district to its destruction of the Naumkeag mills. Over 253 acres of land were destroyed; 251 acres were in the burned district and the others were destroyed by the fire in North Salem that started later in the evening. Over 1,600 buildings were destroyed, 6 people lost their lives and more than 1,500 families were homeless. In mythology, fire is both a creative, cleansing force and a destructive, punishing one. Initially the city of Salem viewed the fire as destructive. The flowers in these ruins on Lafayette Street provide a glimpse of hope that the Great Fire of 1914 could also be a creative force for the city. (*Neg #4618*)

According to the report of the Rebuilding Commission, the first permit for reconstruction in the burned area was granted to George W. Pickering at the end of July; the first permit to erect a residence in the burned area was granted to Miss Hannah Elwell at 4 Holly Street. In the three years between when the first permits were granted and the writing of the Rebuilding Commission's report, 828 permits were granted with an estimated value of over $6,827,675. According to Blackall, 1,792 buildings were destroyed by the fire; as of June 1917, 828 of them had been rebuilt. Land value fell little after the fire and rose slowly during the reconstruction process. The new building code prevented land speculation, which might have raised prices beyond that which families displaced by the fire could afford. Although only 65 percent of the burned area had been rebuilt by the date the report was filed, the assessed value of the burned area was 36 percent greater than before the fire. In the report Blackwell stated, "When all is rebuilt Salem will be more than compensated for her loss, financially."

Public buildings were also rebuilt after the fire. Two schools went up in flames, the old Saltonstall School on Holly Street and the Brown School on Ropes Street. They were combined into a new elementary school erected on Lafayette Street, named the Saltonstall School. A new firehouse was built on Essex Street in Ward 3 on the site of the old firehouse destroyed in the fire. Another new firehouse was built on Loring Street. Public parks were also created as part of the reconstruction: Lafayette Park was built in a triangle bordering Lafayette, Washington and Harbor Streets. Two small parks were created on Highland Avenue at the end of New Broad Street and an open space was created at Endicott Street. A new playground was opened in a square between High and Endicott Streets. Two churches were destroyed in the fire: St. Joseph's Church on Lafayette Street and the Society of Friends on South Pine on the corner of Warren Street. St. Joseph's Church was not rebuilt for many years; parishioners worshipped in the renovated basement. The Society of Friends was rebuilt within three years.

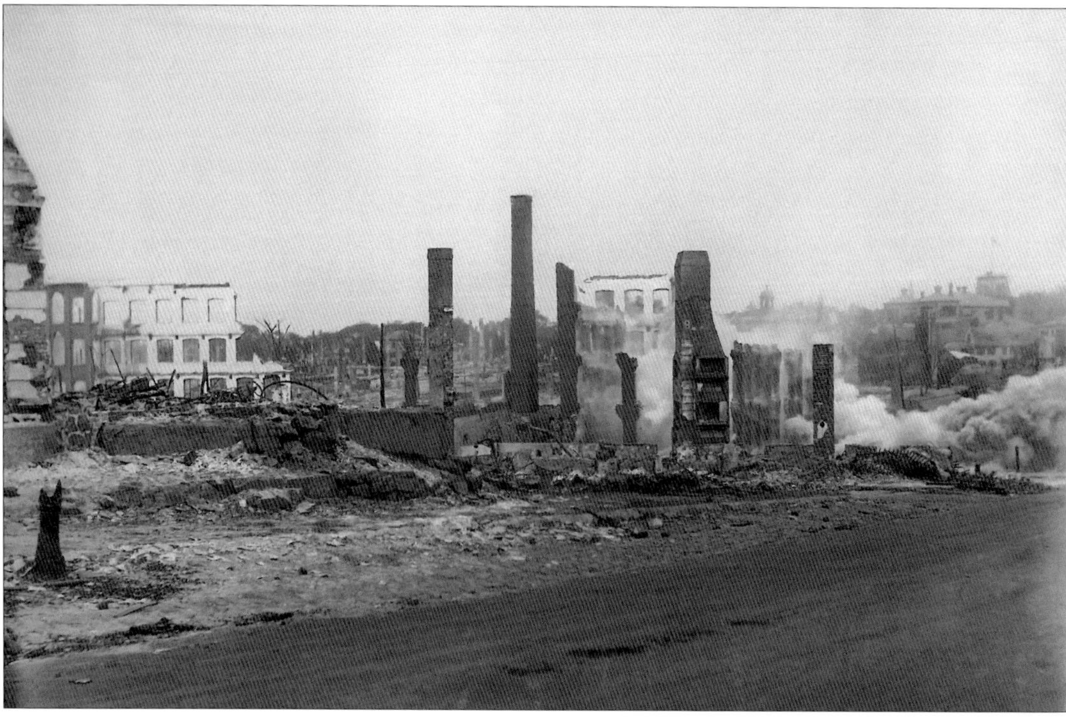

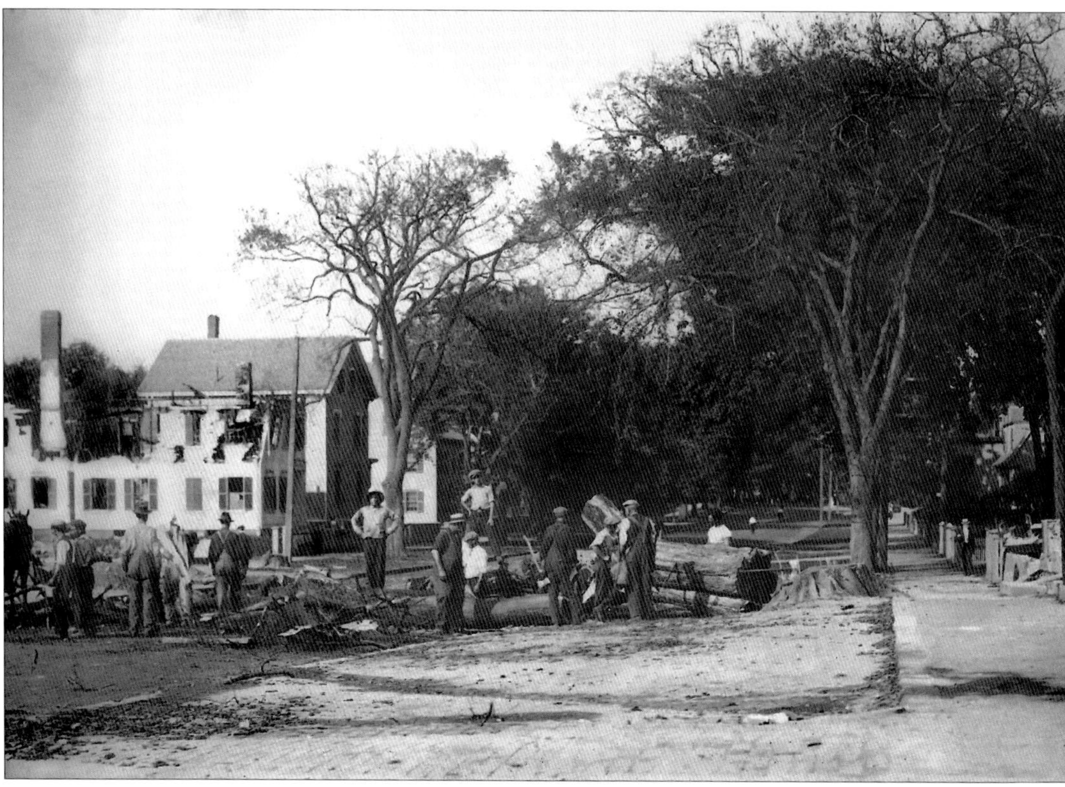

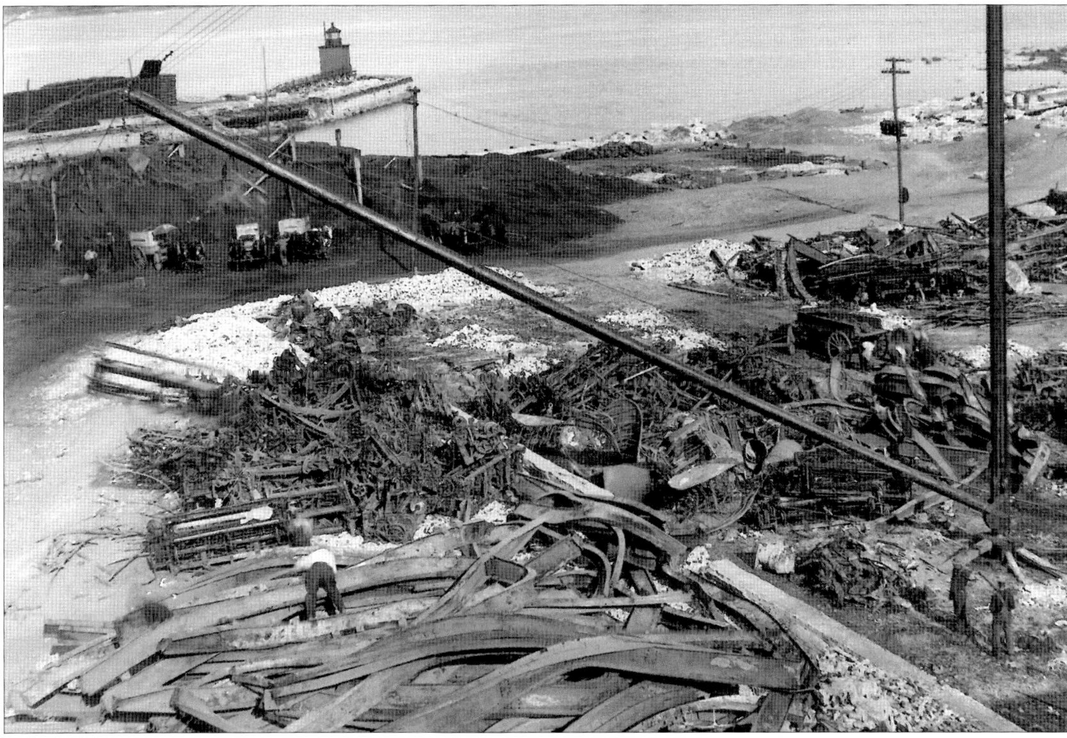

The periodical *Fire and Water Engineering* published many articles about the Salem Fire. In several of them the use of dynamite to stem the fire was discussed. Over ten buildings on Boston Street had been dynamited to create open spaces to slow the fire's spread. Under the direction of P.J. Kelley, the director of Public Works, dynamite was put into the walls of the buildings to be leveled. The periodical called this "a hopeless expedient" because the debris added to the embers, which actually spread the fire more quickly. Kelley and his crew did use dynamite during reconstruction to remove ruins and prevent injury from buildings as they continued to collapse, as shown in the photo on the top of the previous page. The first building to be blown up was the P.A. Field Shoe Factory on the corner of Mill and Canal Streets. (*Neg #32815*) The bottom photo was taken on the corner of Broad and Hathorne Streets. Burned trees needed to be removed before reconstruction could begin. An unidentified news clipping, dated July 23, reported that the Shade Department was responsible for removing the stumps of every burned tree on every side street along Lafayette Street. The steam road roller had "been fitted with spikes which are regularly fitted to the wheels to break the pavement, but in this case they are necessary to give the machine a grip to the ground. Without them the machine…slips with the tremendous weight of the timber, although at the same time that the good work is being done the roadbed is torn up badly and so in many instances is the sidewalk." (*Neg #15528*) The image on this page shows the debris being organized for removal from the Naumkeag mill complex. The harbor lighthouse can be seen in the background. (*Neg #32797*)

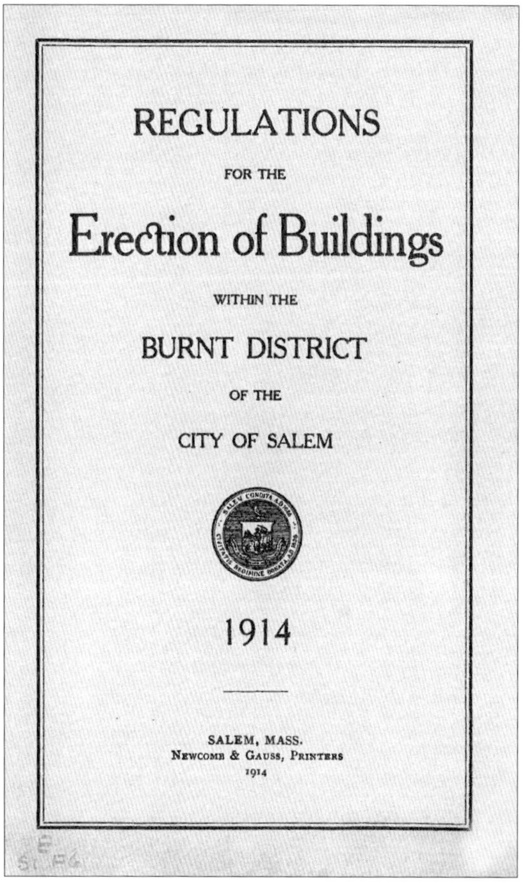

REGULATIONS

FOR THE

Erection of Buildings

WITHIN THE

BURNT DISTRICT

OF THE

CITY OF SALEM

1914

SALEM, MASS.
NEWCOMB & GAUSS, PRINTERS
1914

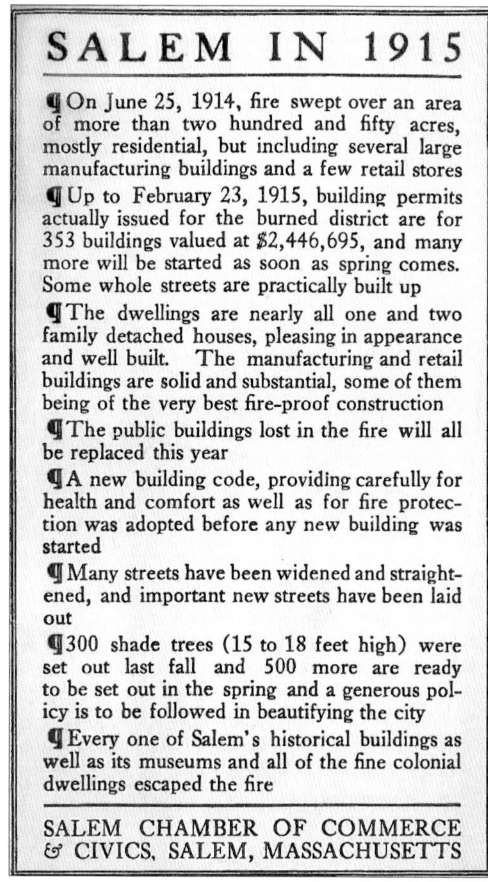

SALEM IN 1915

¶ On June 25, 1914, fire swept over an area of more than two hundred and fifty acres, mostly residential, but including several large manufacturing buildings and a few retail stores

¶ Up to February 23, 1915, building permits actually issued for the burned district are for 353 buildings valued at $2,446,695, and many more will be started as soon as spring comes. Some whole streets are practically built up

¶ The dwellings are nearly all one and two family detached houses, pleasing in appearance and well built. The manufacturing and retail buildings are solid and substantial, some of them being of the very best fire-proof construction

¶ The public buildings lost in the fire will all be replaced this year

¶ A new building code, providing carefully for health and comfort as well as for fire protection was adopted before any new building was started

¶ Many streets have been widened and straightened, and important new streets have been laid out

¶ 300 shade trees (15 to 18 feet high) were set out last fall and 500 more are ready to be set out in the spring and a generous policy is to be followed in beautifying the city

¶ Every one of Salem's historical buildings as well as its museums and all of the fine colonial dwellings escaped the fire

SALEM CHAMBER OF COMMERCE & CIVICS, SALEM, MASSACHUSETTS

As the summer progressed, temporary shelters were built in the burned district to house the homeless. On August 20, forty-four days after the appointment of the Salem Rebuilding Commission, a new building code was published for this district. The temporary buildings would need to be torn down and any new building was required to adhere to the code. Building codes and housing ordinances adopted later by the city for areas outside the burned district were more drastic than the one adopted by the Rebuilding Commission within the burned area; this was because the commission wanted to encourage rebuilding and hesitated to place insurmountable restrictions on those who had lost everything. By 1917, a single building code was adopted by the city. (*Neg #32853*) In 1915, the Salem Chamber of Commerce printed a small poster to publicly affirm the progress of reconstruction and recovery from the fire. (*Neg #32812*

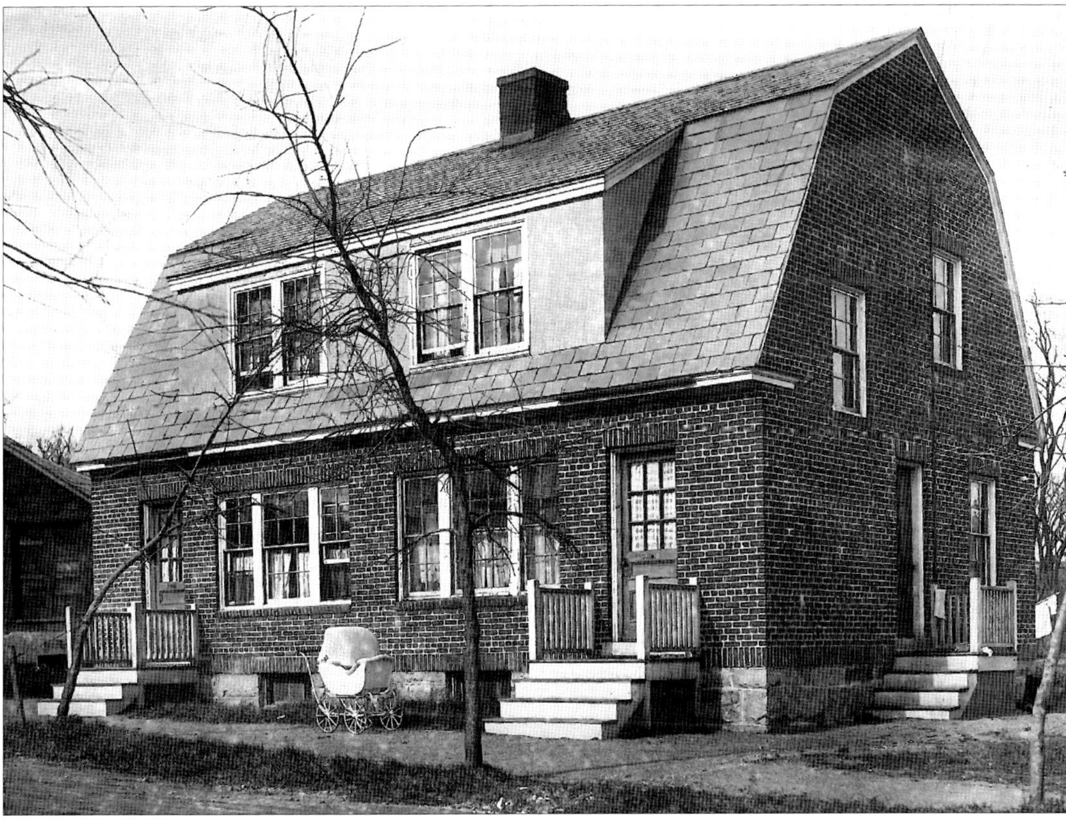

In September, the Salem Rebuilding Trust was organized to assist homeowners who wanted to rebuild. The Committee of Fourteen met in the Now and Then Hall on September 2 to allocate $100,000 from the fire relief fund for the work of the trust. Loans were not to exceed $500 for any individual; all loans were secured by mortgages to return the monies to the city. The trust expanded its mandate to develop building plans for low-cost housing. The Salem Rebuilding Commission had held a contest for architects around the state to develop plans to cost no more than $3,000 for a single dwelling and no more than $5,000 for a two-family house. Mrs. Elmma S. Almy offered a prize of $500 for the winning architect; the prize was awarded to Edwin Sherrill Dodge of Salem. The trust used these plans to build demonstration houses on Winthrop Street. The trust also purchased land in North Salem to erect twenty low-cost cottages for workmen; through community planning and low-rental home development, the trust sought to change the physical and economic landscape of the city. The architectural firm of Kilham & Hopkins in Boston, which had participated in the Rebuilding Commission's lectures on housing options for the city, was hired to build these homes. The cost of land and building costs was just over $2,000 per family unit, less than the typical three-decker home the workmen lived in before the fire. The rent for each was estimated to be $15 per month. The house shown above is typical of what was built on the land purchased by the trust. It was a gambrel, double-house design done in brick, with a green slate roof and a pent roof dormer. (*Neg #32876*) Floor plans for the house can be found in the *Winterthur Portfolio* (Spring 1987). *The Salem Evening News*, dated June 25, 1915, states that twelve homes had been completed in North Salem. Members of the trust at that time included Robert W. Hill, Harry P. Gifford, Samuel H. Batchelder, Josiah H. Gifford, John J. Mack, Morris Newmark, Emile Poirier, Alva P. Thompson and Christian Lantz. The trust continued its work after the fire relief committee disbanded in August.

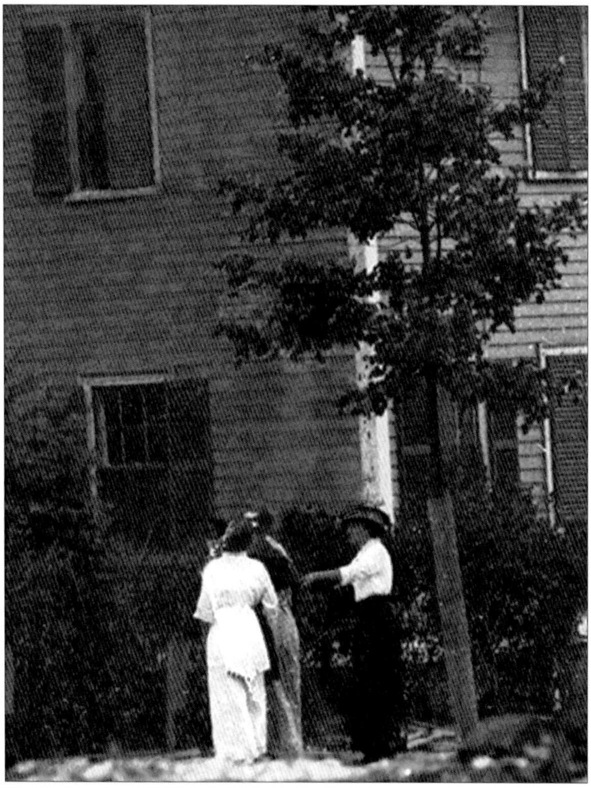

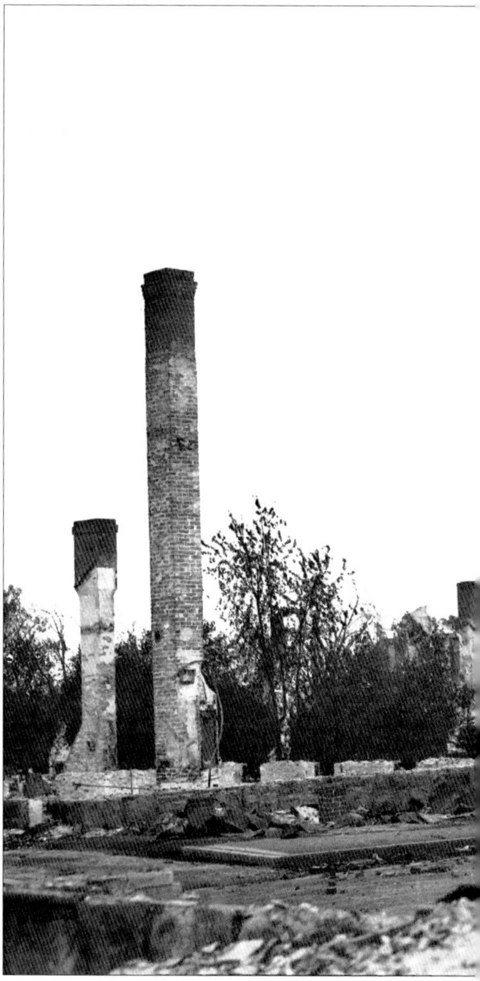

These photographs were taken by Frank Cousins. The negative found in the Phillips Library collection did not identify the location of the house or the names of the women shown in the images. It is possible that one of these women may be the owner of the home. Two women were among the first granted building permits. Miss Georgianna R. Kehew, a teacher at the Saltonstall Grammar School, was granted a building permit to erect a two-and-a-half-story, one-family dwelling at 6 Holly Street. Local contractors Morrison & Young were awarded the contract to build the house at the cost of $5,000. According to the Salem Rebuilding Commission Report, Miss Hannal Elwell was granted a permit to build at 4 Holly Street. (*Neg #32934 and Neg #32934Detail*)

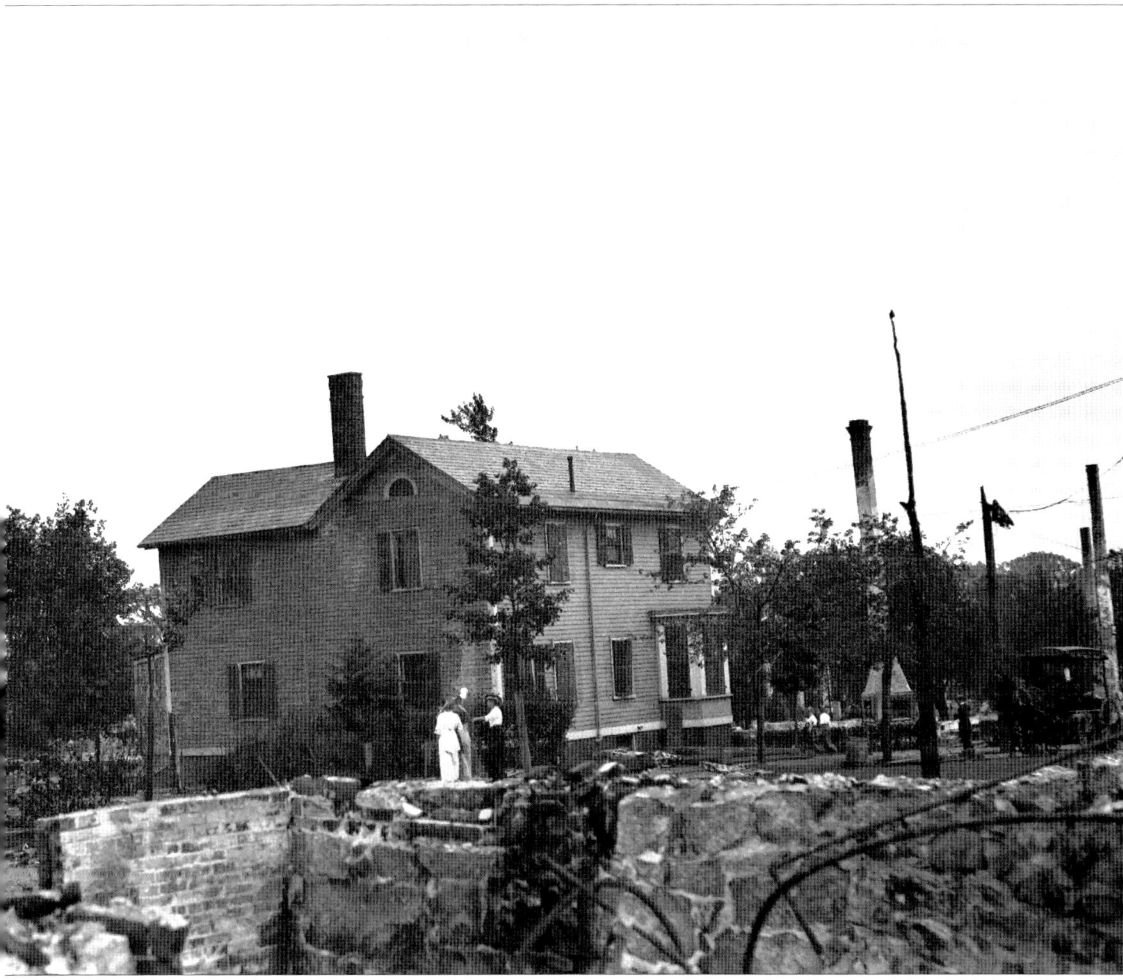

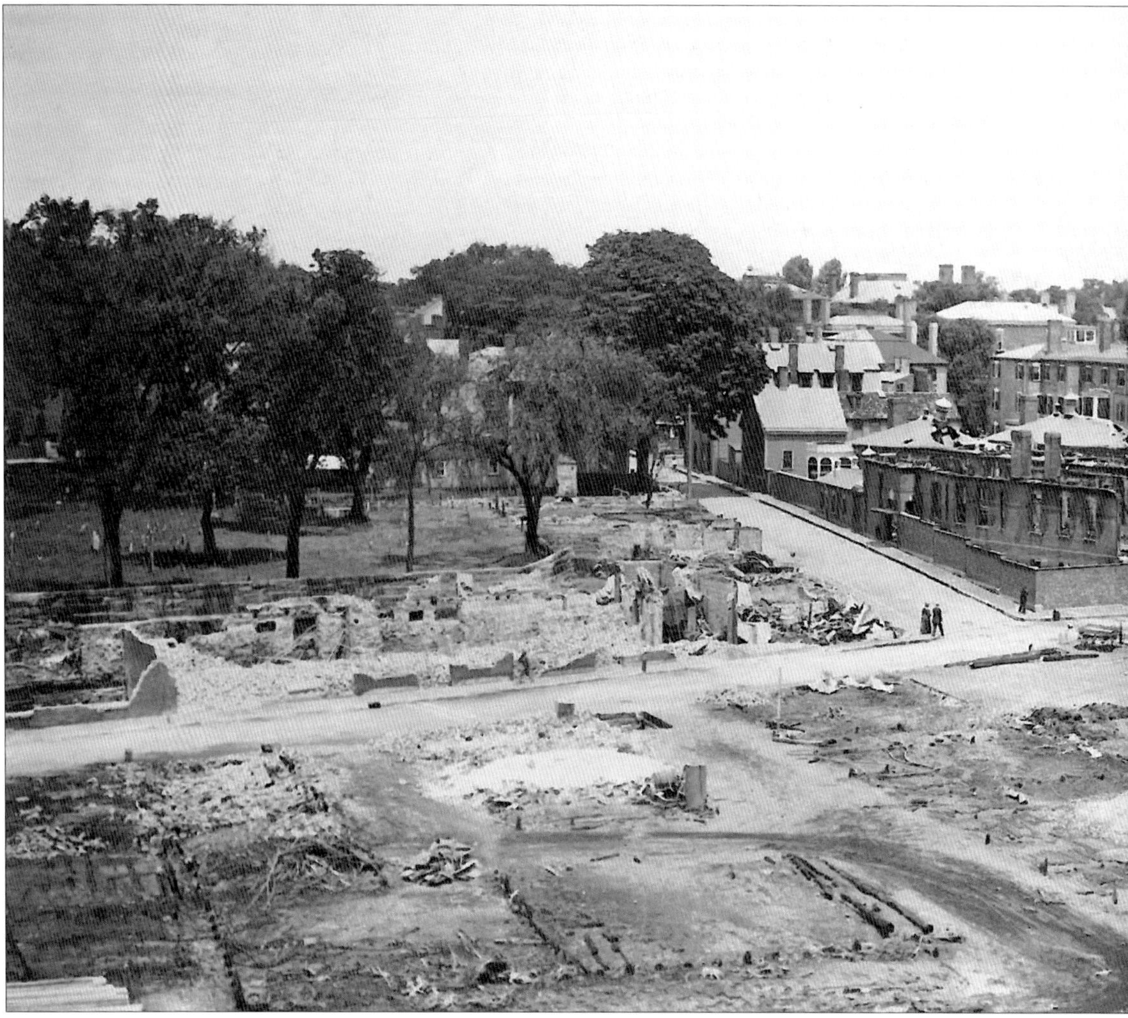

The exposed foundations shown in each of the photographs were all that remained of the burned homes after the debris from the fire had been removed. The larger image that spans both pages was taken from the roof of the Salem Electric Light Power Plant looking toward the damaged Salem Hospital. The Immaculate Conception Church is on the far right side of the picture. (*Neg #4012*) The smaller photograph was taken from the vantage point of the Salem High School. It shows Jackson Street and Upper Essex Street. The brick building standing behind the vacant lots is the remains of the Steamer 4 Engine House on Essex Street. (*Neg #4017, a Robb photo*) *The Salem Evening News*, dated July 3, reported that the director of Public Works, P.J. Kelley, had instructed city workers to begin the removal of the chimneys that had survived the fire. "The modus operandi is simple, but effective. Hitched to a tailboard of a two-horse team is a wire cable. The other end is made fast around a chimney about four or five feet from the ground. The horses are started and down comes the chimney with a rush. The cable is long enough so as to place the team beyond the danger of being hit."

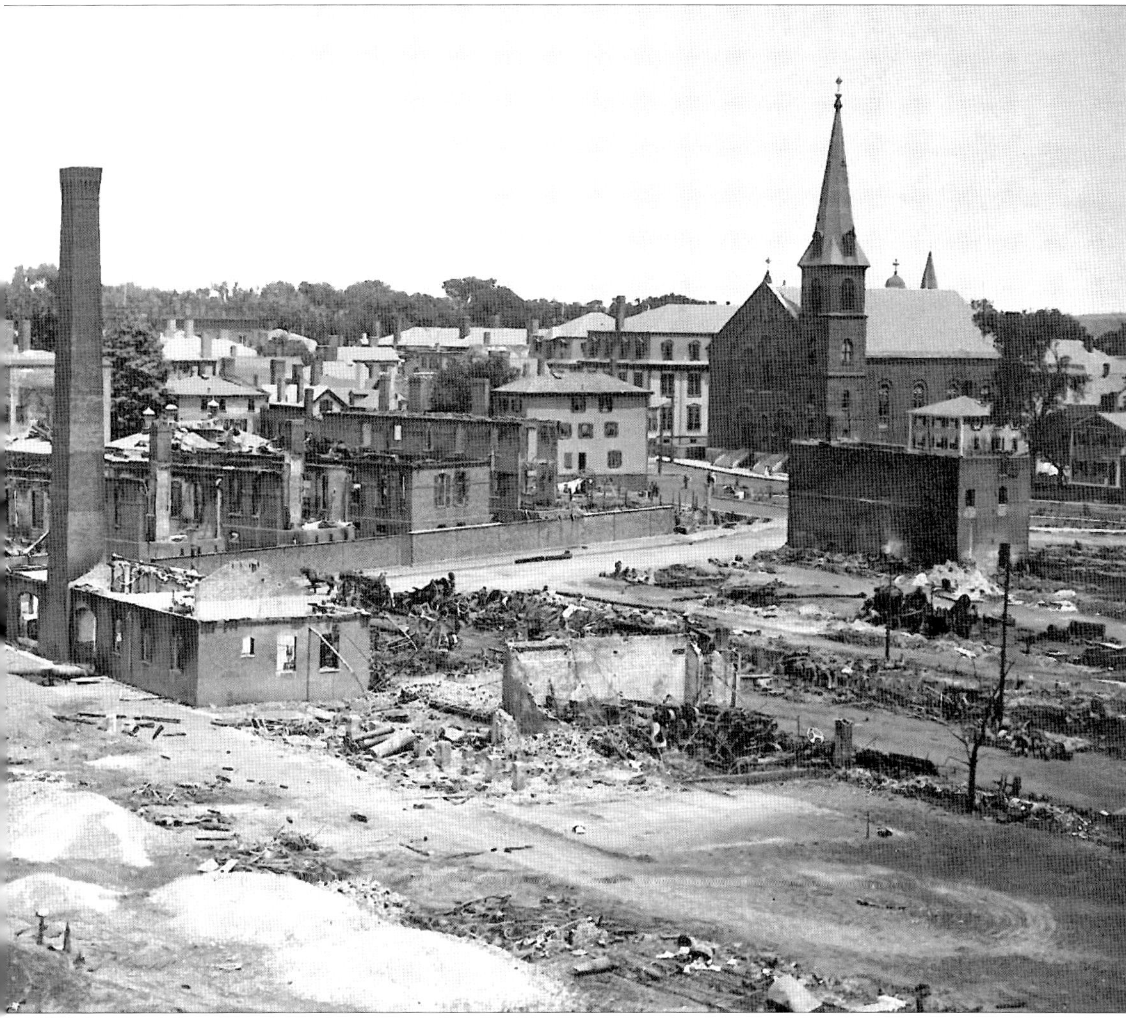

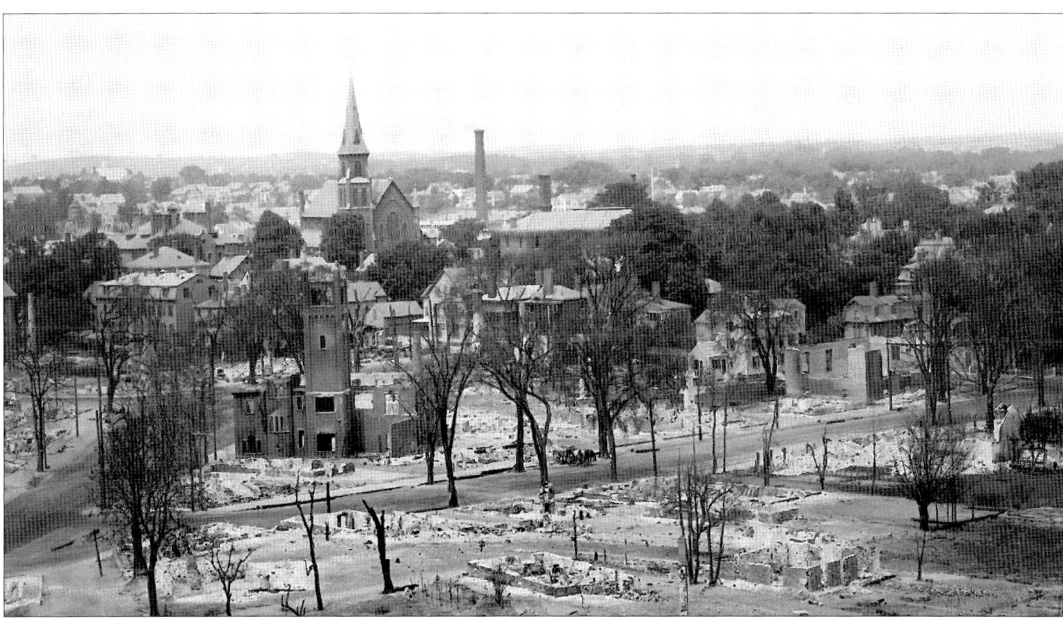

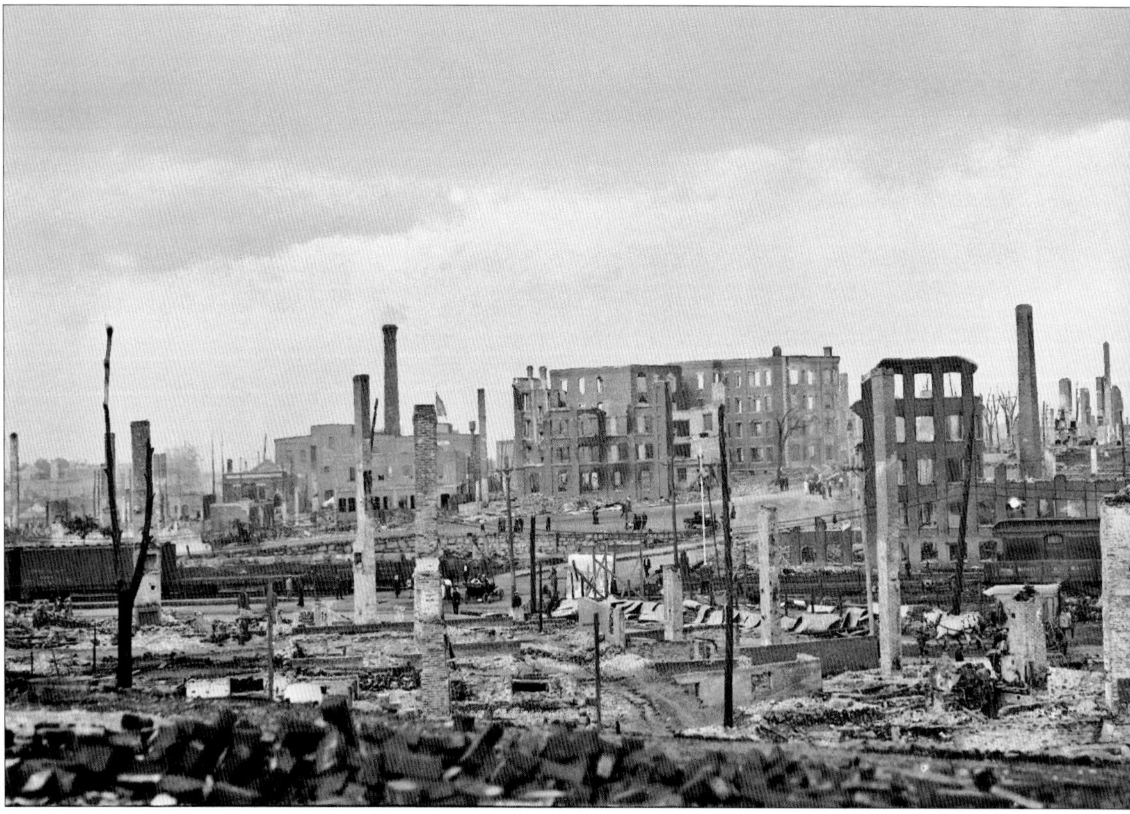

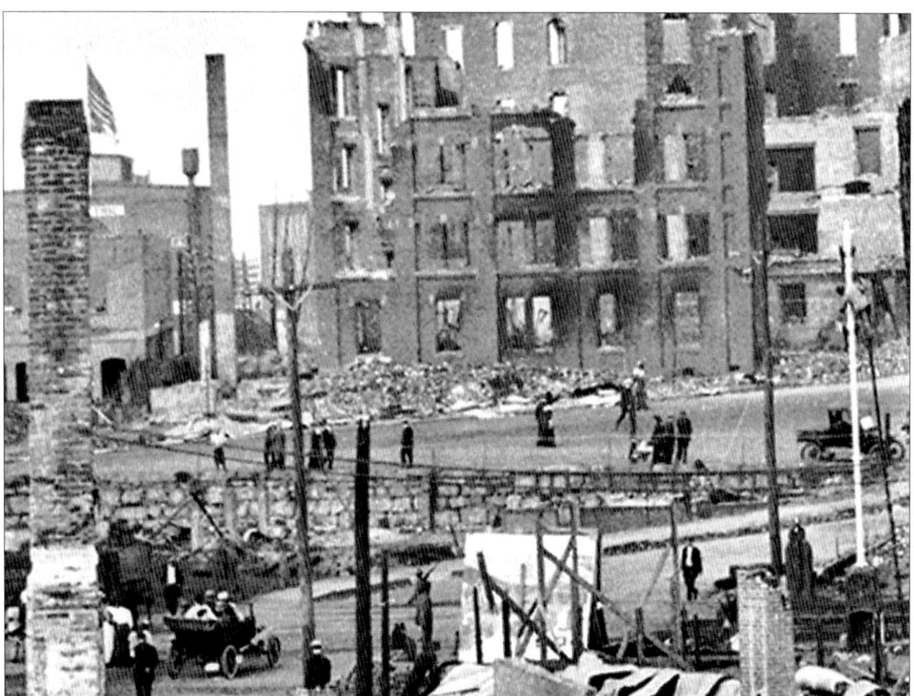

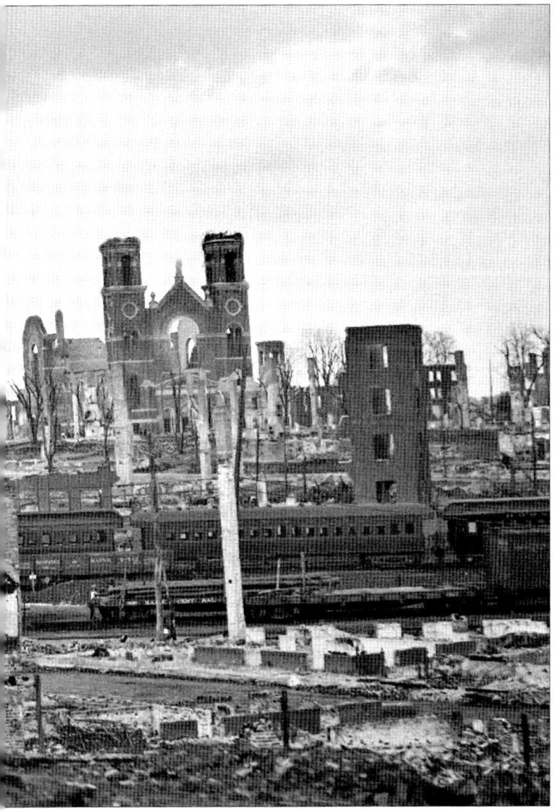

Frank Cousins photographed the reconstruction efforts taking place along the Boston and Maine Railroad. St. Joseph's Church is shown behind the railroad cars; the Salem Electric Light Company's Power House is the building with the American flag in the far left side of the photograph. (*Neg #32933*) Once digitized, it was possible to extract details from the photograph. The bottom image on the previous page shows electrical linemen repairing the damaged electrical lines; a lineman can be seen in the right side of the photograph reaching for the electrical line that is hanging on the ground. (*Neg #32933DetailA*) The railroad car in the bottom image on this page appears to be bringing electrical poles to the area so that the electrical lines can be repaired. (*Neg #32933DetailC*)

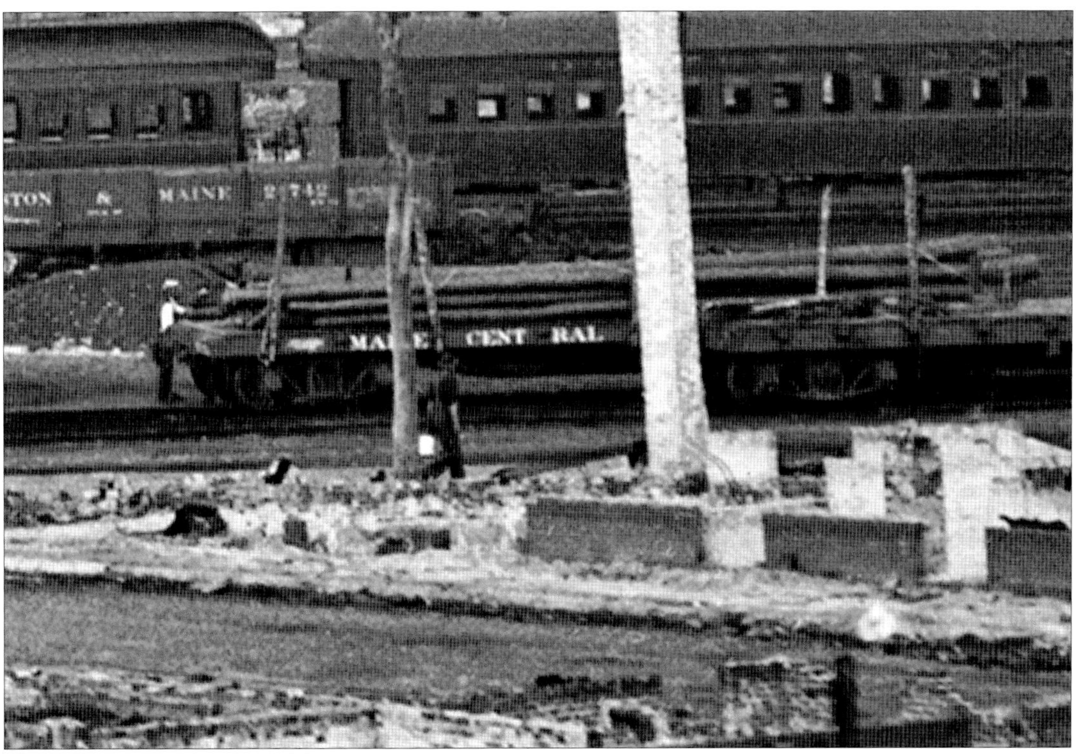

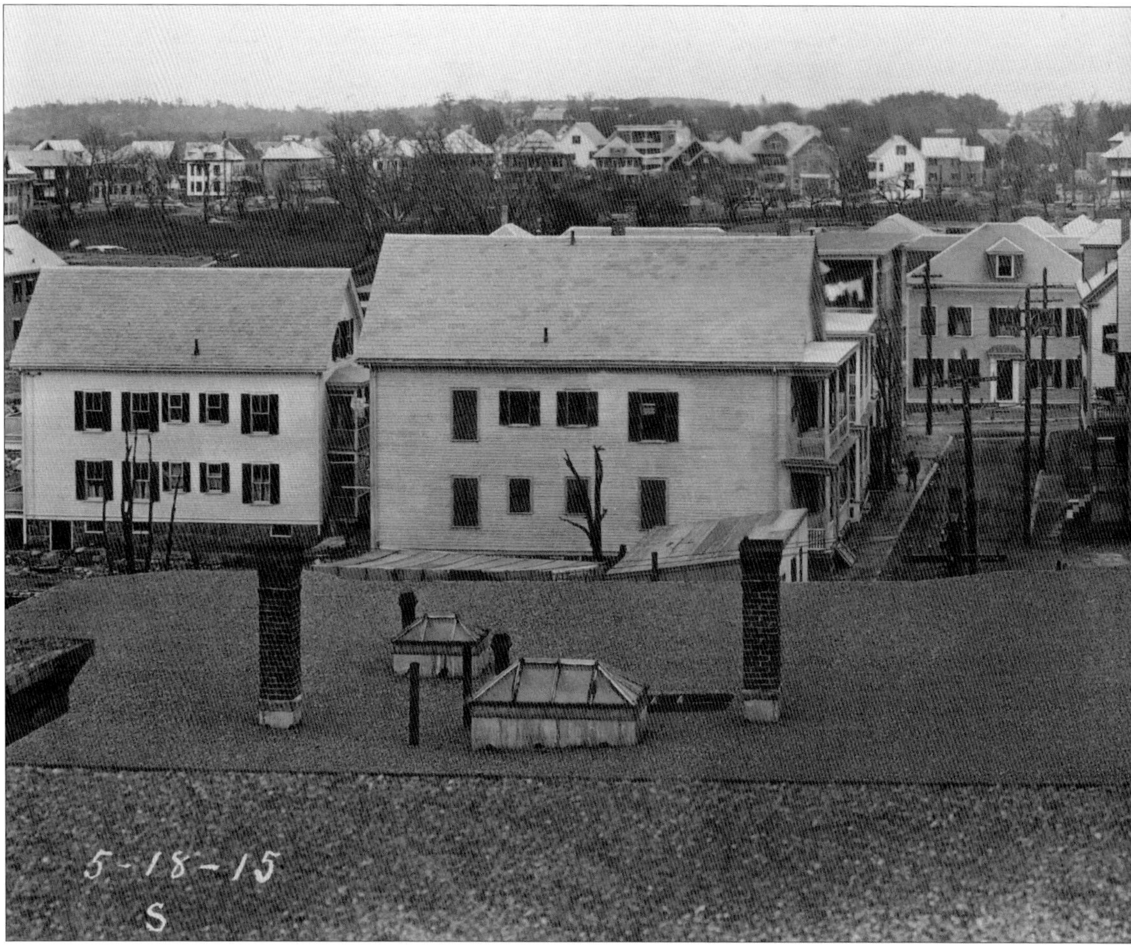
5-18-15
S

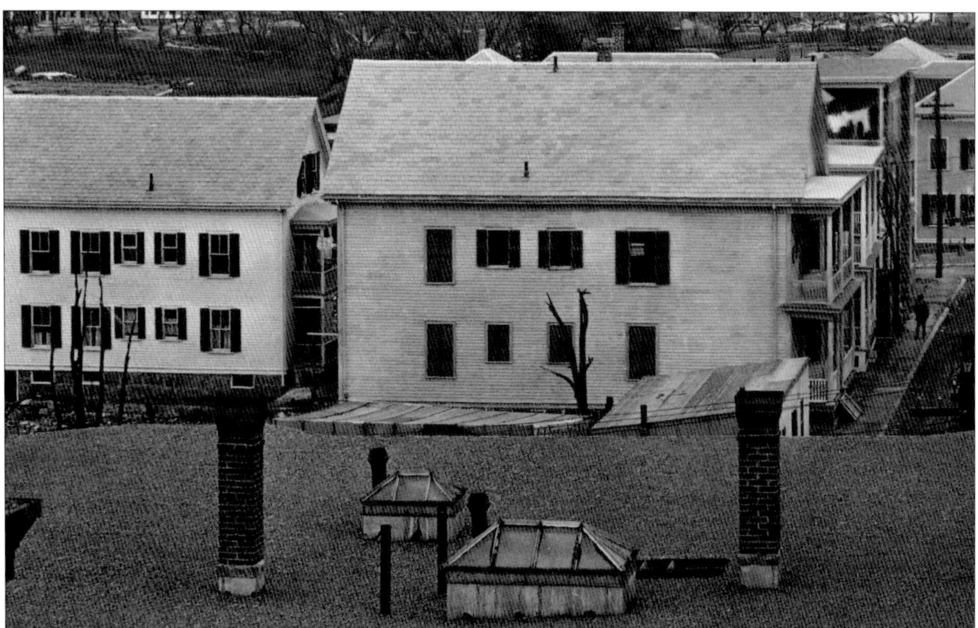

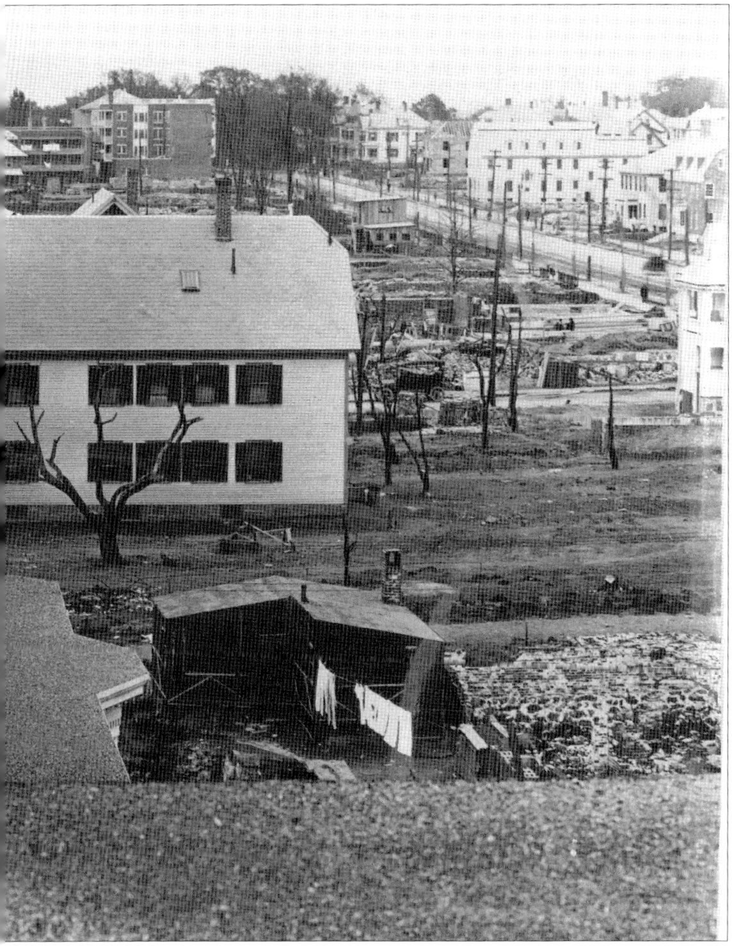

The date stamp on the photograph shows that this section of Salem had been rebuilt within eleven months of the fire. The street on the right side of the image is Tremont Street; the cross street is Grove Street. (*Neg #32936*) The extracted detail on the bottom of the previous page shows that the homes were occupied, as evidenced by the laundry hanging in the second-story back porch. (*Neg #32936DetailB*) The bottom image shows that reconstruction was still in progress. (*Neg #32936DetailA*)

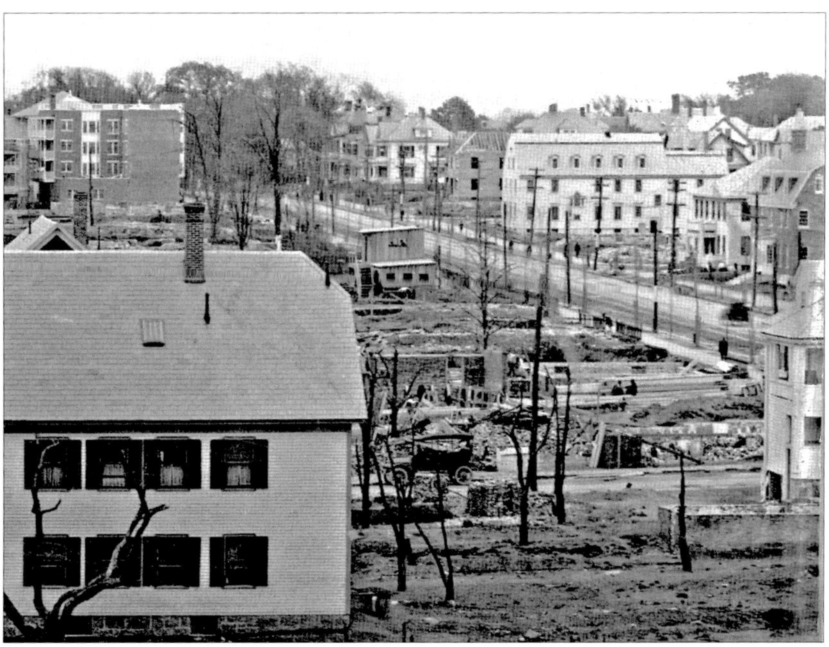

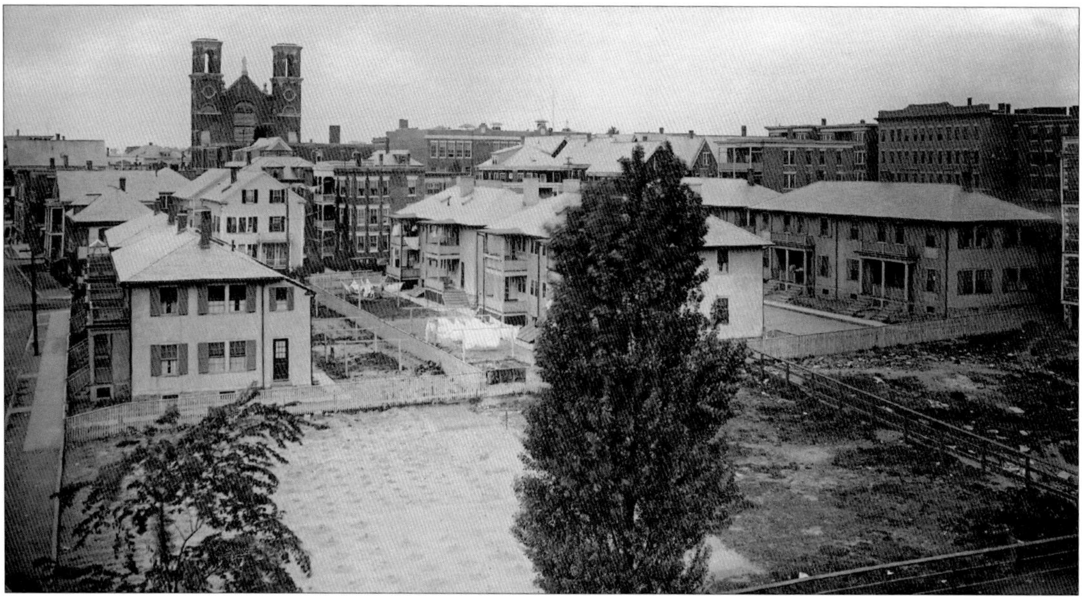

The Point section of Salem suffered severe damage in the fire. The top image on the next page was taken from the Electric Light Company building looking south toward St. Joseph's Church; much of the neighborhood surrounding the church was destroyed. Reconstruction workers can be seen in the foreground as they remove debris from the fire. (*Neg #3281, a Robb photo*) The bottom photograph on the next page depicts the area after reconstruction (*Neg #32875*), as do the two images on this page. The detail shot portrays the statue of St. Joseph watching over the residents of the neighborhood. These images, photographed by Tilford, were in the William Abbott photograph album in the Photo Archives collection of the library. (*Neg #32926 and Neg #32926Detail*)

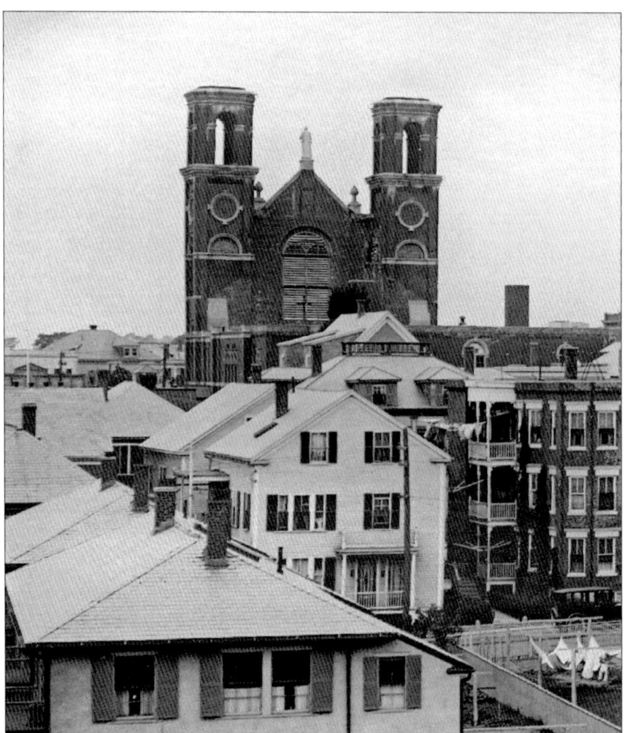

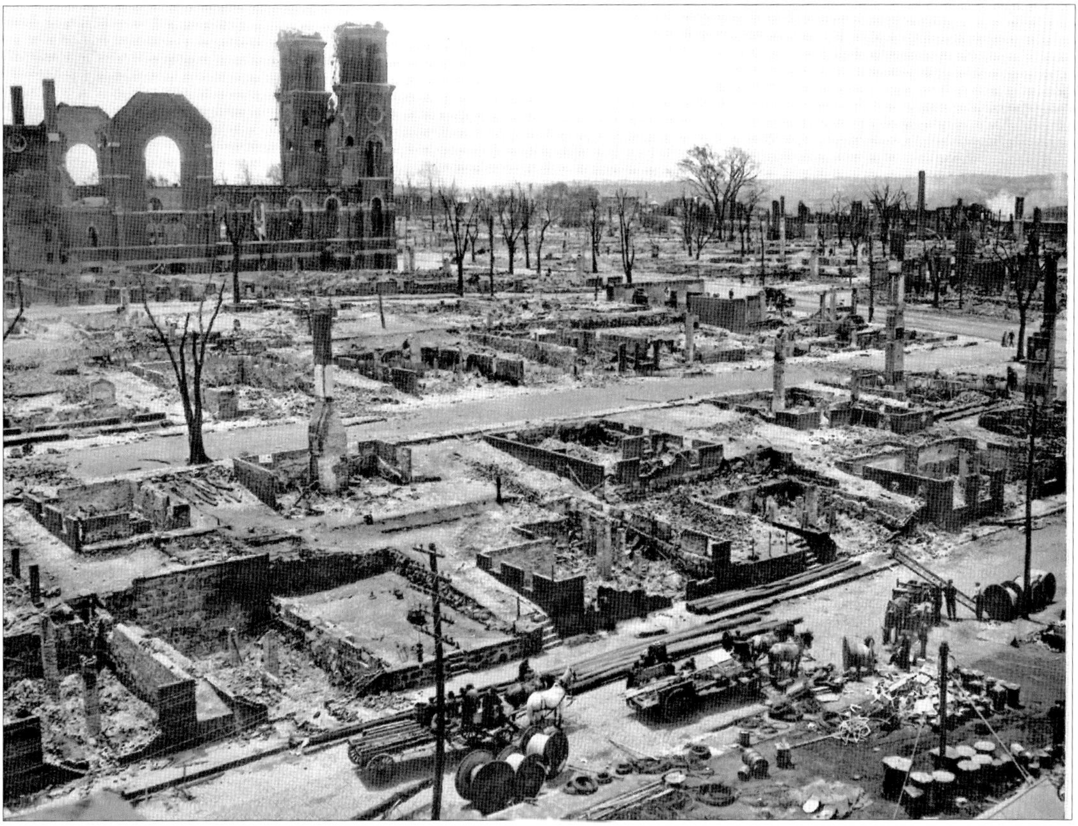

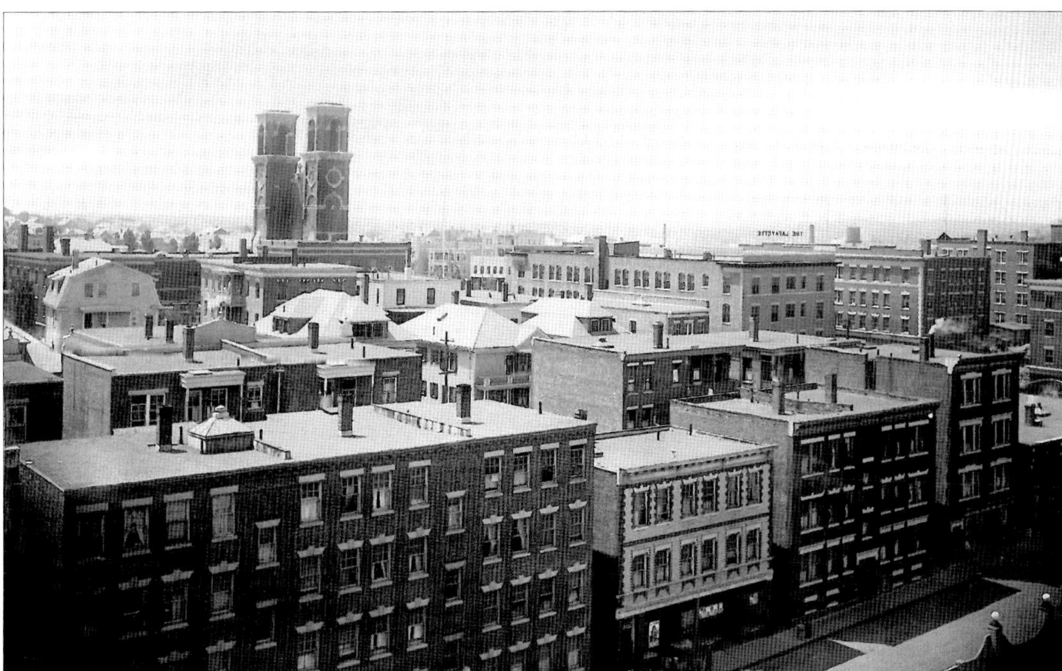

As a local contractor, William Abbott was interested in the redevelopment of the city after the fire. His photograph album has many pictures chronicling the process, three of which are shown on these two pages. The photo to the right depicts another view of the destruction along the Point. The photographer, probably Tilford, captured the process of installing new poles to hold telephone and electrical wires, one of which looks like a large cross standing outside of the church. (*Neg #32874*) The bottom image on the next page shows the completed reconstruction efforts in that area of the Point. (*Neg #32873*) The photograph on the bottom of this page portrays the rebuilt St. Joseph's Parish House. (*Neg #32878*)

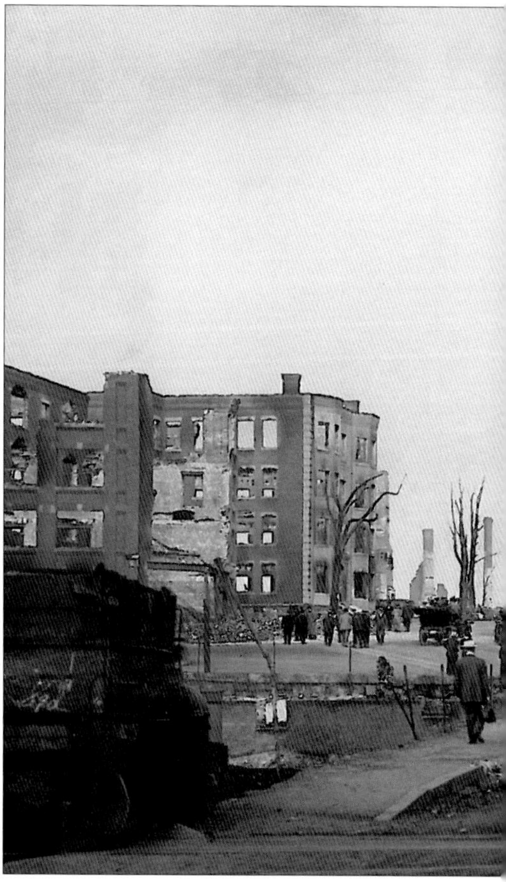

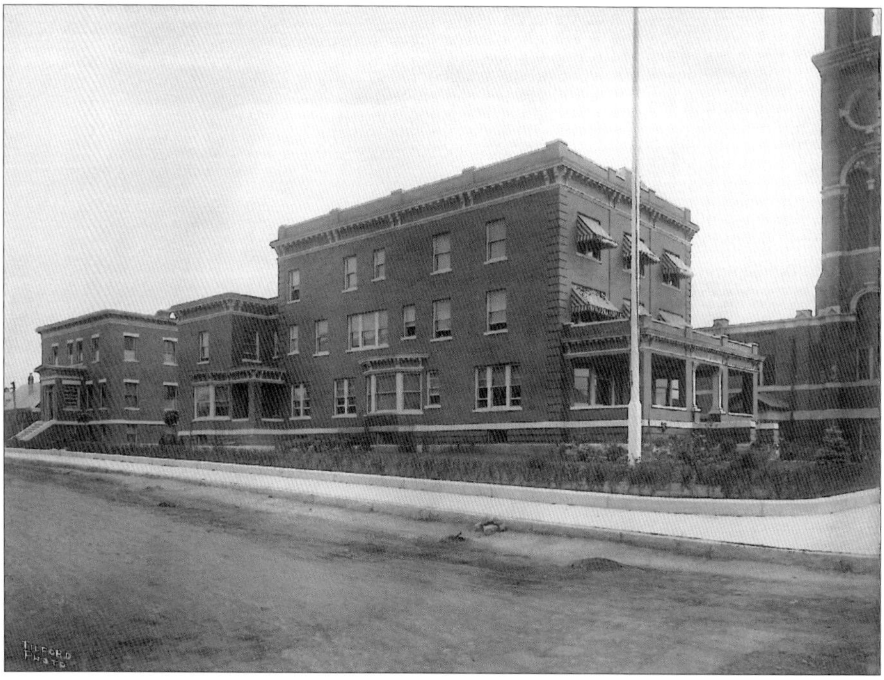

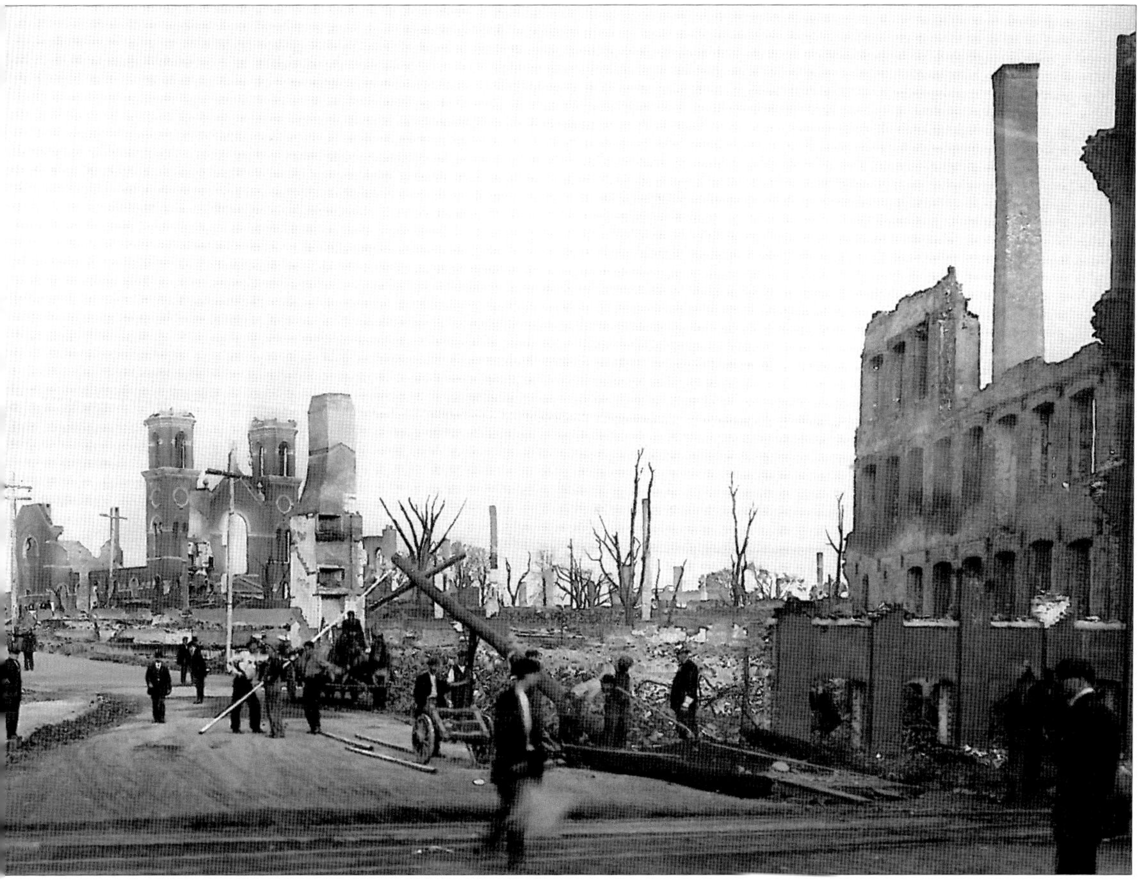

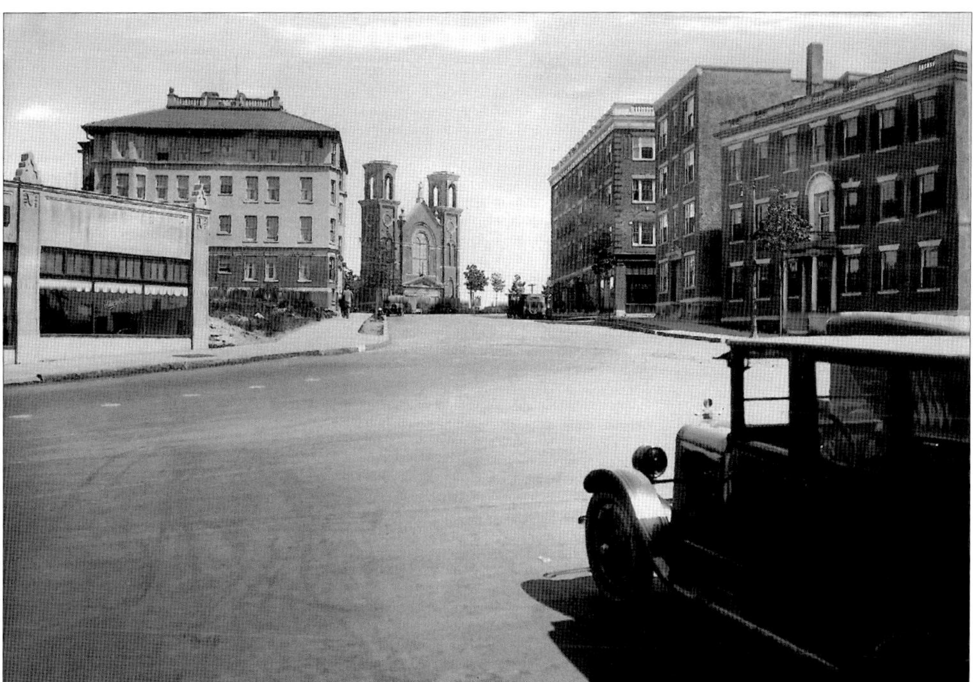

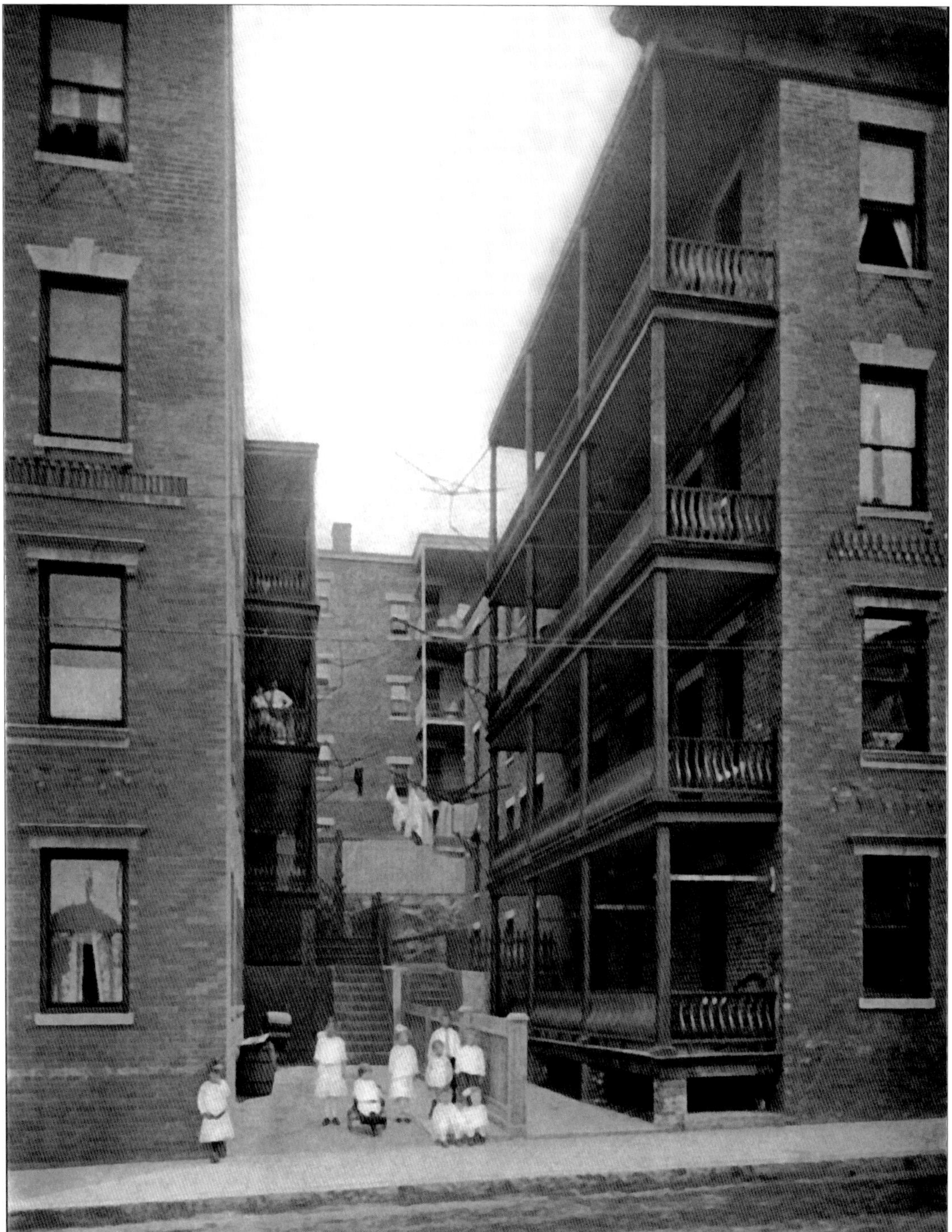

Walter Kilham, one of the principal architects of Kilham & Hopkins, had visited Salem the day of the fire. He called the industrial tenements in Blubber Hollow "conflagration breeders." The Naumkeag Steam Cotton Company hired Kilham & Hopkins to rebuild the boardinghouse for single workers and to build four-family dwellings, similar to those shown in the bottom image on the next page. (*Neg #32839*) The gardens are planted between the new homes built at 34 and 38 Park Street. (*Neg #32838*) The top image on this page shows the new brick tenements built on the Point between 52 and 56 Ward Street. (*Neg #32837*)

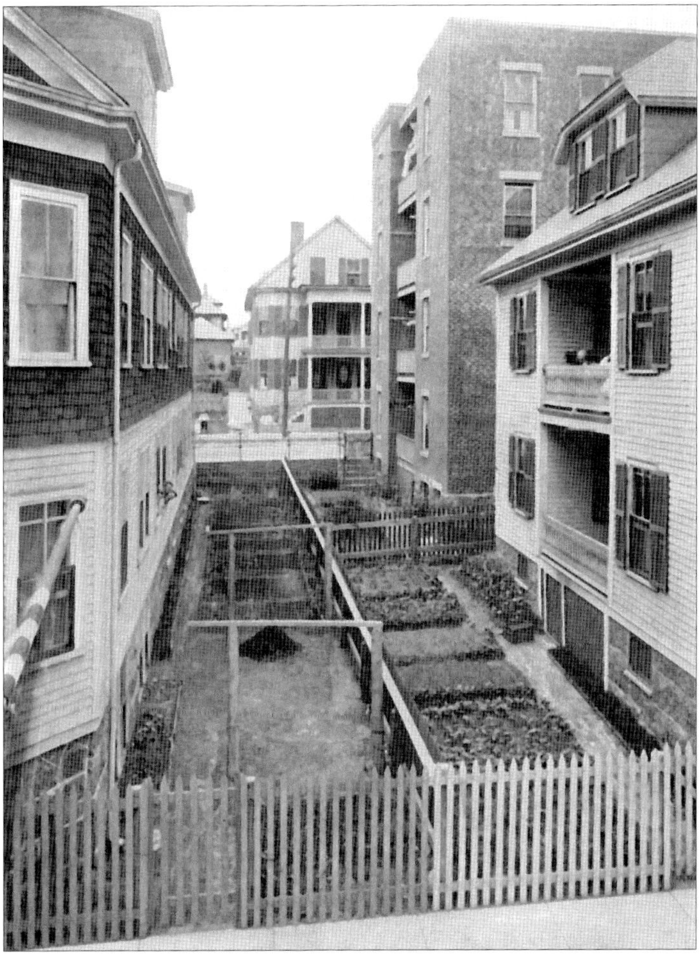

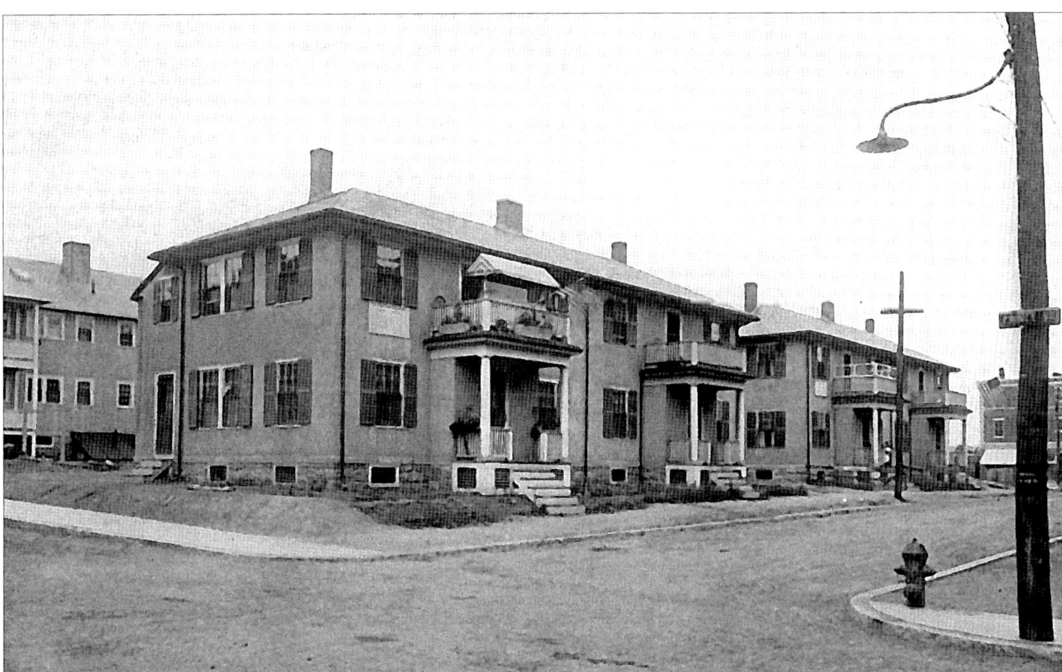

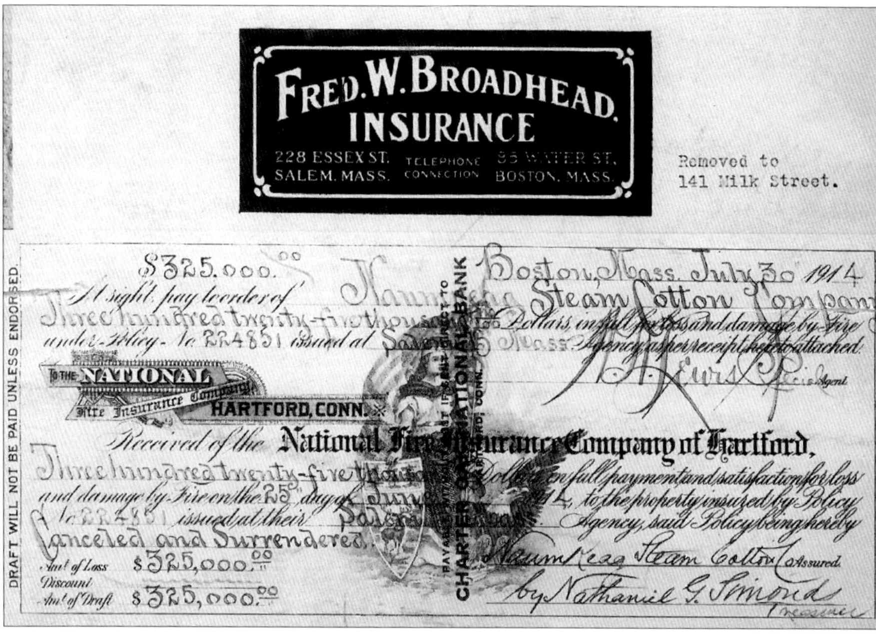

FRED. W. BROADHEAD. INSURANCE
228 ESSEX ST. SALEM, MASS. TELEPHONE CONNECTION 85 WATER ST. BOSTON, MASS.

Removed to 141 Milk Street.

DRAFT WILL NOT BE PAID UNLESS ENDORSED.

$325,000.00 Boston, Mass. July 30 1914

At sight pay to order of Naumkeag Steam Cotton Company

Three hundred twenty-five thousand Dollars in full for loss and damage by Fire under Policy No. 224851 issued at Salem Mass. Agency as per receipt hereto attached.

Agent

To THE NATIONAL Fire Insurance Company HARTFORD, CONN.

PAYABLE THROUGH CHARTER OAK NATIONAL BANK

Received of the National Fire Insurance Company of Hartford, Three hundred twenty-five thousand Dollars in full payment and satisfaction for loss and damage by Fire on the 25th day of June 1914, to the property insured by Policy No. 224851 issued at their Salem Agency, said Policy being hereby Canceled and Surrendered.

Am't of Loss $325,000.00
Discount
Am't of Draft $325,000.00

Naumkeag Steam Cotton Co Assured.
by Nathaniel G. Simonds Treasurer

The Naumkeag Steam Cotton Company was insured for its facility. It received $3,500,000 for property damages from the Boston Manufacturers Mutual Fire Insurance Company of Boston. New to the insurance industry was the option for use and occupancy insurance, which was written for the total or partial interruption of business due to fire. Naumkeag was insured under this option; it received insurance benefits to cover salaries, taxes and loss of profit due to the interruption of business for a maximum of three hundred days. The check from the National Insurance Company of Hartford paid the dividends for this option, allowing Naumkeag to continue to pay 10 percent dividends to its investors. The cost for such insurance was less than $200 per year. (*Neg #32833*) The image to the right shows the rebuilding of the No. 6 Weave Shed. The steeple-like device shown in the right side of the picture was a Lakewood lay-out, a steel tower 150 feet high, constructed by the New England Concrete Construction Company; half of the concrete floor, an area of about seven acres, was poured in twenty-one days because of the device. (*Neg #32895*) Hundreds of new looms had been ordered for the company before the fire; they were not delivered until after the fire and were stored in the surviving warehouse. The completed weave shed is shown next to the No. 1 Storehouse, which survived the fire. (*Neg #8153a*)

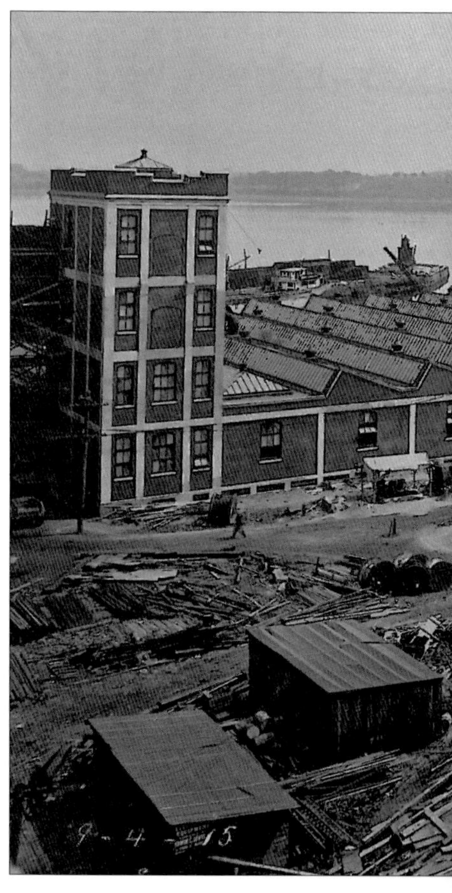

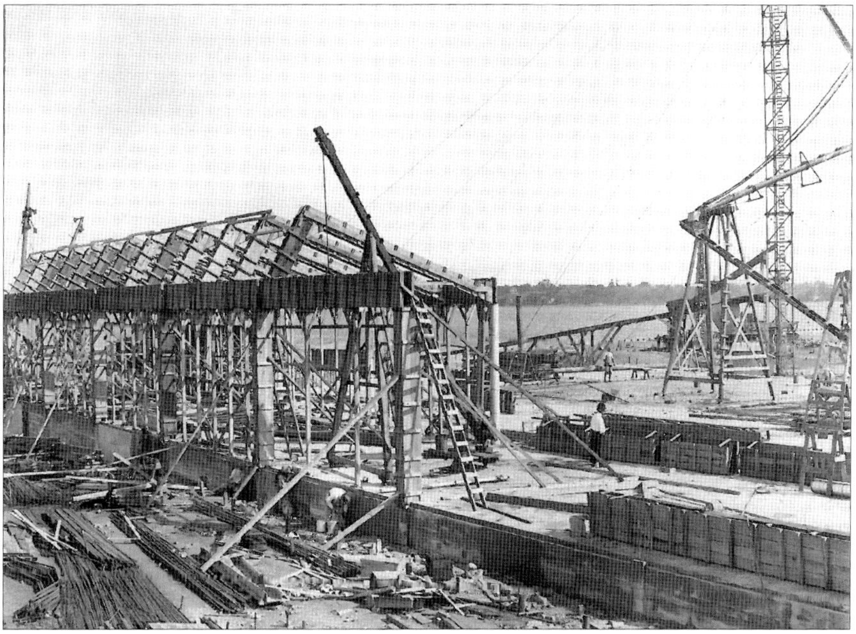

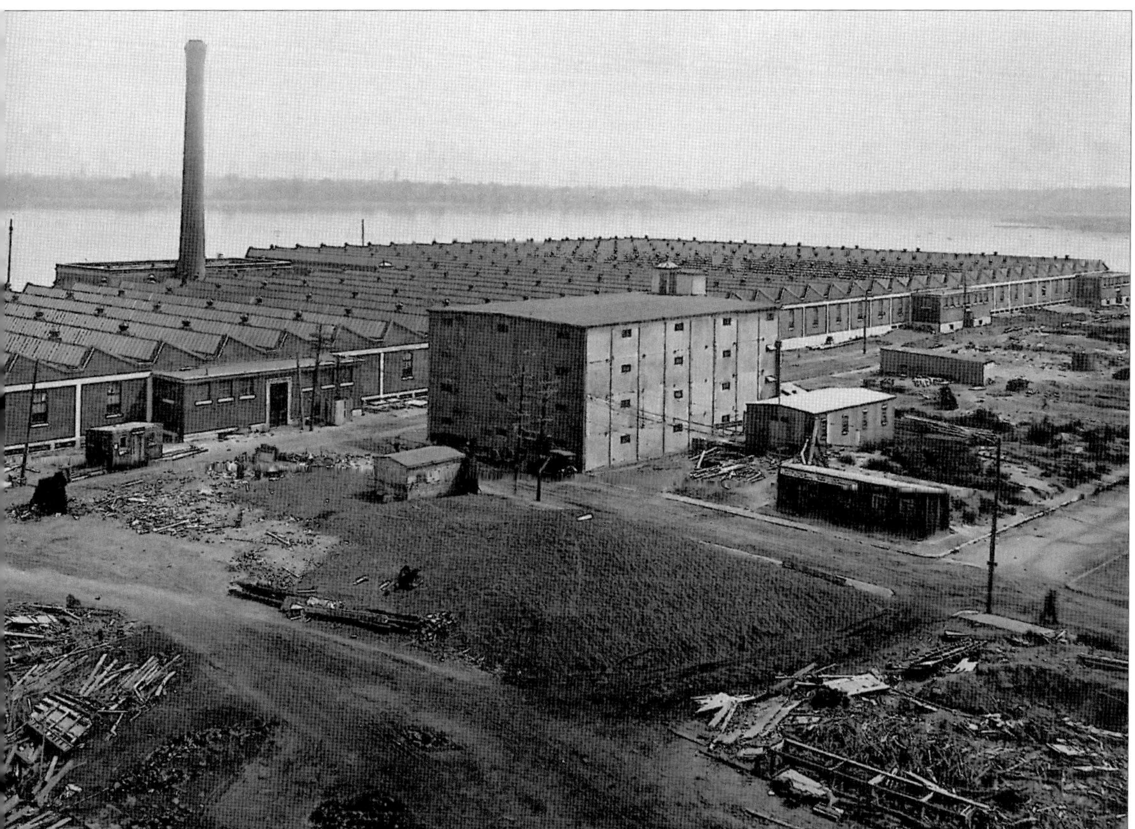

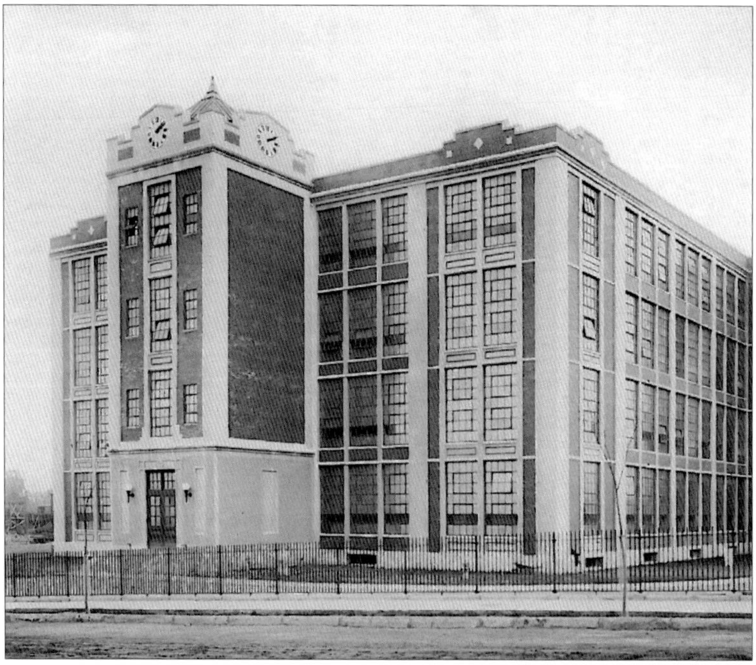

The Naumkeag Steam Cotton Company rebuilt on a large scale. The weave shed opened in December 1915. By February of the following year, the entire plant had been rebuilt and was working at capacity production. The manufacturing buildings were replaced by four new buildings: the weave shed, a storehouse, a spinning mill and a boiler house. All were made of reinforced concrete construction with brick panels, metal sash and wired-glass windows. The buildings were designed and engineered by Lockwood Green Engineers, Inc. Steam was no longer used as a source of power; electric motors powered all the machinery. The design of the new buildings paid attention to the issues of light, ventilation, sanitation and positioning of machinery so that mass production of sheetings would be effective and working conditions would be safe for employees. The company remained in operation until 1953, when it moved to South Carolina to take advantage of cheaper labor. In 1958, Robert Lappin, owner of the Shetland Company, which manufactured home appliances, purchased the property; it was renamed the Shetland Industrial Park, known today as Shetland Park since manufacturing firms have moved out of the facility. (*Neg #32877 and Neg #8154*)

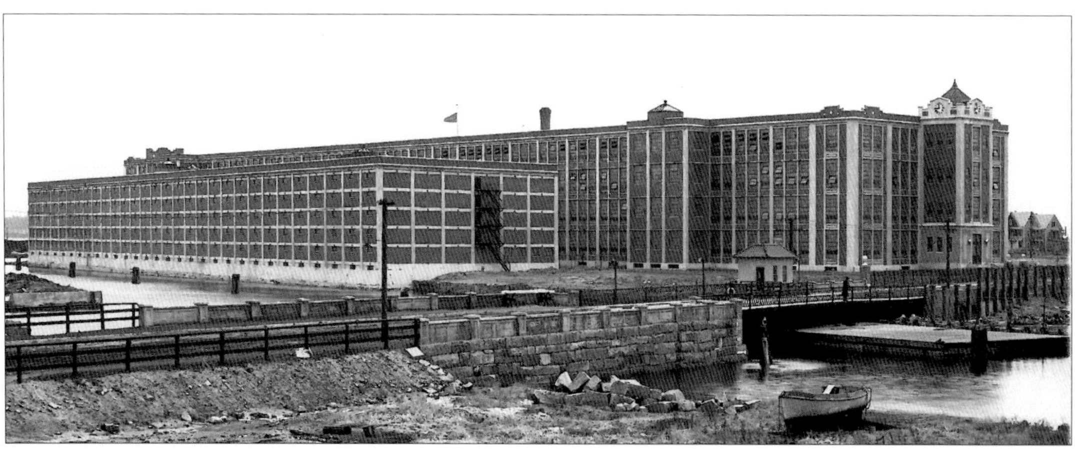

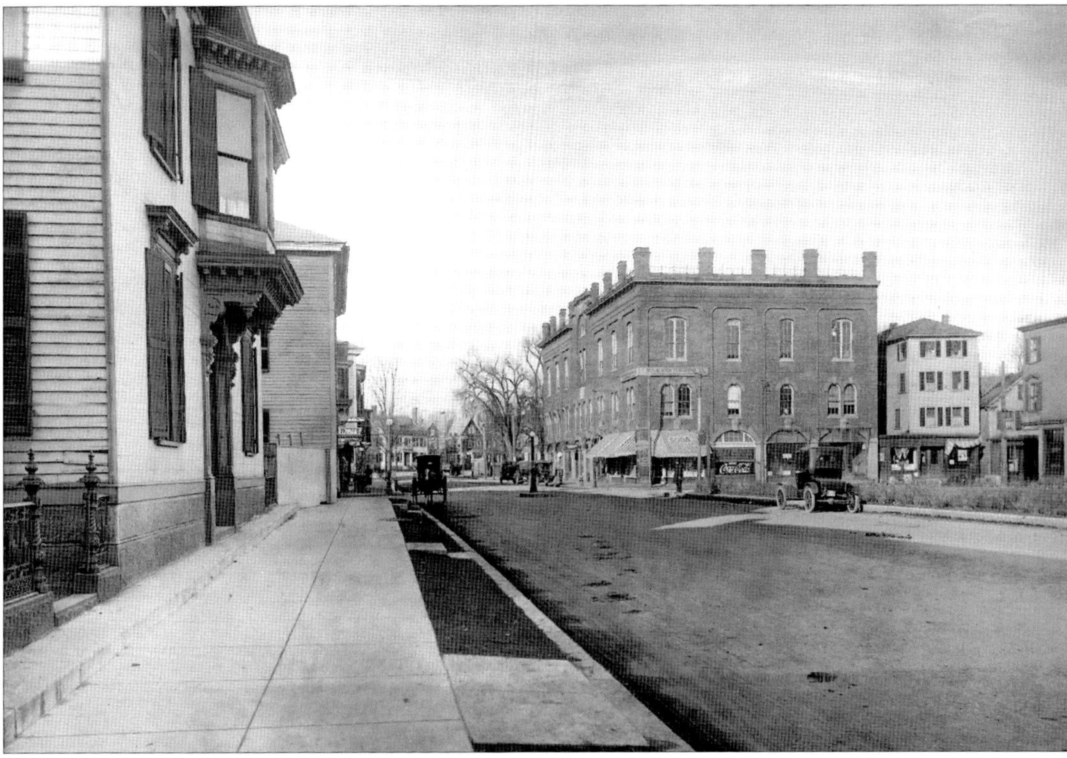

These two photographs show the newly built Hawthorne Boulevard from two directions. The top image looks toward the Salem Common and the large brick building known today as the Hawthorne Hotel. (*Neg #32924*) The bottom image looks toward the Immaculate Conception Church. The street becomes Congress Street as it passes the rebuilt Naumkeag Mills. (*Neg #32881*)

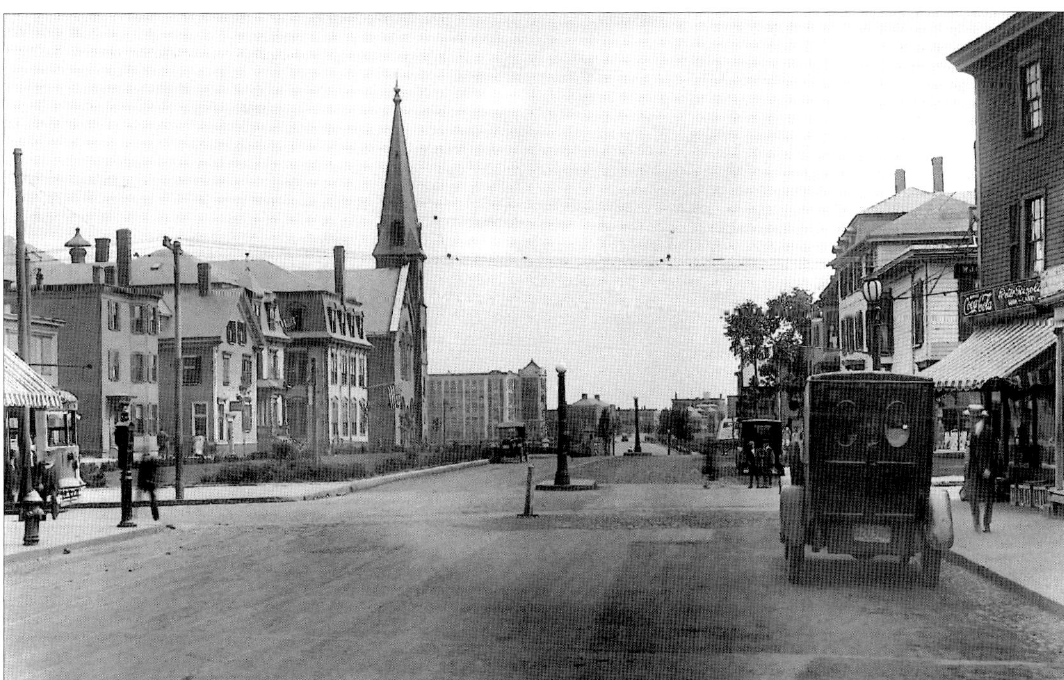

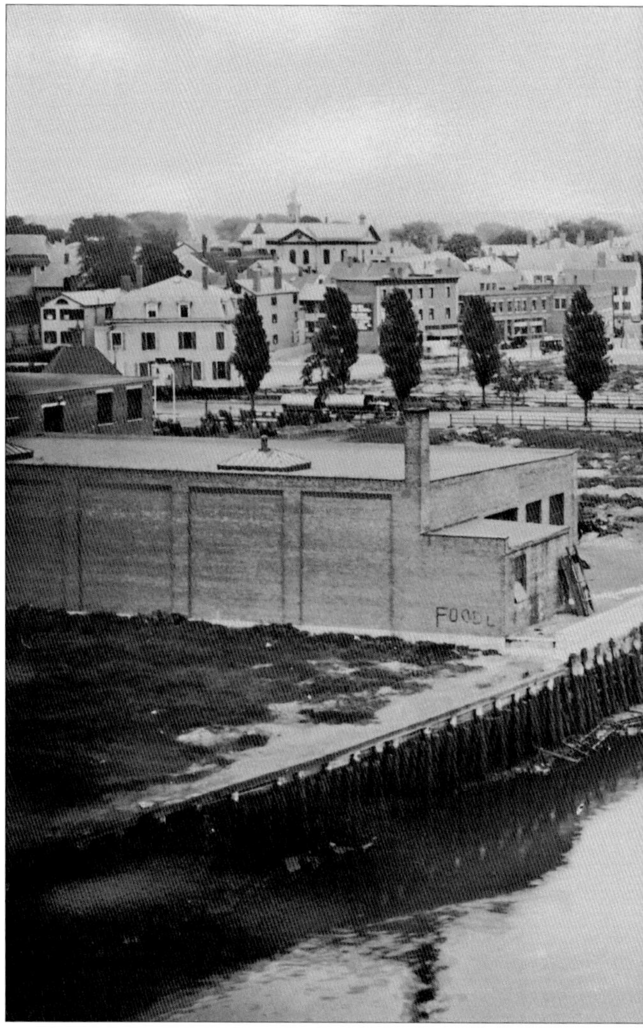

These three images show the Congress Street section of Salem. The top photograph shows the reconstruction after the fire and the newly built Congress Street Bridge (*Neg #32923*), while the bottom image shows the area before reconstruction began. The bottom image is a view from the Electric Light Company building looking east across the South River, depicting the intersection of Derby and Congress Streets. (*Neg #4479*) The smaller image to the right provides a closer look at the rebuilt Congress Street as it passes the reconstructed Naumkeag mill facility. (*Neg #32922*)

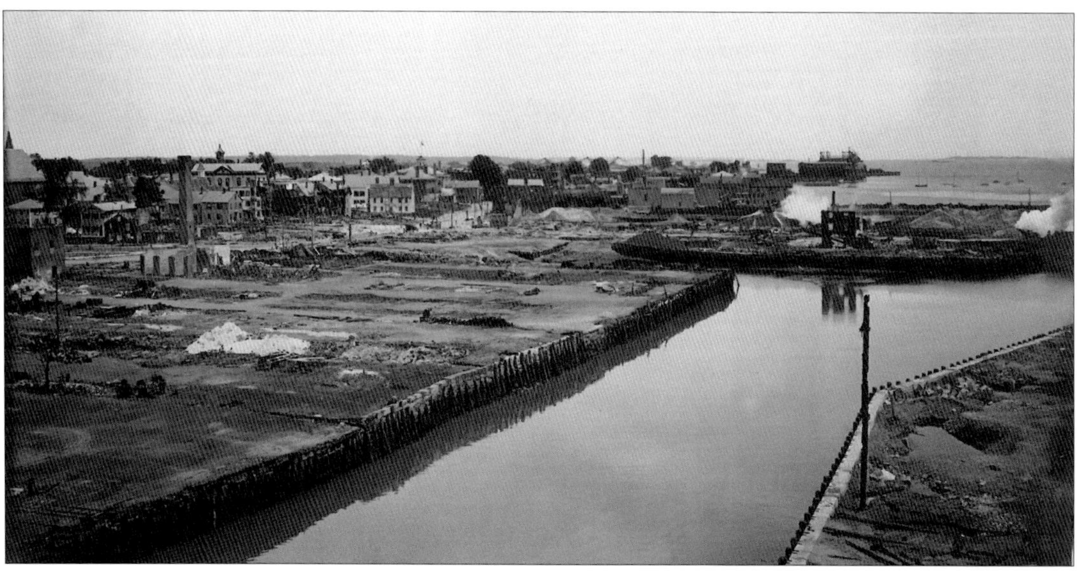

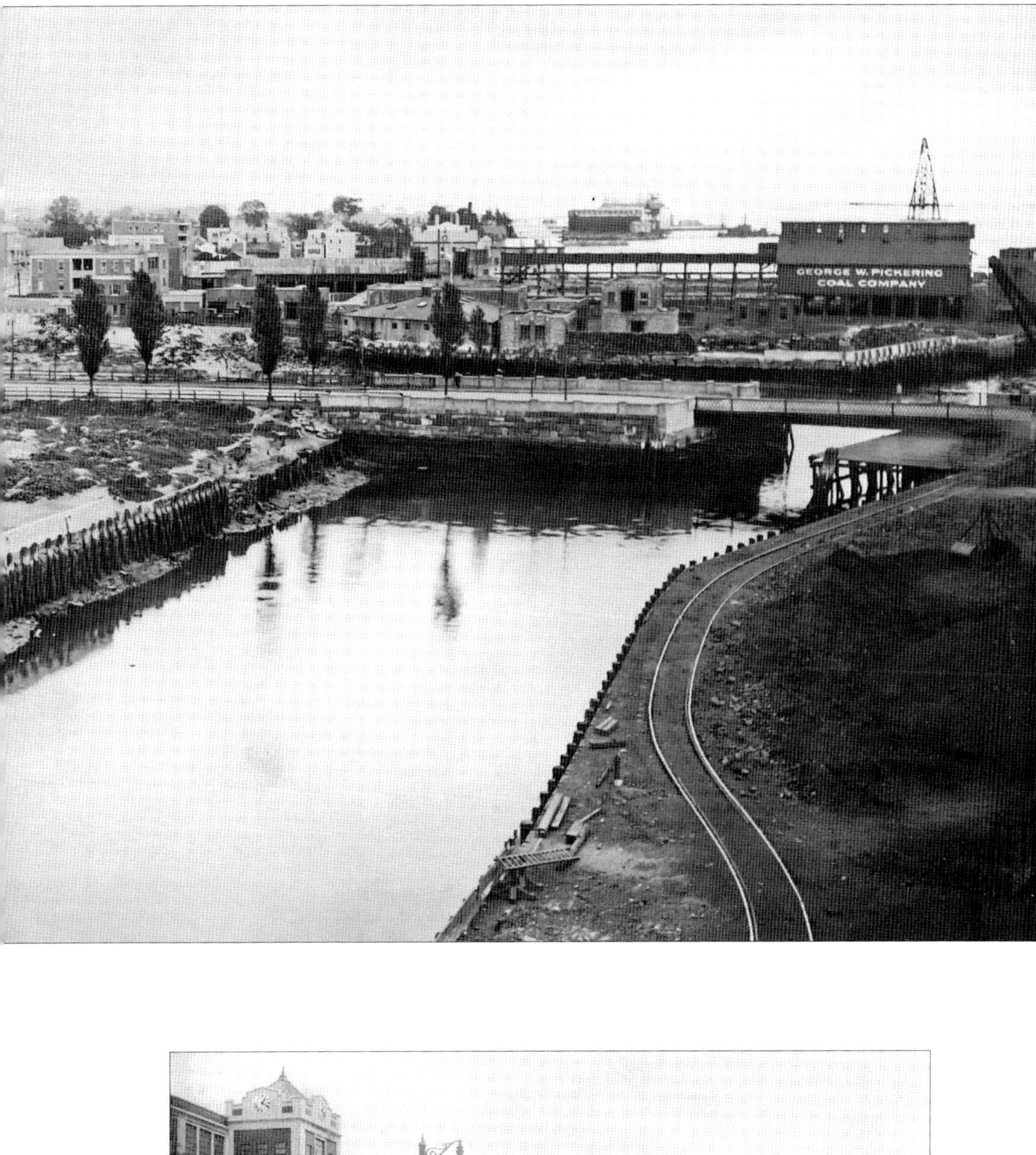
GEORGE W. PICKERING
COAL COMPANY

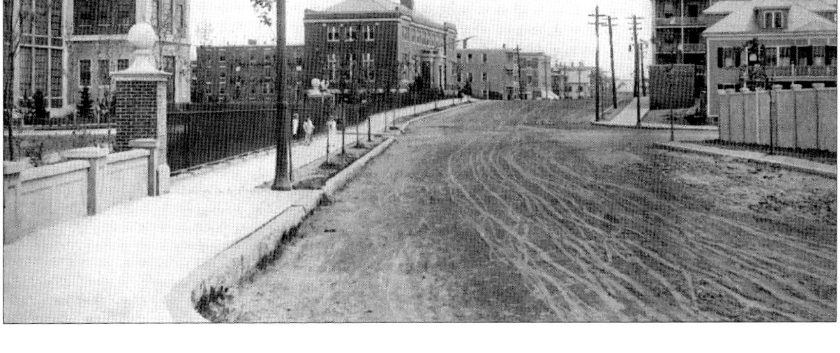

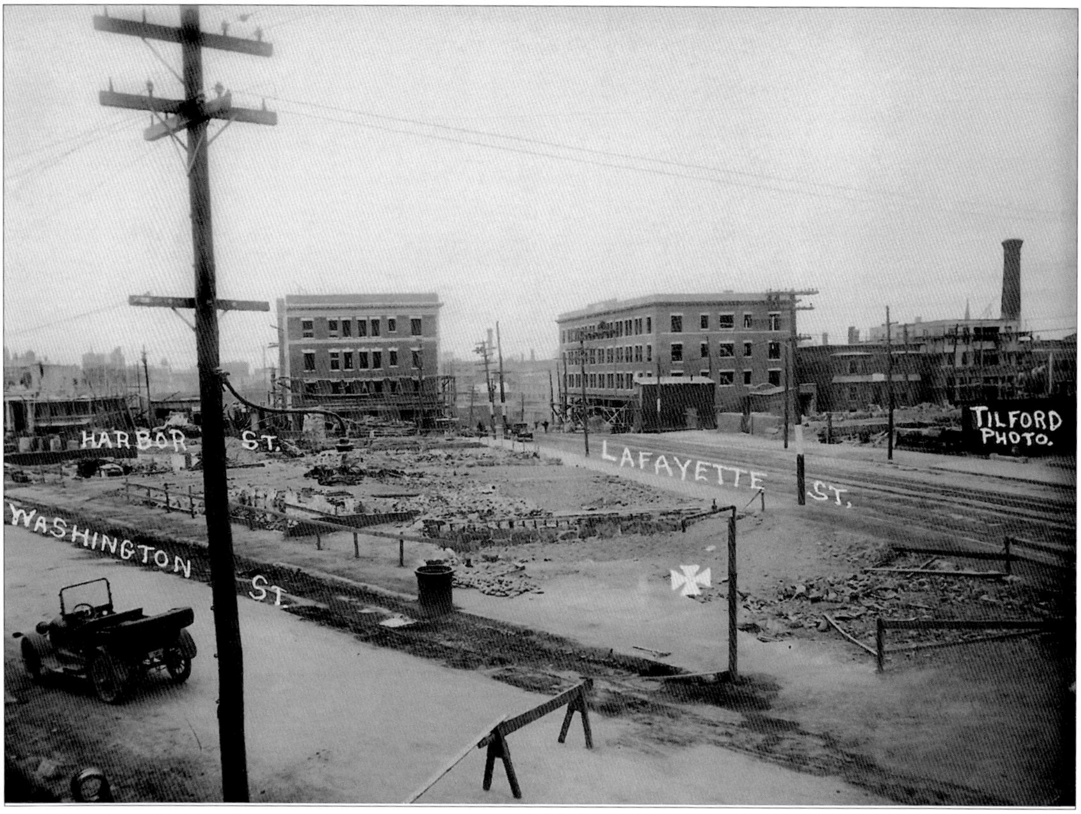
HARBOR ST.
LAFAYETTE ST.
WASHINGTON ST.
TILFORD PHOTO.

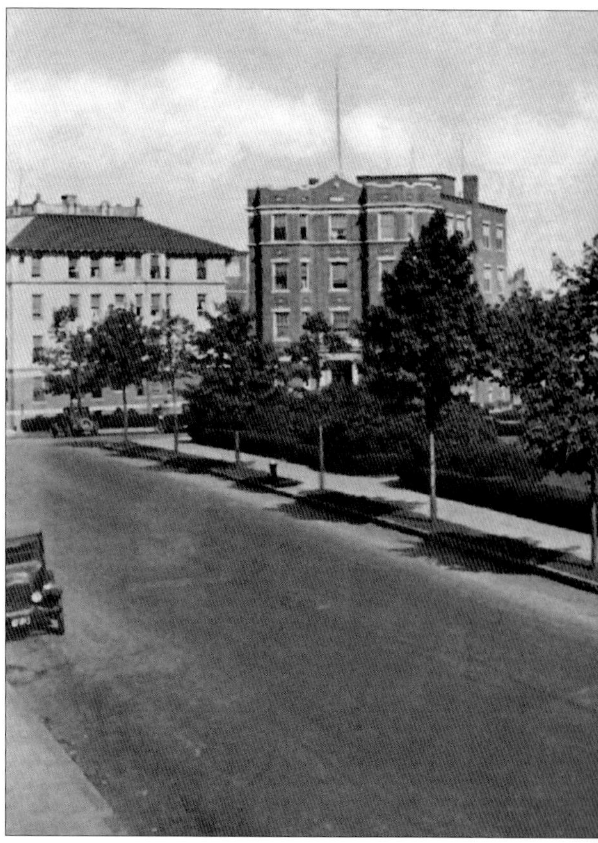

In 1915, the Salem Rebuilding Commission planned a memorial for the fire. It decided to create a public park on a triangle of land bordered by Washington, Harbor and Lafayette Streets. The top photo outlines the park before reconstruction; it was taken from the roof of one of the houses on Upper Washington Street. On the extreme right, in the foreground, is city property on which Engine House 1 originally stood at the intersection of Washington and Lafayette Streets. The standpipe for watering carts and the old watering trough can be seen on the Washington side of the triangle. At the rear are the Poirier and Levesque Hotels on each corner of Lafayette and Harbor Streets. (*Neg #32793, a Tilford photo*) Reverend James L. Hill contacted Dr. J. Ackerman Coles of New York City for a donation to provide a flagpole and a bronze commemorative tablet for the park. Coles said he would bear the expense "to record the civic pride and heroism of the people of Salem." The flagpole was placed facing the residential side of Lafayette Street. The tablet, bent to fit the flagpole, commemorated "the enterprise and resolute spirit with which Salem arose from her ashes, looked calamity in the face and rebuilt her walls which were broken down." Protest arose when additional land needed for the park was requested from abutting landowners. News reports indicated that the proposed taking of land by eminent domain created more reaction than any other act of the Rebuilding Commission outside of the new building code issued in 1914. Citizens who protested the building of the park felt that taxable property should not be used for such a purpose; money spent to create the park was better spent on schools, sewers, water pipes and streetlights. In mid-May, the Salem Rebuilding commission voted to take the land for the park, awarding a total of $19,250.20 in damages to the abutting landowners, which was a rate of $0.75 per foot for the land taken. The total amount of land taken was 23,019 square feet. The park was dedicated in December 1915. The finished park is depicted in the bottom photograph. (*Neg #32826*)

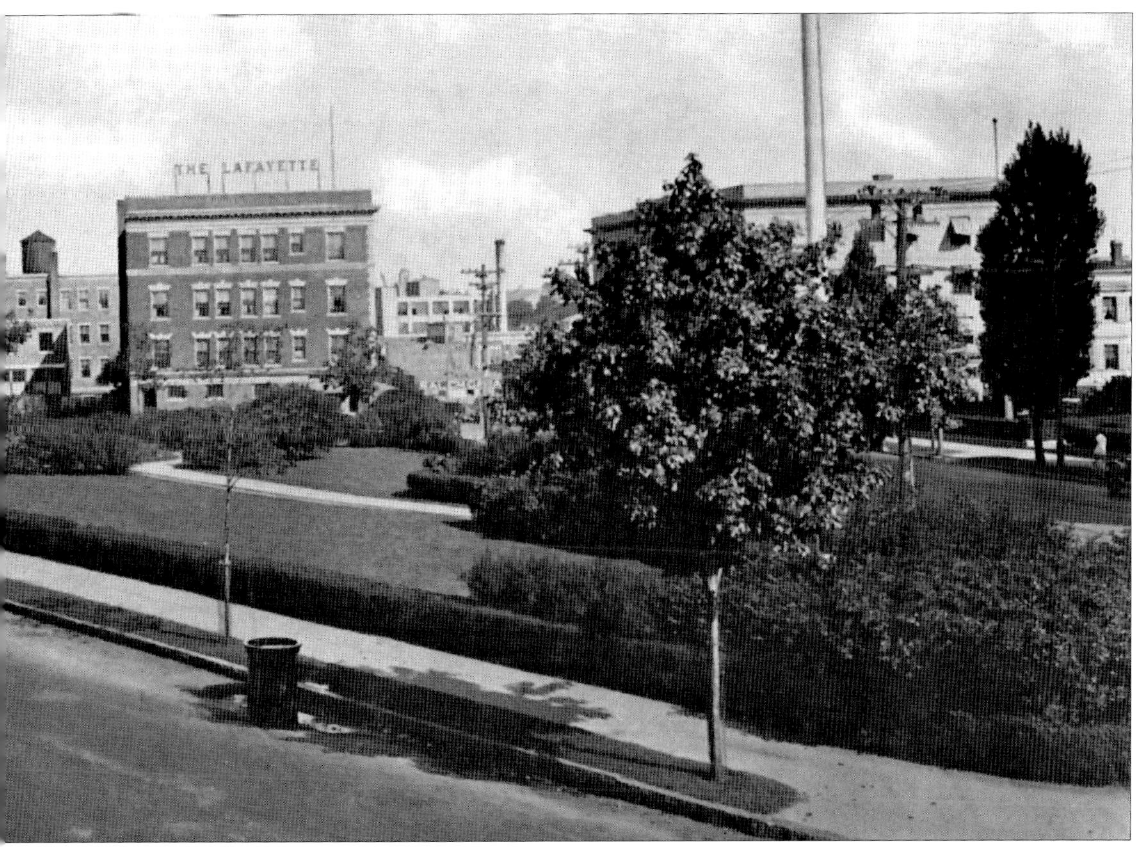

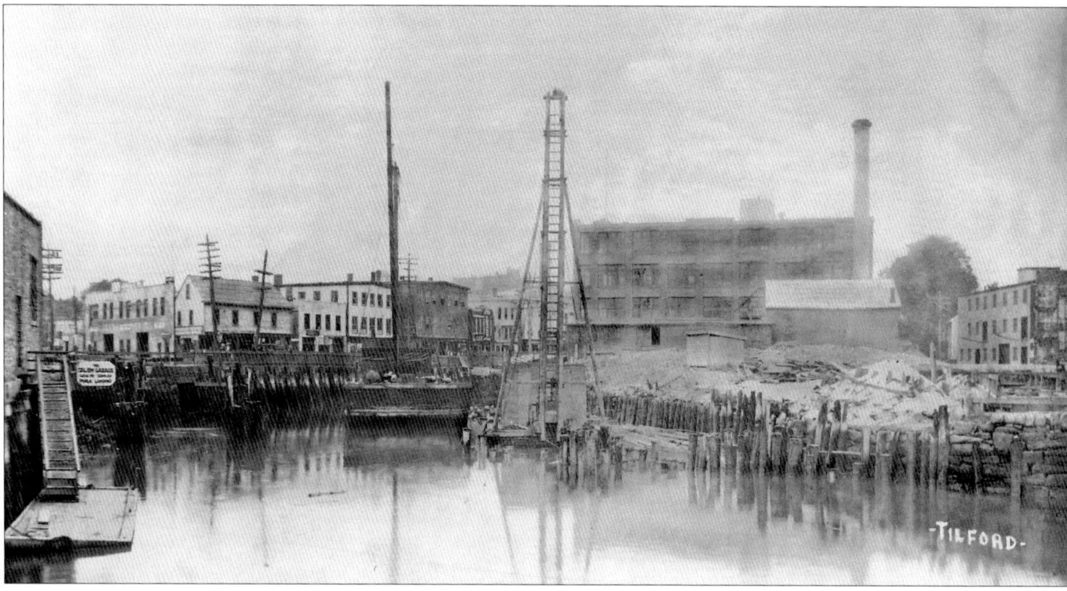

The Fairfield Dock property was conveyed to the city by Charlotte Fairfield, a coal dealer in Salem. Miss Fairfield leased the dock for many years but contended that sewerage filled up her dock so that coal barges could not unload at her wharf. After winning a lawsuit against the property owner, Spencer T. Williams, Miss Fairfield conveyed the property to the city. The property consisted of twenty thousand feet of land, a portion of which would be filled in to extend Derby Street to an intersection with a new boulevard intended to connect Washington Square with Congress Street and South Salem; the north portion of the property was a probable site for the new Central Fire Station. The image above shows the dock before the reconstruction process began, while the photograph below depicts the process of filling in the dock. (*Neg #32929, above; Neg #32930, below; both Tilford photos*)

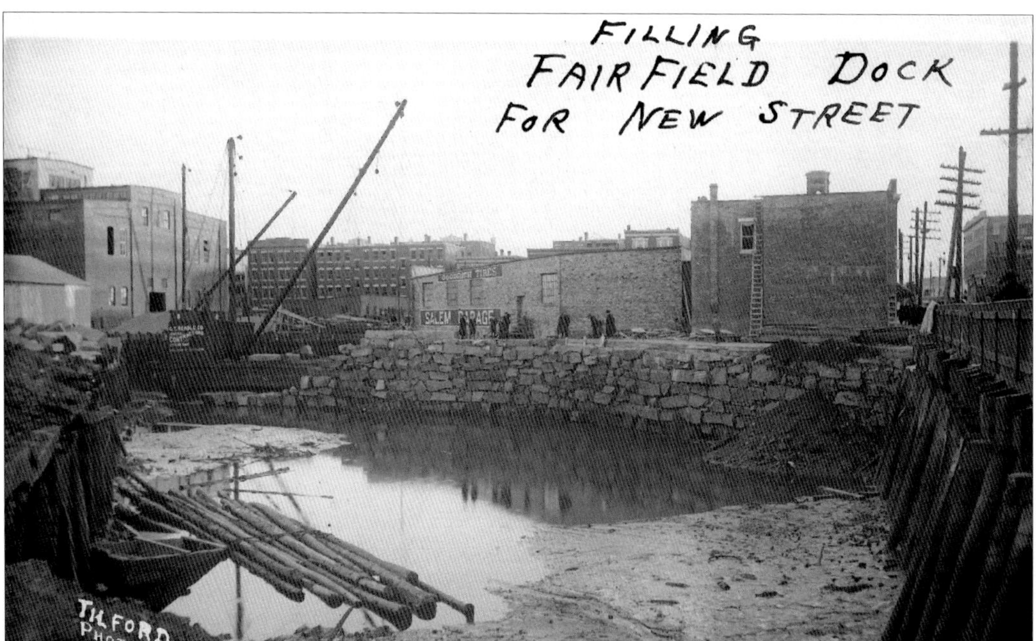

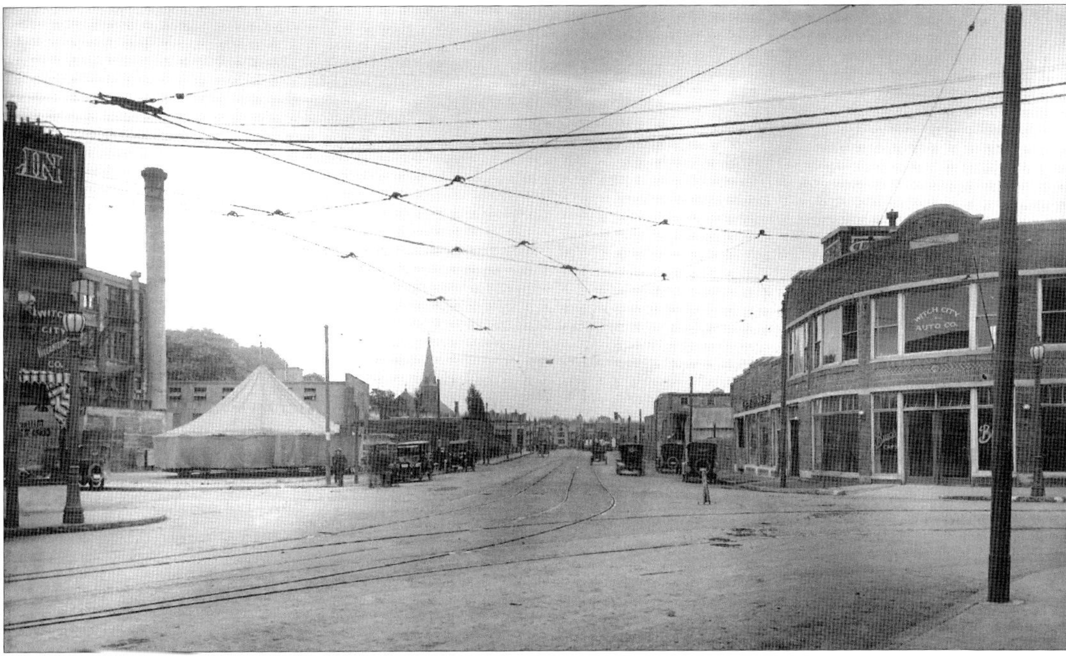

The top image shows the new Derby Street extension connecting with Lafayette Street after the Fairfield Dock was filled in. (*Neg #32925*) The image below shows the completed reconstruction on Lafayette Street; one of the towers of St. Joseph's Church can be seen through the trees at the end of the street. (*Neg #32931*)

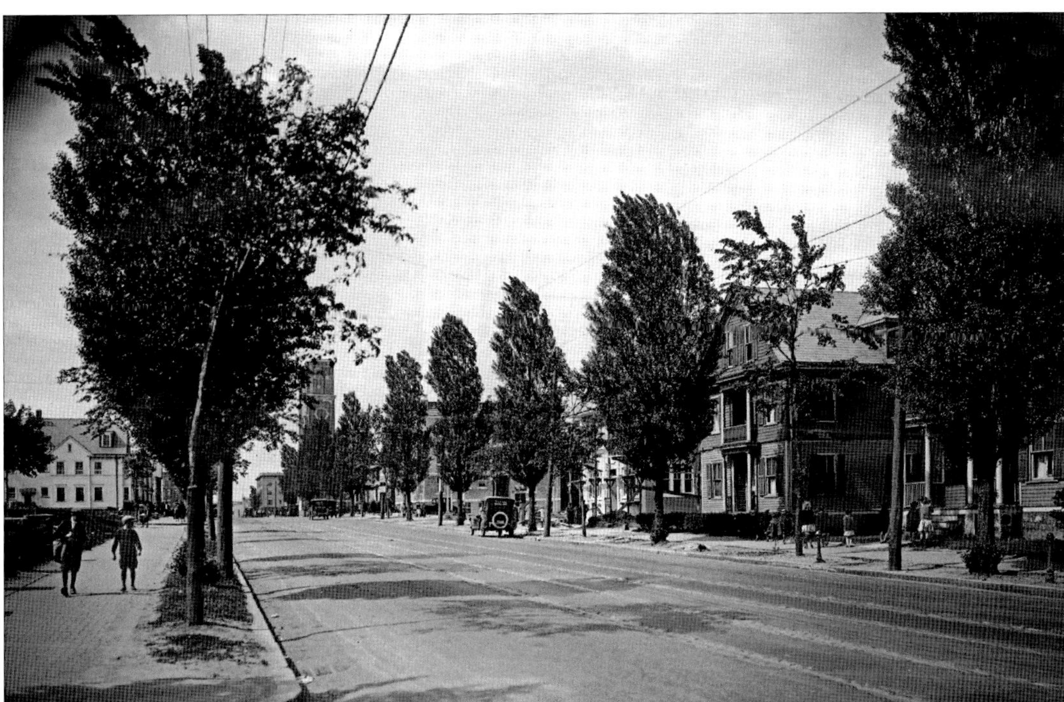

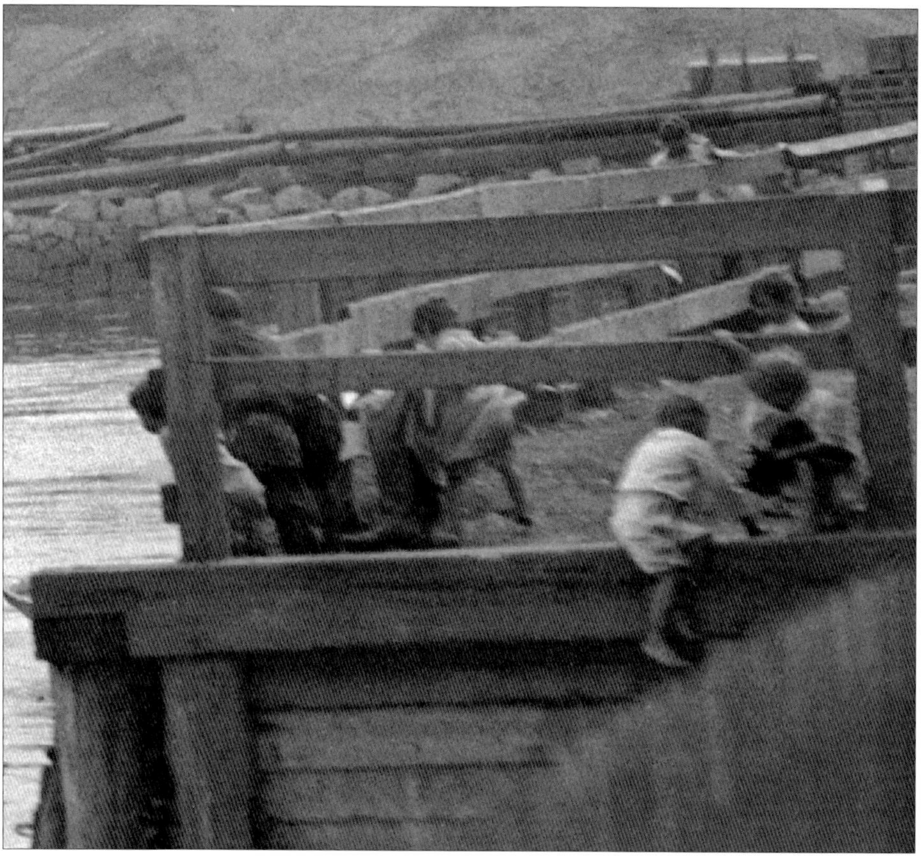

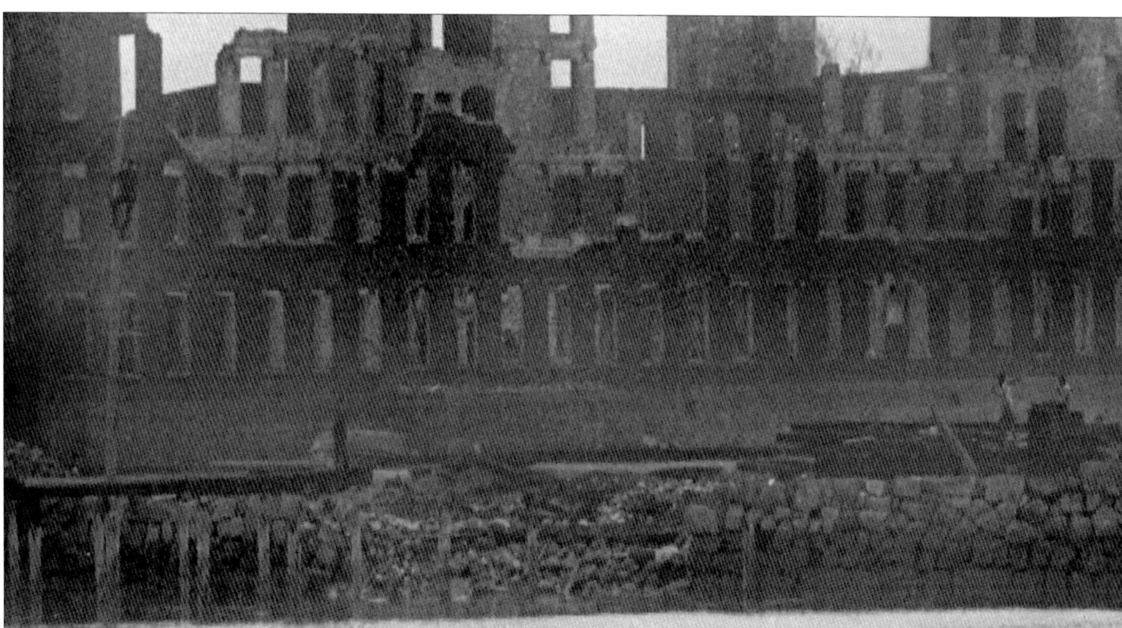

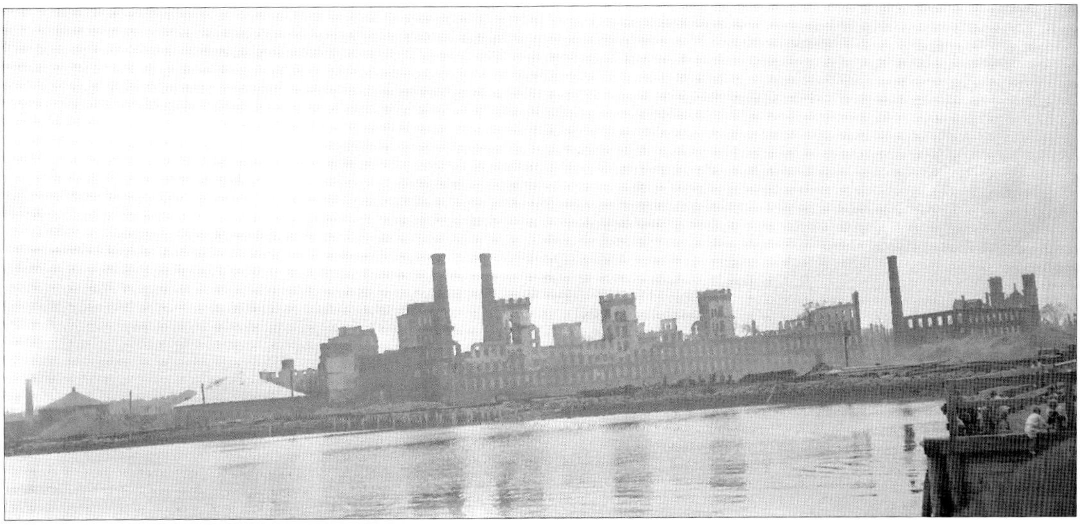

These prints were made from a negative found in the Frank Cousins collection. The size of the negative was very similar to the small photo above. Cousins photographed the ruins of the Naumkeag mill property from across the harbor. (*Neg #32935*) Both of the larger photographs were extracted after the negative was digitized. The bottom photograph depicts workers beginning the reconstruction process. (*Neg #32935DetailB*) The image to the left shows the resiliency of children as they played on a neighboring dock. (*Neg #32935DetailA*) After the fire, many news stories appeared about the generous acts of the citizens of Salem. In one story, the generous citizens were two young boys. Before the fire, the two boys, unnamed in the article, had made plans to attend summer camp at the Salem Fraternity Camp in Rowley, Massachusetts. The two boys decided that since their homes were not burned out, their places should be taken by two boys whose homes had been destroyed by the fire.

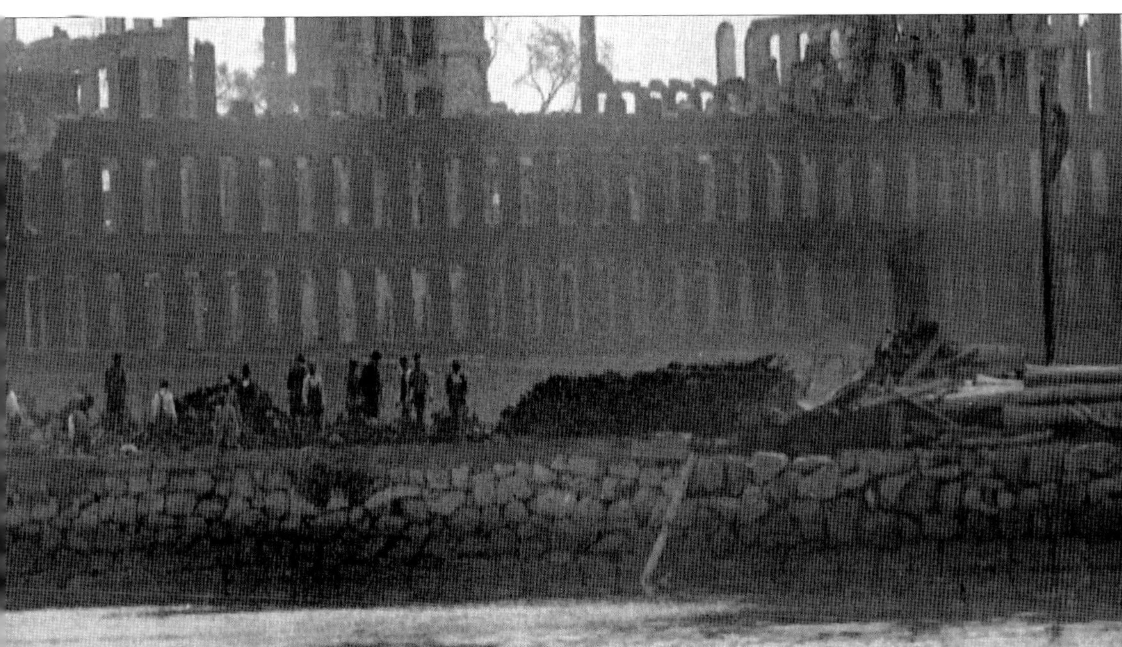

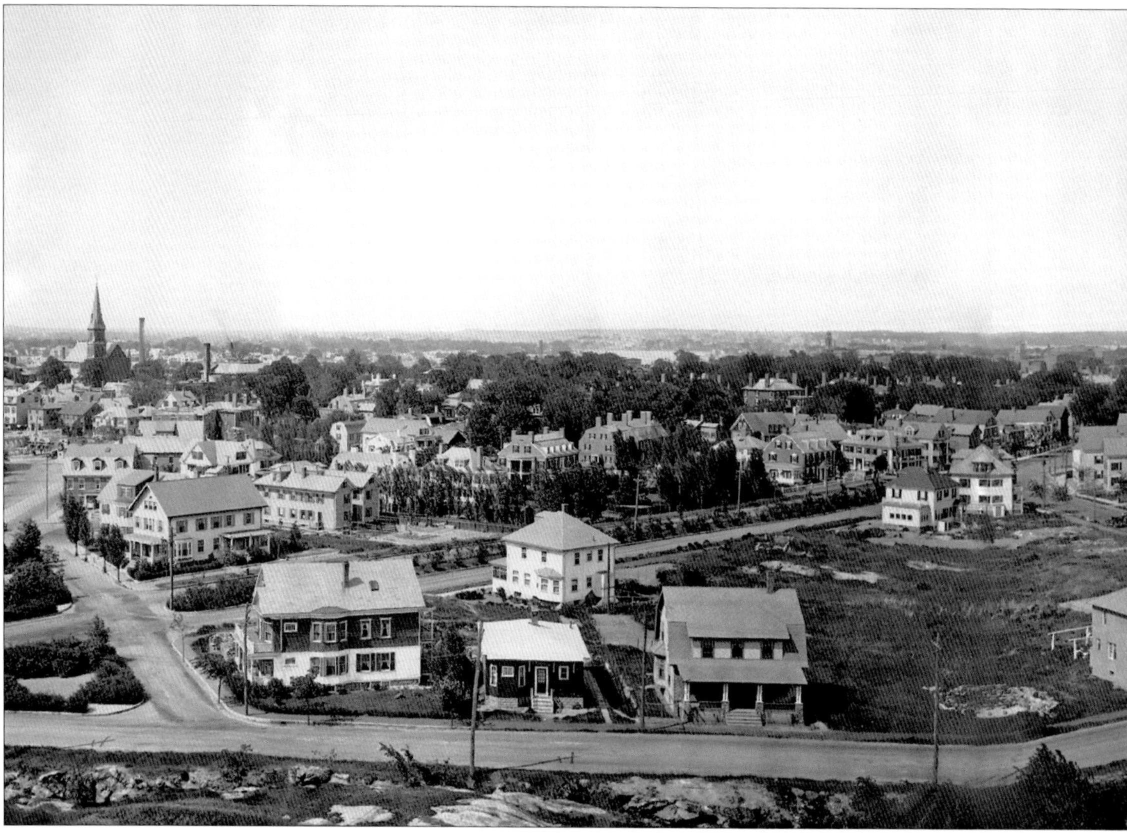

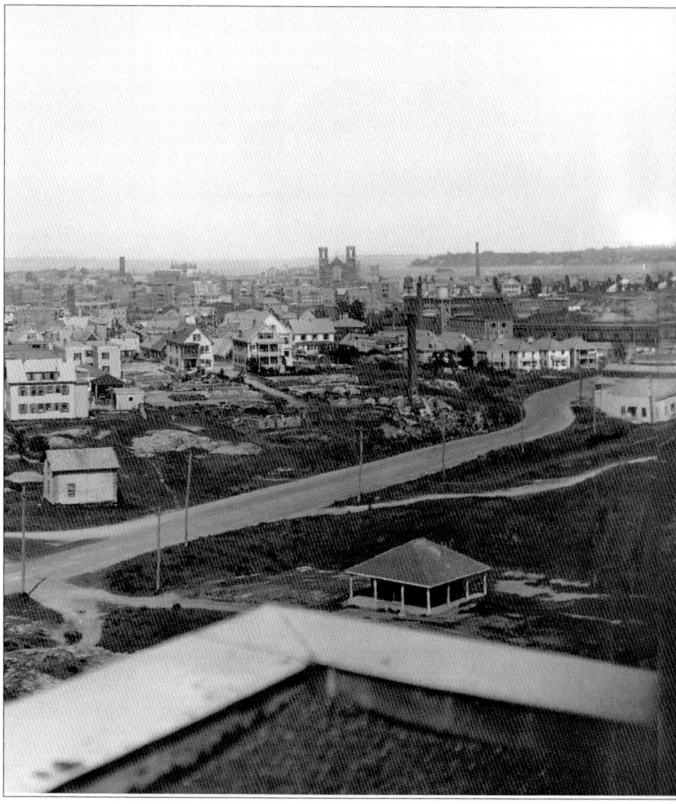

This split panoramic view is a trifold image from the William Abbott photograph album. Probably photographed by Tilford, as were many of the images in the Abbott album, this photograph was taken from the vantage point of the Salem High School, showing that much of the city had been rebuilt. (*Neg #32920A and Neg #32920B*)

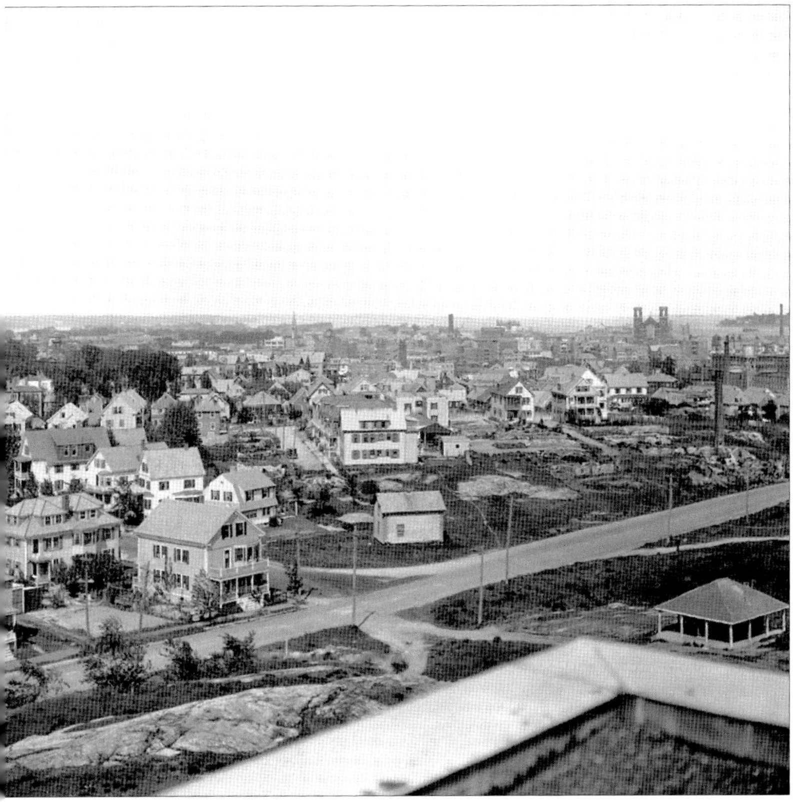

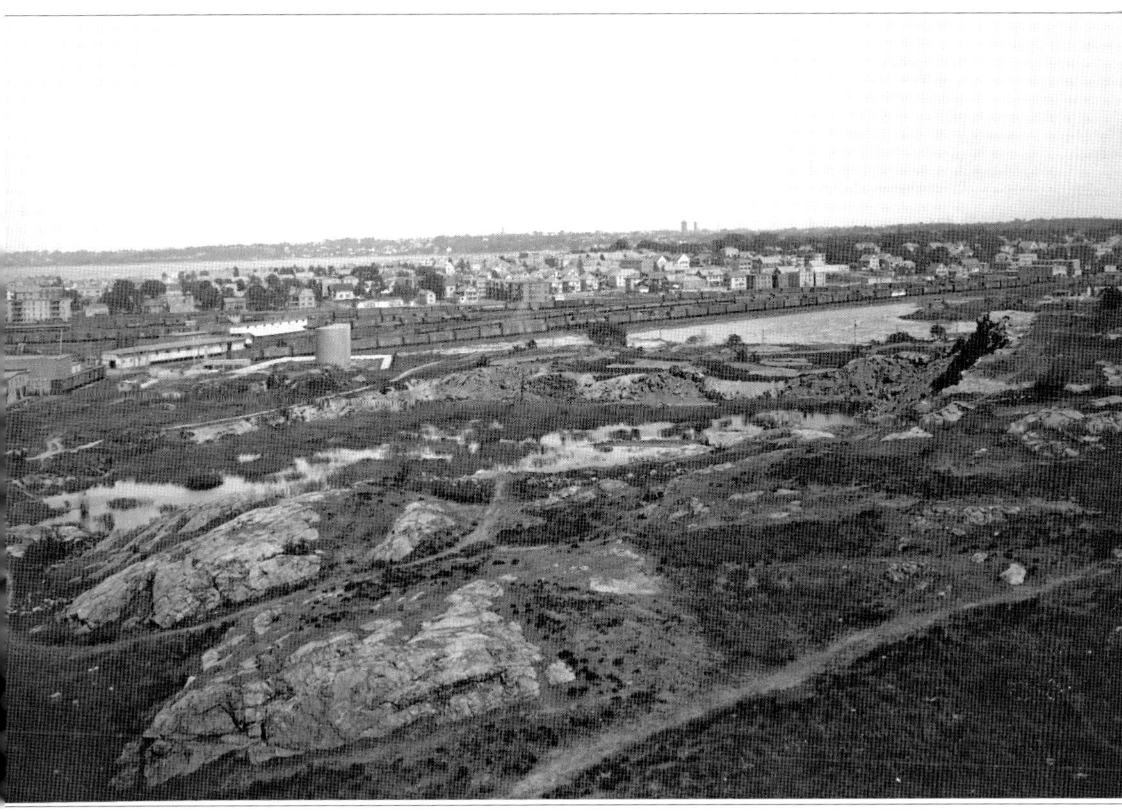

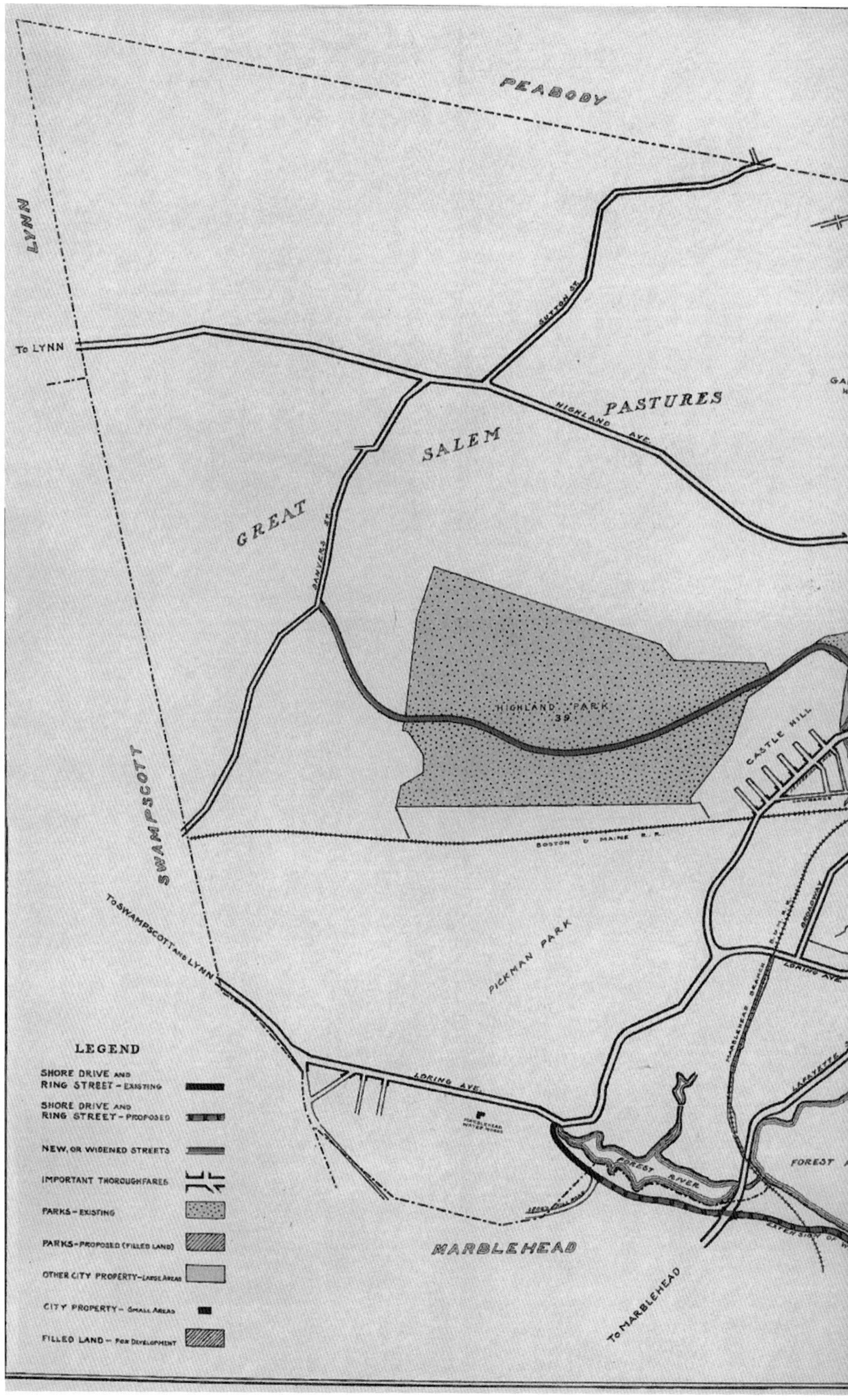

Salem was the first city in the Commonwealth to have a City Plans Commission. Appointed in April 1911, the commission presented a plan to the city for improvements in December 1912. The chief engineer of the city, George Ashton, helped to formulate suggested areas of improvement: widening city

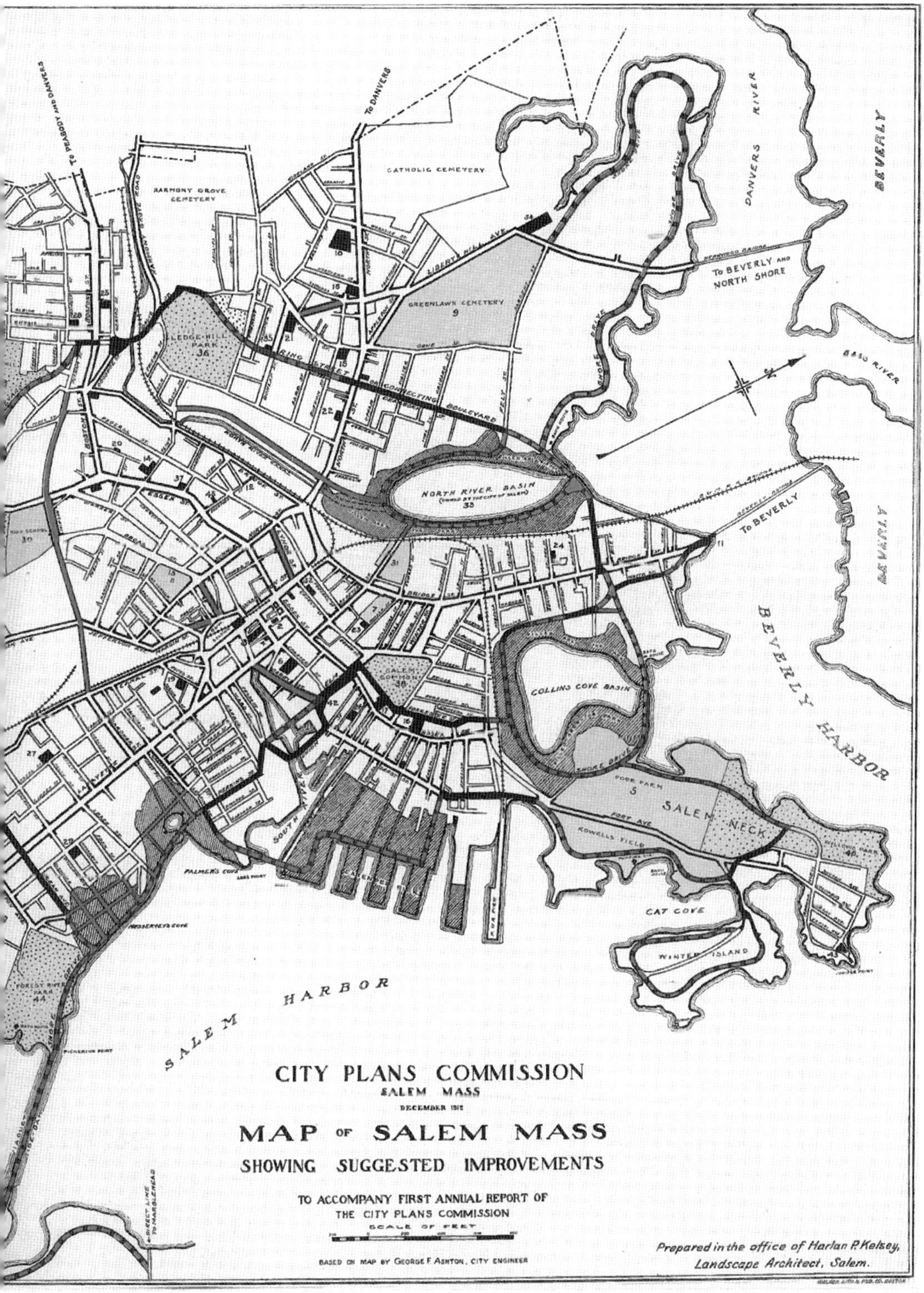

streets, extensions of certain streets and improvements to specified streets. The map above was published by the City Plans Commission in 1912; it facilitated the efforts of the Salem Rebuilding Commission in 1914. (*Neg #32903*)

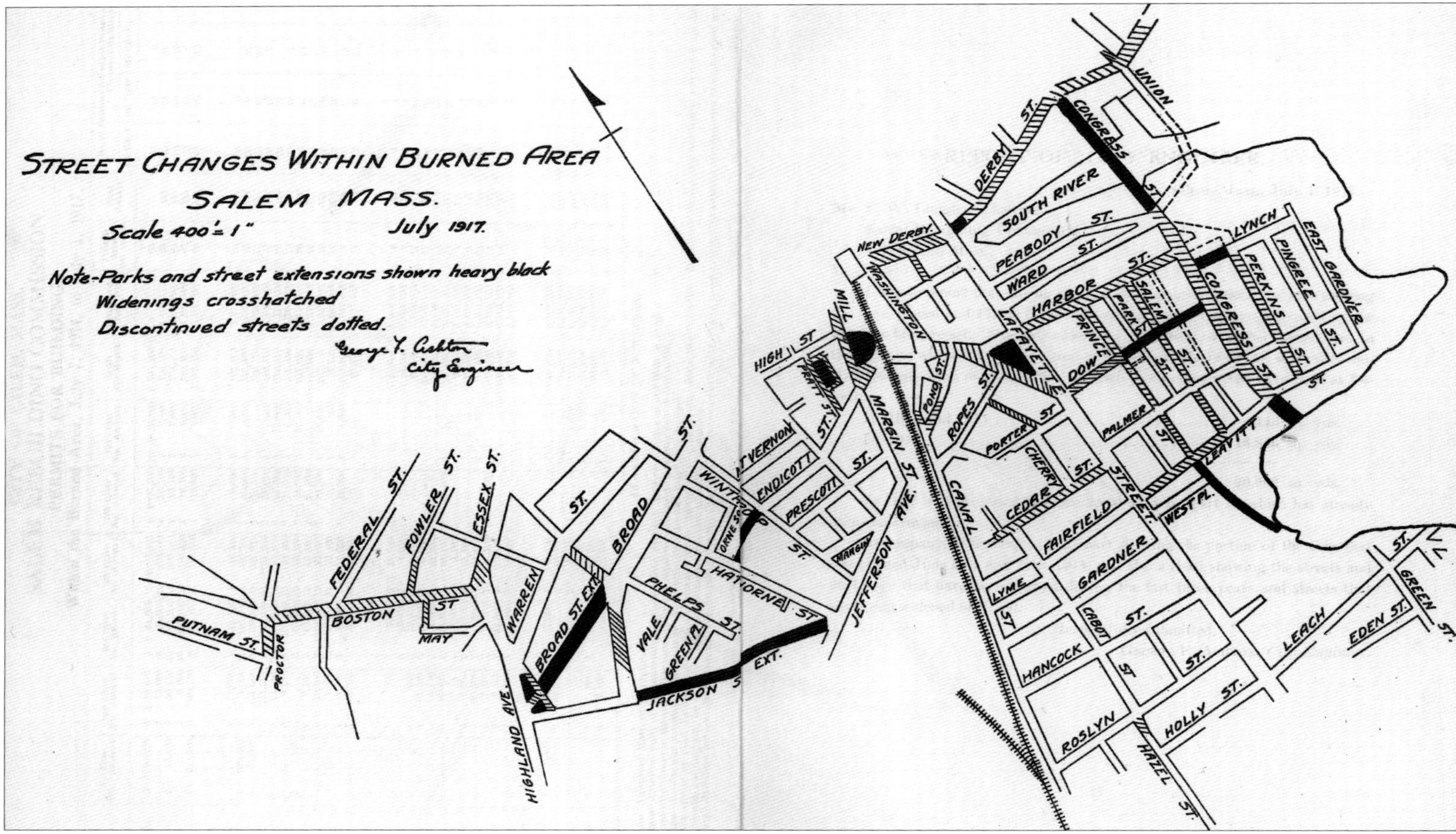

This map was created in 1917 by the city engineer, George Ashton, to show the changes made to streets within the burned area of Salem. Newly created parks and street extensions are identified by heavy black coloring; discontinued streets are represented with dotted lines, while streets that were widened are crosshatched. Prior to the changes, many of the streets had been very narrow and curved, which impeded fire apparatus in their attempts to fight the fire. (*Neg #32921, published in the Salem Rebuilding Commission Report, July 1917*)

Chapter 8

The Impact of the Fire

Beyond the obvious destruction of the fire, what was its actual impact other than the changes to the streets, the new building code, the improvement to the fire department and the new water system already discussed? How would the impact of the fire have been different if the historic sections of town and the mercantile district rather than the immigrant district had been destroyed? One can surmise that the impact would have been quite different and the fire may have held more importance in the collective memory of Salem and its residents. The editors of *Salem: Place, Myth, and Memory*, Dane Morrison and Nancy Schulz, indicate that the fire does not hold a prominent place in Salem's collective memory. It is possible that this is because of the juxtaposition of the fire and world events or the fact that many of the residents burned out by the fire moved away from the city and did not return. Reading first-person reminiscences of the fire indicates that it was a pivotal event for a family and that the fire was remembered by subsequent generations.

The front page of the June 29, 1914 *Boston Journal* reminds the reader that the fire occurred when there was major concern about an impending world war. Archduke Franz Ferdinand, heir to the Austria-Hungarian throne, and the Duchess of Hohenberg, his wife, were assassinated on the streets of Sarajevo, Bosnia. His assassination precipitated Austria-Hungary's declaration of war against Serbia. In the context of such an event, does the impact of the fire have the same importance?

The Phillips Library collection includes several poems written about the fire. Alexander McCabe thought about the fire for many years, but did not write his poem until he was eighty. Donated to the library in 1999, the poem was written many years after the fire, indicating that the fire did have an impact on Mr. McCabe. Below is an excerpt from his poem, "The Salem Fire."

It was on June the Twenty-Fifth
In the Nineteenth-Fourteenth year.
The air was soft and balmy.
The day was bright and clear.
At One Thirty-Seven Post Merid
Whistles heard afar
Brought to homes more terror
Than rumors of the war.

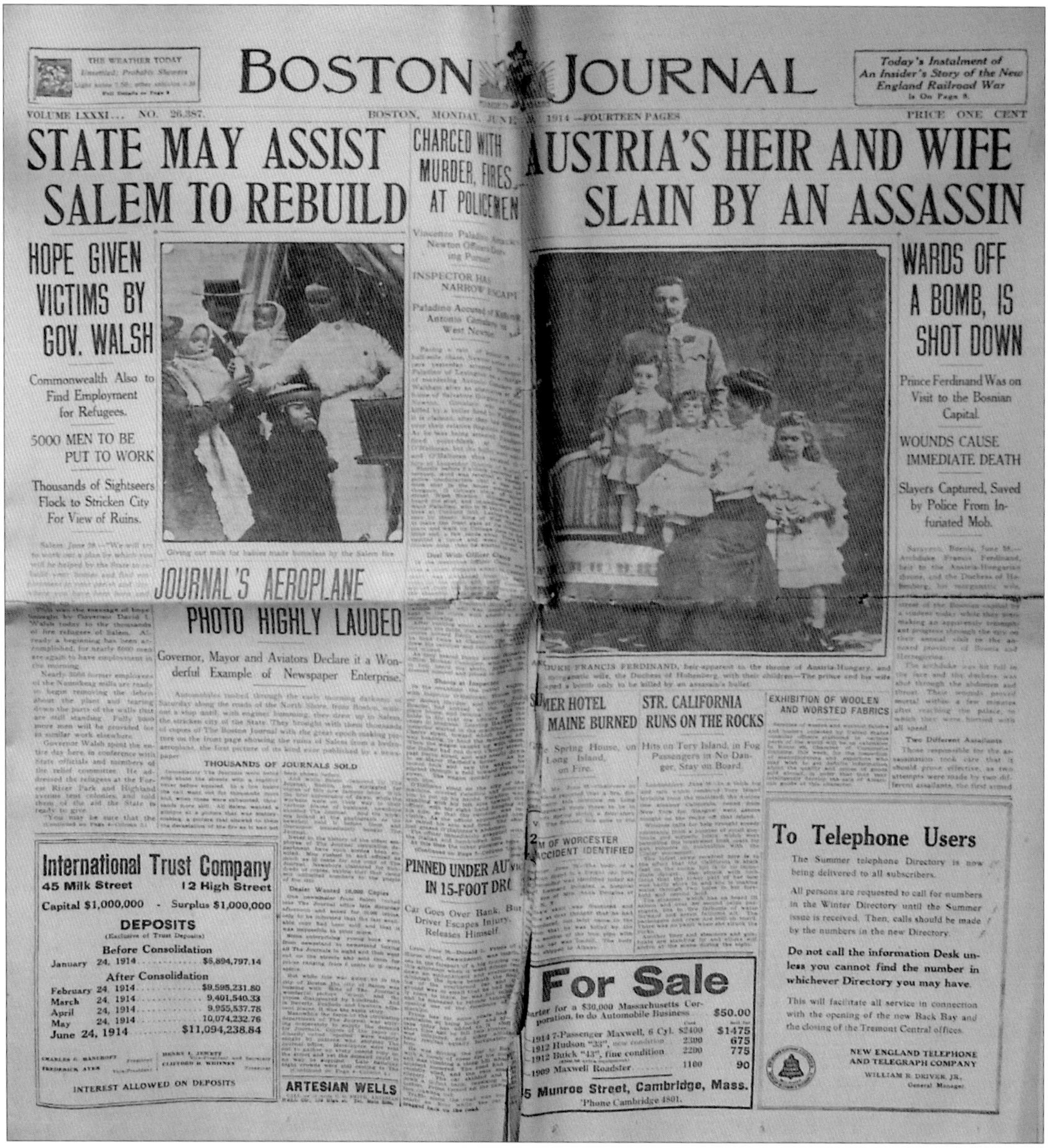

THE WEATHER TODAY

BOSTON JOURNAL

Today's Instalment of An Insider's Story of the New England Railroad War Is On Page 9.

VOLUME LXXXI... NO. 26,387. BOSTON, MONDAY, JUNE 29, 1914 —FOURTEEN PAGES PRICE ONE CENT

STATE MAY ASSIST SALEM TO REBUILD

HOPE GIVEN VICTIMS BY GOV. WALSH

Commonwealth Also to Find Employment for Refugees.

5000 MEN TO BE PUT TO WORK

Thousands of Sightseers Flock to Stricken City For View of Ruins.

Giving out milk for babies made homeless by the Salem fire

JOURNAL'S AEROPLANE PHOTO HIGHLY LAUDED

Governor, Mayor and Aviators Declare it a Wonderful Example of Newspaper Enterprise.

THOUSANDS OF JOURNALS SOLD

CHARGED WITH MURDER, FIRES AT POLICEMEN

INSPECTOR HAS NARROW ESCAPE

AUSTRIA'S HEIR AND WIFE SLAIN BY AN ASSASSIN

WARDS OFF A BOMB, IS SHOT DOWN

Prince Ferdinand Was on Visit to the Bosnian Capital.

WOUNDS CAUSE IMMEDIATE DEATH

Slayers Captured, Saved by Police From Infuriated Mob.

DUKE FRANCIS FERDINAND, heir-apparent to the throne of Austria-Hungary, and his morganatic wife, the Duchess of Hohenberg, with their children—The prince and his wife escaped a bomb only to be killed by an assassin's bullet.

Two Different Assailants

SUMMER HOTEL MAINE BURNED

STR. CALIFORNIA RUNS ON THE ROCKS

Hits on Tory Island, in Fog. Passengers in No Danger, Stay on Board.

EXHIBITION OF WOOLEN AND WORSTED FABRICS

M OF WORCESTER ACCIDENT IDENTIFIED

PINNED UNDER AUTO IN 15-FOOT DROP

Car Goes Over Bank, But Driver Escapes Injury, Releases Himself

ARTESIAN WELLS

International Trust Company

45 Milk Street — 12 High Street

Capital $1,000,000 - Surplus $1,000,000

DEPOSITS

(Exclusive of Trust Deposits)

Before Consolidation

January	24, 1914	$8,894,797.14

After Consolidation

February	24, 1914	$9,595,231.80
March	24, 1914	9,401,540.33
April	24, 1914	9,955,537.78
May	24, 1914	10,074,232.76
June	24, 1914	$11,094,238.84

INTEREST ALLOWED ON DEPOSITS

For Sale

Charter for a $30,000 Massachusetts Corporation, to do Automobile Business — $50.00

	Cost	Sell for
1914 7-Passenger Maxwell, 6 Cyl.	$2400	$1475
1912 Hudson "33", new condition	2300	675
1912 Buick "43", fine condition	2200	775
1909 Maxwell Roadster	1100	90

5 Munroe Street, Cambridge, Mass. 'Phone Cambridge 4801.

To Telephone Users

The Summer telephone Directory is now being delivered to all subscribers.

All persons are requested to call for numbers in the Winter Directory until the Summer issue is received. Then, calls should be made by the numbers in the new Directory.

Do not call the information Desk unless you cannot find the number in whichever Directory you may have.

This will facilitate all service in connection with the opening of the new Back Bay and the closing of the Tremont Central offices.

NEW ENGLAND TELEPHONE AND TELEGRAPH COMPANY

WILLIAM R. DRIVER JR., General Manager

Found in a box of folded newspapers, the image of the front page of the *Boston Journal*, dated June 29, shows the juxtaposition of the timing of the Great Fire in Salem with that of the assassination of the Archduke of Austria. The world was on the brink of World War I as Salem was recovering from the devastating effects of the fire. The *Boston Journal* also used the front page to report the response to the historic aerial photographs taken of the devastation created by the fire. (*Neg #32948*)

Well, Salem has been built anew
Through years elapsed since then;
But it's grand old quaint appearance
We'll never see again.

Elizabeth Bishop, an internationally known poet who spent part of her childhood in New England, attributes her alcoholism to the Salem fire. In her poem "The Drunkard" she says that she was three years old when the fire occurred. Living on Marblehead Neck at the time, she remembers watching the fire across the harbor and feeling an intense thirst. Was the fire truly responsible for her alcoholism?

I do not live in Salem and have no relatives who have experienced the fire. However, doing the research for this book had a significant impact on me. I was amazed at the strength of the residents and the resilience they had to move forward and rebuild their lives and their homes. The images in the library collection are powerful in their starkness after the fire and in their sense of hope as the city rebuilt itself. As Edward D. Johnson writes in his reminiscences of the fire,

> *From her experience Salem learned new duties. She started on toward the last years of her third century strong and quietly resolute. Not forgetting the things that were behind, the city pressed on toward a mark of new accomplishment. She sought and occupied a worthy place in the midst of new conditions which the passing years had brought for which a great disaster had not disqualified her.*

Selected Bibliography

Newspapers of the time, professional periodicals for firemen, insurance reports, first-person accounts, letters and an oral history were among the resources used to research the content for this text. A selection of the titles is listed below. Additional works are available at the Phillips Library.

Ames, Bertram E. "Report No. 150 on Conflagration, June 25, 1914." Salem, MA: Underwriters' Bureau of New England. (Phillips Library call number: E S1 F6 1914_{15})

Haywood, Charles F. *General Alarm: A Dramatic Account of Fire and Firefighting in America.* New York: Dodd, Mead & Co., 1967. (E H42702 1967)

Inspector Department. "Salem Conflagration and Destruction of the Mills of Naumkeag Steam Cotton Co., June 25, 1915." Boston. (E S1 B6 N1 1917)

Jones, Arthur B. *The Salem Fire.* Boston: The Gorham Press, 1914. (E J76.3 1914)

O'Keefe, Joseph A. "Salem's Conflagration of 1914." *Fireman* 21, no. 16. Boston: National Fire Protection Association, 1964. (E Si F6 1914_{2})

Perry, Montayne. *The Salem Fire Relief.* Salem, MA: Milo A. Newhall & Co., Printers, 1915. (974.462 S1 P5 c2)

Philbric, Charles M. "Our Story of June 25, 1915: The Day of the Fire in the Electric Light Office." Salem, MA: self-published. (E S1 F6 1914_{8})

Salem Rebuilding Commission. Report of the Salem Rebuilding Commission. Salem, MA: Newcomb & Gauss, Printers, 1914. (E S2 B8 1917 c2)

Salem Rebuilding Commission, Salem Rebuilding Trust. Detailed Reports from Secretary. Salem, MA, 1915. (E S1 B8 1915)

Shepperd, F.W., ed. *Fire & Water Engineering.* 56–58 (1914–15) Uncatalogued periodical.

Smith, Charles H. "Fire Report #431, Index 12375, Factory of Naumkeag Steam Cotton Company, Insurance Report." Boston: Assoc. Factory Mutual Fire Insurance Companies. (E S1 F6 1914_{2})

Tufts, Edward R. *A History of the Salem Fire Department.* Salem, MA: Holyoke Mutual Insurance Company, 1975. (TH 9117 T914)